Advances in Organic Corrosion Inhibitors and Protective Coatings

Advances in Organic Corrosion Inhibitors and Protective Coatings

Editor

Ioannis Kartsonakis

MDPI • Basel • Beijing • Wuhan • Barcelona • Belgrade • Manchester • Tokyo • Cluj • Tianjin

MDPI

Editor
Ioannis Kartsonakis
Institute of Nanoscience and Nanotechnology
Greece
National Technical University of Athens
Greece

Editorial Office
MDPI
St. Alban-Anlage 66
4052 Basel, Switzerland

This is a reprint of articles from the Special Issue published online in the open access journal *Applied Sciences* (ISSN 2076-3417) (available at: https://www.mdpi.com/journal/applsci/special_issues/corrosion_inhibitors_coatings).

For citation purposes, cite each article independently as indicated on the article page online and as indicated below:

LastName, A.A.; LastName, B.B.; LastName, C.C. Article Title. *Journal Name* **Year**, *Volume Number*, Page Range.

ISBN 978-3-0365-0640-1 (Hbk)
ISBN 978-3-0365-0641-8 (PDF)

Cover image courtesy of Ioannis Kartsonakis.

Contents

About the Editor

Ioannis Kartsonakis is a chemist from the National and Kapodistrian University of Athens (NKUA), Greece. His master's degree was in polymer science (NKUA) with a focus on the synthesis and characterization of standard polymers. He has a Ph.D. in chemistry (NKUA in collaboration with the National Centre for Scientific Research "Demokritos") with a focus on corrosion protection of metal alloys, synthesis, and characterization via a combination of the sol–gel method, electrochemistry, and the polymer science of hybrid organic–inorganic coatings, as well as micro- and nanocontainers. He has been involved in the technical/scientific development, implementation, and project administration of several European and national research projects as a chemist researcher. His scientific interests are focused on the synthesis and characterization of materials, such as nanocomposites, nanocontainers, and coatings (polymeric, inorganic, hybrid organic–inorganic), and the study of corrosion processes on materials.

Preface to "Advances in Organic Corrosion Inhibitors and Protective Coatings"

Metallic components consisting of magnesium, aluminum, steel, and their alloys are widely used in a plethora of applications because of their exceptional mechanical properties and their overall durability. However, they are often damaged by mechanical or chemical processes. Mechanical damage that is caused by the impact of solid particles is called erosion, while electrochemical damage that results from chemical degradation is termed corrosion. The corrosion phenomena, to which every metallic substrate is subjected to, eventually result in the degradation of the metal and the deterioration of its properties. The interaction of a metal with its environment that results to its chemical alteration is called metallic corrosion. According to the literature, corrosion is classified into two types: uniform and localized corrosion. The intervention of either in the alloy environment or in the alloy structure can influence the corrosion protection of metallic materials. Furthermore, interference in the metal alloy environment can be conducted with the utilization of cathodic or anodic protection via the corresponding inhibitors. Therefore, the most common categorization is cathodic, anodic, and mixed-type inhibitors, taking into account which half-reaction they suppress during corrosion phenomena. The majority of the organic inhibitors are of mixed type and perform through chemisorption. The corrosion control of metallic structures is an important task in technical, economic, environmental, and safety terms. Several types of corrosion inhibitors are being employed to prevent metallic dissolution in corrosive media, for which the use of organic inhibitors is one of the most frequent and economic methods. Heteroatoms (O, S, N, and P) and π-electrons in the conjugated form act as excellent corrosion inhibitors for metals and alloys in aggressive solutions [1]. These inhibitors can be incorporated into corrosion-protective coatings. Coatings considered for corrosion inhibition must offer an effective physical barrier, impeding the access of violent materials to the metal surface. In order to update the field of corrosion protection of metal and metal alloys with the use of organic inhibitors, a Special Issue entitled "Advances in Organic Corrosion Inhibitors and Protective Coatings" is introduced. This book gathers and reviews a collection of ten contributions (nine articles and one review), from authors from Europe, Asia, and Africa, that were accepted for publication in this Special Issue of *Applied Sciences*.

Ioannis Kartsonakis
Editor

applied sciences

Editorial

Special Issue on "Advances in Organic Corrosion Inhibitors and Protective Coatings"

Ioannis A. Kartsonakis

Laboratory of Advanced, Composite, Nanomaterials and Nanotechnology, R-Nano Lab, School of Chemical Engineering, National Technical University of Athens, 9 Heroon Polytechniou str., Zografou Campus, 15773 Athens, Greece; ikartso@chemeng.ntua.gr; Tel.: +30-210-772-3310

Received: 3 December 2020; Accepted: 21 December 2020; Published: 24 December 2020

Abstract: Featured Application**This Special Issue collects new findings and recent advances in the development, synthesis, and structure–activity relationships of organic corrosion inhibitors and protective coatings**.

---

1. Introduction

Metallic components consisting of magnesium, aluminum, steel, and their alloys are widely used in a plethora of applications because of their exceptional mechanical properties and their overall durability. However, they are often damaged by mechanical or chemical processes. Mechanical damage that is caused by the impact of solid particles is called erosion, while electrochemical damage that results from chemical degradation is termed corrosion. The corrosion phenomena, to which every metallic substrate is subjected, eventually result in the degradation of the metal and the deterioration of its properties.

The interaction of a metal with its environment that results in its chemical alteration is called metallic corrosion. According to the literature, corrosion is classified into two types; uniform and localized corrosion. The intervention of either in the alloy environment or in the alloy structure can influence the corrosion protection of metallic materials. Furthermore, the interference in the metal alloy environment can be conducted with the utilization of cathodic or anodic protection via the corresponding inhibitors. Therefore, the most common categorization is among cathodic, anodic and mixed-type inhibitors, taking into account which half-reaction they suppress during corrosion phenomena. The majority of the organic inhibitors are of mixed-type and perform through chemisorption.

Corrosion control of metallic structures is an important task in technical, economic, environmental, and safety terms. Several types of corrosion inhibitors are being employed to prevent metallic dissolution in corrosive media, for which the use of organic inhibitors is one of the most frequent and economic methods. Heteroatoms (O, S, N, and P) and π-electrons in the conjugated form act as excellent corrosion inhibitors for metals and alloys in aggressive solutions [1]. These inhibitors can be incorporated into corrosion-protective coatings. Coatings considered for corrosion inhibition must offer an effective physical barrier, impeding the access of violent materials to the metal surface.

In order to update the field of corrosion protection of metal and metal alloys with the use of organic inhibitors, a Special Issue entitled "Advances in Organic Corrosion Inhibitors and Protective Coatings" has been introduced. This editorial manuscript gathers and reviews the collection of ten contributions (nine articles and one review), with authors from Europe, Asia and Africa accepted for publication in the aforementioned Special Issue of Applied Sciences.

2. Corrosion Inhibitors

Steel has gained an increasing utilization and is widely used for machinery parts, building construction, and pipelines, because of its low price as well as its machinability and weldability.

Three articles were published in this Special Issue related to corrosion protection of steel in the presence of inhibitors. The corrosion resistance of mild steel under the conditions used for industrial applications was investigated by Kartsonakis et al. [2]. Their study investigated the application of organic 2-mecraptobenzothiazole (MBT) and inorganic Na_2HPO_4 as corrosion inhibitors for the protection of API 5L X42 pipeline steel. The electrochemical characterizations proved that both compounds can be considered as corrosion inhibitors of steel and the corrosion protection mechanisms can be assigned to the protective layers created onto the metal surface through chemisorption of the inhibitors, which prevent chloride's penetration. The synergistic effect of the MBT and Na_2HPO_4, in a molar ratio of 1:1, revealed the effective performance of the additives with corrosion inhibition efficiency above 90%.

Lahrour et al. [3] investigated the use of glycerin-grafted starch as a bio-copolymer working as a corrosion inhibitor for the protection of C-Mn steel, in the presence of HCl, by electrochemical techniques and the weight loss method. The obtained results revealed that inhibition efficiency enhances with increasing bio-copolymer concentration, reaching a maximum of 94%. Moreover, it was concluded that the corrosion inhibition mechanism might be not only a simple physisorption process on the steel surface, but can also be explained in terms of chemisorption of polymer on the steel surface. The corrosion behavior of steel reinforcement incorporated in mortar specimens, in which both the cement and aggregate are partially substituted by ladle furnace slag (LFS) in respect of the presence of chloride ions, was investigated in the work of the Prieto et al. [4]. Electrochemical techniques were used to estimate the corrosion rate as well as the symptoms created in steel rebars in the presence of LFS. Considering the obtained results, it was concluded that in mortars with the presence of mixed-in chlorides, the embodiment of LFS in the mortar at the time of kneading does not negatively influence the corrosive process of the steel rebar.

In recent years, studies on copper alloy corrosion have gained much attention in the industrial sector due to the fact that copper and its alloys create several industrially important materials. Many important properties ranged from good thermal properties, formability, electrical conductivities and visual appearance to good corrosion resistance influenced by the alloying elements of copper. Two articles based on copper alloy corrosion protection were published in this Special Issue. In the work of Kozaderov et al. [5], the corrosion inhibition efficiency of 3-mercaptoalkyl derivatives of 5-amino-1H-1,2,4-triazole for the a-brass (copper > 65 wt.% and zinc < 35 wt.%) in a chloride media was investigated using electrochemical techniques. Taking into account the obtained result, it was estimated that the degree of protection achieved for all inhibitors can reach a maximum of over 99%. Moreover, it was proven that the protective effect enhances with the length of the alkyl chain. The corrosion protective mechanism is based on the formation of a protective film on the a-brass surface that contains oxides as well as complex compounds of copper and zinc with the inhibitor molecules.

In the work of Shinato et al. [6], the protection role of cysteine for the copper-based alloys Cu-5Zn-5Al-1Sn, containing zinc (5 wt.%), aluminum (5 wt.%) and tin (1 wt.%) corrosion in sodium chloride solutions, was studied. The electrochemical results proved that the inhibition efficiency improved with increasing cysteine concentration. Furthermore, the results of the surface analysis techniques indicated that the corrosion inhibition was based on the adsorption of the inhibitor molecules onto the copper alloy surface.

Aluminum alloy is one of the most widely used materials in the subsea industry due to its excellent thermal conductivity, low cost, acceptability for short-term development, good strength, and low density. One article related to corrosion protection of aluminum alloy 7075 was published in this Special Issue. The study of Seo et al. [7] examined the corrosion resistance of oil impregnated anodic aluminum oxide surfaces of aluminum 7075 for subsea application. Several experiments were conducted based on contact angle measurements together with salt spray and pressure tests. The obtained results revealed that the corrosion resistance of aluminum 7075 could be improved by oil impregnation on the aluminum alloy surface, and therefore can be used in the subsea industry.

3. Corrosion Protective Coatings

In general, three different types of coatings exist for corrosion protection of metal alloys. Barrier coatings prevent the aggressive elements from coming together with the substrate to onset the corrosion process. Inhibitive coatings actively block the electrochemical reaction from happening by interfering with the electrolytes needed to start the corrosion process. Sacrificial coatings are a type of metal coatings that become more oxidated than the metal surface they protect. Three articles and one review based on corrosion protective coatings were published in this Special Issue. In the work of Shi et al. [8], smooth coatings of TiNb and TiNbN were deposited on 316L stainless steel by magnetron sputtering in order both to improve the corrosion resistance and the electronic conductivity of bipolar plates for proton exchange membrane fuel cells.

In the study by D'Elia et al. [9], poly(phenylene methylene)-based coatings were developed for corrosion protection of aluminum alloy AA2024. These copolymer-based coatings contained n-octyloxy side chains and their anti-corrosion behavior was estimated via several electrochemical methods. The obtained results revealed that these coatings exhibited good corrosion protection of the metal surface towards a sodium chloride solution. Thin multifunctional coatings of cerium oxide nanoparticles with anti-reflective properties were developed onto glass and silicon substrates by Romasanta et al. [10]. These coatings could be used as an intermediate layer between air and substrates for providing anti-reflection features. Finally, in the review paper of Fu et al. [11], the advances and challenges of corrosion and topology detection of grounding grid are discussed. This paper presents the research status of grounding corrosion and topological detection in detail and introduces the basic principles, research difficulties and existing problems of the methods such as the electromagnetic field method, electric network method, ultrasonic detection method, electrochemical method and electromagnetic imaging method.

4. Future Strategies

Although the Special Issue has been closed, more in-depth research in the field of corrosion protection technologies with the use of organic inhibitors together with coatings development is expected. It can be anticipated that more friendly applications will be demanded in large numbers in the future for the protection of metal alloys against corrosion.

Funding: This research received no external funding.

Institutional Review Board Statement: Not applicable.

Informed Consent Statement: Not applicable.

Data Availability Statement: Data sharing not applicable.

Acknowledgments: The Guest Editor would like to thank all the authors and peer reviewers for their fruitful and valuable contributions to this Special Issue. The confluence of the editorial team of Applied Sciences is highly appreciated. Finally, special thanks to Tamia Qing, Section Managing Editor of Applied Sciences from the MDPI Branch Office, Wuhan, China.

Conflicts of Interest: The author declares no conflict of interest.

References

1. Kartsonakis, I.A.; Charitidis, C.A. Corrosion Protection Evaluation of Mild Steel: The Role of Hybrid Materials Loaded with Inhibitors. *Appl. Sci.* **2020**, *10*, 6594. [CrossRef]
2. Kartsonakis, I.A.; Stamatogianni, P.; Karaxi, E.K.; Charitidis, C.A. Comparative Study on the Corrosion Inhibitive Effect of 2-Mecraptobenzothiazole and Na_2HPO_4 on Industrial Conveying API 5L X42 Pipeline Steel. *Appl. Sci.* **2019**, *10*, 290. [CrossRef]
3. Lahrour, S.; Benmoussat, A.; Bouras, B.; Mansri, A.; Tannouga, L.; Marzorati, S. Glycerin-Grafted Starch as Corrosion Inhibitor of C-Mn Steel in 1 M HCl solution. *Appl. Sci.* **2019**, *9*, 4684. [CrossRef]

4. Prieto, M.; González, M.; Rodríguez, Á.; Cobo, A. The Influence of Replacing Aggregates and Cement by LFS on the Corrosion of Steel Reinforcements. *Appl. Sci.* **2019**, *9*, 683. [CrossRef]
5. Kozaderov, O.; Shikhaliev, K.; Prabhakar, C.; Tripathi, A.; Shevtsov, D.; Kruzhilin, A.; Komarova, E.; Potapov, A.; Zartsyn, I.; Kuznetsov, Y. Corrosion of α-Brass in Solutions Containing Chloride Ions and 3-Mercaptoalkyl-5-amino-1H-1,2,4-triazoles. *Appl. Sci.* **2019**, *9*, 2821. [CrossRef]
6. Shinato, K.W.; Huang, F.; Xue, Y.; Wen, L.; Jin, Y. The Protection Role of Cysteine for Cu-5Zn-5Al-1Sn Alloy Corrosion in 3.5 wt.% NaCl Solution. *Appl. Sci.* **2019**, *9*, 3896. [CrossRef]
7. Seo, Y.; Jung, J.-Y.; Chung, J.; Lee, S. Enhancement of Corrosion Resistance of Aluminum 7075 Surface through Oil Impregnation for Subsea Application. *Appl. Sci.* **2019**, *9*, 3762. [CrossRef]
8. Shi, K.; Li, X.; Zhao, Y.; Li, W.-W.; Wang, S.-B.; Xie, X.-F.; Yao, L.; Jensen, J.O.; Li, Q.-F. Corrosion Behavior and Conductivity of TiNb and TiNbN Coated Steel for Metallic Bipolar Plates. *Appl. Sci.* **2019**, *9*, 2568. [CrossRef]
9. D'Elia, M.F.; Magni, M.; Trasatti, S.P.; Schweizer, T.B.; Niederberger, M.; Caseri, W.R. Poly(phenylene methylene)-Based Coatings for Corrosion Protection: Replacement of Additives by Use of Copolymers. *Appl. Sci.* **2019**, *9*, 3551. [CrossRef]
10. Romasanta, L.J.; D'Alençon, L.; Kirchner, S.; Pradère, C.; Leng, J. Thin Coatings of Cerium Oxide Nanoparticles with Anti-Reflective Properties. *Appl. Sci.* **2019**, *9*, 3886. [CrossRef]
11. Fu, Z.; Wang, X.; Wang, Q.; Xu, X.; Fu, N.; Qin, S. Advances and Challenges of Corrosion and Topology Detection of Grounding Grid. *Appl. Sci.* **2019**, *9*, 2290. [CrossRef]

Publisher's Note: MDPI stays neutral with regard to jurisdictional claims in published maps and institutional affiliations.

applied sciences

MDPI

Article

Comparative Study on the Corrosion Inhibitive Effect of 2-Mecraptobenzothiazole and Na_2HPO_4 on Industrial Conveying API 5L X42 Pipeline Steel

Ioannis A. Kartsonakis, Panagiota Stamatogianni, Evangelia K. Karaxi and Costas A. Charitidis *

Research Unit of Advanced, Composite, Nano-Materials and Nanotechnology, School of Chemical Engineering, National Technical University of Athens, 9 Heroon Polytechniou St., Zographos, GR-15773 Athens, Greece; ikartso@chemeng.ntua.gr (I.A.K.); giota_grrock.n.metal@hotmail.com (P.S.); dcar@chemeng.ntua.gr (E.K.K.)
* Correspondence: charitidis@chemeng.ntua.gr; Tel.: +30-2107724046

Received: 1 December 2019; Accepted: 27 December 2019; Published: 31 December 2019

Abstract: Mild or low-carbon steel has an increasing utilization and is widely used for building construction, machinery parts, and pipelines, because it can be machined easily and has enhanced weldability as well as a low price. In any case, the corrosion resistance of mild steel under the conditions in industrial applications or in atmosphere is a thoughtful concern. This study inquires into the application of 2-mecraptobenzothiazole (MBT) and Na_2HPO_4 as corrosion inhibitors for the protection of API 5L X42 pipeline steel in 3.5 wt % NaCl as well as in water from the Athens city supply system. The electrochemical/morphological characterizations of the aforementioned mild steel proved that the corrosion protection mechanisms can be assigned to the protective layers created onto the metal surface because of the presence of the inhibitors, which prevent chloride's penetration. The synergistic effect of the MBT and Na_2HPO_4 corrosion inhibition behavior, in a molar ratio of 1:1, revealed that the additives performed effectively with corrosion inhibition efficiency above 90%.

Keywords: mild steel; EIS; SEM; Raman spectroscopy; pitting corrosion; synergistic effect

1. Introduction

One of the most widespread kinds of steel is low-carbon or mild steel, not only for structural purposes but also for several applications in industry. Therefore, its usage and operation range from machinery parts, building materials, and domestic appliances to cutting tools, conveying tubes, cables, and magnets. Due to its low concentration in carbon (up to 0.29 wt %), mild steel exhibits extreme durability, great affordability, and significant mechanical, thermal, and magnetic properties [1–3]. Due to mild steel's increasing utilization, there has been great concern for the susceptibility of such steels caused by environmental corrosive factors, such as the humidity, acidity, or salinity of the atmosphere [4–6]. Thus, lots of studies have been focusing on low-alloy (weathering) steel corrosion and how to encounter it in a functional way [7–9].

In general, metallic corrosion is associated with any chemical alteration of the metal that stems from interaction with its environment [10]. Although intense research studies have taken place over all these decades in order to face corrosion, it still remains in the foreground since it is a continuous and inevitable process [11]. The main classification of corrosion types emphasized in the literature is between uniform and localized corrosion [12]. Pitting corrosion is a type of localized corrosion in which the metal is corroded in depth via the formation of indistinguishable pits. In the case of water and usually oil-conveying pipes, the corrosion effect often triggers uniform surface degradation internally, in parallel with localized corrosion underneath and oxide film formation [13,14]. Taking into account that pitting may cause the initiation of stress corrosion cracking [15], it is considered as the most detrimental corrosion effect. While uniform corrosion occurs during the subjection of mild steel in various environmental conditions [8,16], localized corrosion, such as pitting and scaling, takes place in the presence of heavy metal ions [17,18] or in locally high pH values [19]. Localized corrosion phenomena may thin locally the inner wall in a pipe, resulting in the creation of areas susceptible to cracking [20].

The protection of metallic materials from the corrosion effect could be achieved by intervening either in the alloy structure or in the alloy environment. The intervention in the alloy structure can be performed by changing the alloying ratio, by creating/applying metallic or organic coatings/films on the surface, by decreasing mechanical operation tensions, by the use of anodic or cathodic protection, and by adding corrosion inhibitors in the corroding solution [21]. In the literature, several studies indicate corrosion inhibition as an effective protection method [22,23]. A conventional classification of corrosion inhibitors is according to their inhibiting action. Therefore, there are adsorption inhibitors that undergo chemisorption in the metallic surface, film-forming inhibitors, which are divided into passivation inhibitors (oxidizing or non-oxidizing) and precipitation inhibitors (deposition of three-dimensional film). However, the most common discrimination is between anodic, cathodic, and mixed-type inhibitors, considering which half-reaction they suppress during corrosion phenomena. Most organic inhibitors are of mixed-type and act as chemisorptive inhibitors. Phosphates act as cathodic inhibitors, whereas benzoates and azelates represent non-oxidizing film-forming inhibitors [24].

Concerning inorganic inhibitors, ideally the central atom of the inhibitor tends to form a complex entering the metallic lattice, without the need of additional energy and with the necessary stability [25]. Refaey et al. compared the inhibition ability of phosphate, chromate, molybdate, and nitrite as potential inhibitors for mild steel, in 0.1 M NaCl near neutral solutions. They concluded that phosphate demonstrated higher inhibition efficiency compared to the other inhibitors. Furthermore, phosphate deposited a strongly adherent layer, consisting of γ-Fe_2O_3 and $FePO_4{\cdot}2H_2O$, from the solution on the metal surface [26]. Regarding organic corrosion inhibitors, a perusal of the literature reveals a variety of organic compounds that have been suggested for encountering pitting corrosion on mild steel. An outstanding investigation was carried out by Marczewska-Boczkowska and Kosmulski using steel samples in aqueous solutions and indicating certain derivatives of azoles and thiazoles. They reported that imidazoles, benzothiazoles, and mercaptobenzothiazoles behaved as significantly efficient corrosion inhibitors against pitting. They claimed that these organic substances are capable of forming self-assembled monolayers, which improve inhibition and confer a protective layer of great stability, due to the spontaneous self-assembly process [27]. Moreover, Wang et al. synthesized and utilized an organic chemical compound, 4-salicylideneamino-3-phenyl-5-mercapto-1,2,4-triazole, as a corrosion inhibitor for mild steel, in electrolytic solution of 1 M HCl solution in several temperatures, which was based on the enhanced corrosion inhibition of the combined N and S elements in heterocyclic organic compounds [26]. They also pointed out that the presence of both elements functions more efficiently instead of the use of substances that contain only nitrogen or sulfur separately. This outcome was attributed to the combination of phenyl, mercapto, and azomethine reactive groups. Taking into account these claims, 2-mercaptobenzothiazole (MBT) was rendered as one of the most preferable mixed-type organic inhibitors for mild steel in aqueous solutions [28].

The self-assembling effect [29] and the chemisorption ability in conjunction with the aromatic nature of the inhibitor, which confers increased stability [25], lead to the need to investigate how to optimize inhibition by the use of MBT. Furthermore, its efficiency in near neutral aqueous solutions was taken into account. A common utilization of MBT was in coolant mixtures as a corrosion inhibitor for low-carbon steel pipes [30]. It exhibited improved inhibition efficiency when mixed with different organic inhibitors, denoting the advanced synergistic inhibition effect through smart combinations. Gunasekaran et al. evaluated the synergistic corrosion inhibition of several phosphonic acids substituting metal ions for azoles, provided that phosphorous inhibition ability could be exploited in chemically different systems [31]. A conventional use of phosphate anions in corrosion inhibitors is in cooling systems that maintain cavitation problems and intense corrosion phenomena [32]. Calmon et al. patented certain mixtures against pitting and the galvanic corrosion of copper and iron surface, consisting of sodium phosphates (Na_2HPO_4) or polyphosphates ($Na_2P_4O_{13}$) synergistically with sodium mercaptobenzothiazole in aqueous solutions emphasizing the immediate connection between inhibition efficiency and inhibitor concentration [33].

The aim of this study is to investigate the susceptibility of mild steel to corrosion in industrial hot/cooling systems after its exposure in 3.5 wt % NaCl as well as in the water of the Athens city supply system. The main goal of the presented experiments is to identify which inhibitors perform best at restricting the aforementioned corrosion process. The importance of these experiments is directly linked to the current urgent need for industrial mild steel pipeline corrosion protection.

In this regard, the protection effectiveness of the complexes, oxides, or salts that are created on the mild steel surface, as well as the morphological conversions that occur on the surface of the metal alloy panels following exposure in the presence or absence of MBT and Na_2HPO_4, were evaluated by three families of techniques: electrochemical, microscopy, and spectroscopy. The electrochemical characterization demonstrated that MBT and Na_2HPO_4 can be considered as corrosion inhibitors of mild steel as they reduced the corresponding anodic and/or cathodic corroding reactions. Moreover, the synergistic effect of the corrosion inhibition behavior of MBT and Na_2HPO_4 in a molar ratio 1:1 at several concentrations was studied. The analysis revealed that the admixtures performed effectively with inhibition efficiency above 90%. The presence of both aforementioned inhibitors into the corrosive environment exhibited the highest impedance modulus ($|Z|$) and polarization resistance (R_p) value as the exposure time elapsed.

Regarding the X-ray diffraction (XRD) measurements and scanning electron microscopy (SEM) characterization, it was disclosed that the exposure of mild steel to a corrosive environment in the presence or absence of inhibitors resulted in the creation of several oxide, hydroxide, and hydroxide–phosphate compounds on the steel surface. According to the aforementioned characterization techniques, the corrosion protection mechanisms of steel can be ascribed to the protective films created onto the metal surface because of the inhibitors' presence, which prevent chlorides' insertion. The performed experiments shed light on the corrosion mechanisms of mild steel in industrial hot/cooling systems.

2. Materials and Methods

2.1. Reagents and Solutions

All the compounds and reagents were of analytical reagent grade. Sodium phosphate dibasic dihydrate ($Na_2HPO_4 \cdot 2H_2O$, Sigma-Aldrich) and 2-mercaptobenzothiazole (MBT, Sigma-Aldrich) were used without further purification. Hot-rolled black (non-galvanized) mild steel panels were manufactured by TMK-ARTROM S.A. and accompanied with all certifications required (ASTM A568/A568M-09 [34]). Electrochemical measurements were performed on mild steel grade API 5L X42 conveying a pipeline with the following chemical composition (wt %): C: 0.15, Mn: 0.56, S: 0.002, P: 0.12, Si: 0.21, Ni: 0.07, Cr: 0.04, Mo: 0.01, Cu: 0.22, Al: 0.020, N: 0.009, V+Ti+Nb: 0.004, and Fe as remainder.

2.2. Preparation of Substrates

The mild steel panels were abraded with SiC paper up to 5 μm grain size (P4000) and then cleaned in agreement with ASTM D6386-99 (reapproved 2005) [35] prior to being used in the conducted experiments. This procedure is necessary because the exposed steel interacts with the environment to form several iron oxides such as α-FeO(OH), γ-FeO(OH), β-FeO(OH), and Fe_3O_4 [36]. According to the aforementioned standard, the cleaning procedure includes the degreasing of steel panels with their subjection in a mixture of acetone and ethanol (50:50) of purity 96% *v*/*v* for about 20 min, and then their exposure to NaOH solution of pH 11 for 5 min at 60 °C. Finally, the panels are rinsed with distilled water and dried in a desiccator in order to avoid the formation of new corrosion products, as uncoated black mild steel is susceptible to corrosion by humidity, as analyzed above.

2.3. Characterization

Several imaging techniques providing information at macro- and microscopic scales were used in order to evaluate the surface morphology of the mild steel panels. The microscale information was collected with SEM imaging performance using a Hitachi Tabletop Microscope TM3030 Scanning Electron Microscopy equipped with an energy dispersive X-ray spectrophotometer (EDS) system (QUANTAX 70), and with an ultra-high resolution scanning electron microscopy (UHR-SEM) using NOVA NANOSEM 230 (FEI Company). The macroscale information with respect to optical details was acquired using a Samsung Galaxy A7 auto focus triple camera 24 MP + 5 MP + 8 MP. The chemical analysis was conducted via micro-Raman measurements using a Renishaw inVia spectrometer working in backscattering configuration and equipped with a near-infrared diode laser emitting at 532 and 785 nm. The spectra were recorded by focusing the laser beam on the sample surface and adjusting the light power so that 1 mW was provided for a spot of about 1 μm diameter. The compounds that are created on the subjected mild steel panels were studied via XRD measurements. The crystal structure was identified by powder X-ray diffraction using an X Bruker D8 Advance Twin Twin, employing Cu–Kα radiation (λ = 1.5418 Å).

The electrochemical characterizations based on linear polarization resistance (LPR) and potentiodynamic polarization (PP) were conducted using a VersaSTAT 3 Potentiostat/Galvanostat/Frequency Response Analyzer instrument (Princeton Applied Research, AMETEK). Concerning the PP measurements, a three-electrode electrochemical cell was used, consisting of a saturated silver chloride electrode [Ag/AgCl,$KCl_{(sat)}$] as reference, a working electrode (≈1.0 cm^2 of exposed area), and a platinum foil as counter electrode. The assessment of the polarization curves on the bare alloy was accomplished after different exposure intervals. Cathodic and anodic branches of polarization curves were recorded separately with the potential scan rate of 1 mVs^{-1} in the cathodic and in the anodic directions, starting from the open circuit potential (OCP, E_{OC}, corrosion potential), at room temperature. Regarding the LPR measurements, the scan rate was 0.1 mVs^{-1} and the potential range was ± 2.5 mV versus OCP. The R_p (polarization resistance) is the definition of the ratio of the applied potential to the applied current ($\Delta E/\Delta I$). The slope of the potential versus current plot is used for the calculation of the R_p [37]. The R_p denotes the degree of the specimen susceptibility to be corroded [38,39]. The corrosion protection performance of the inhibitors was estimated via electrochemical impedance spectroscopy (EIS). The EIS measurements were conducted using the aforementioned potentiostat instrument with the arrangement of the three-electrode electrochemical cell, in a frequency range from 10 μHz to 1 MHz. All spectra were recorded at open circuit potential, applying a 10 mV sinusoidal perturbation (rms signal) at room temperature. During all measurements, the electrochemical cell was placed in a Faraday cage. The Z-view Software (Scribner Associates Incorporated, SAI) was used for the analysis of the obtained spectra utilizing the adequate equivalent electric circuits. Two different types of electrolytic solutions (300 mL) were used: (i) water from the supply system of Athens city (WSS: water of the Athens city supply system after a softening process, Table 1), and (ii) 3.5 wt % NaCl solution prepared in distilled water [ASTM G44-99 (2013)] [40].

Table 1. Ion composition of the water of the Athens city supply system after a softening process (WSS) estimated via atomic absorption spectroscopy (mg/L).

K	Ca	Mg	Fe	Mn	Zn
1.7733	0.346	0.09755	1.36	0.005	0.0052

Moreover, the corrosion phenomena on a metallic uncoated surface was further evaluated with respect to the influence of corrosion inhibitors. Consecutively, 6 mM electrolytic solutions of the impending corrosion inhibitors were prepared for each compound as well as certain equimolecular 6 mM solutions of both Na_2HPO_4 and MBT for the investigation of their synergistic inhibition effect (Table 2). The sample *Steel-blank* is ascribed to mild steel panel prior to subjection in electrolytic solutions.

Table 2. Composition of the solutions for testing the effectiveness of the inhibitors onto mild steel. Reagents concentration 6 mM, reagents solubility in water (mg/100 g, 20 °C) 1.7×10^3 Na_2HPO_4, 1.9×10^1 MBT.

Composition	pH	Sample
-	-	*Steel–blank*
WSS	7.41–7.52	*Steel–WSS–blank*
WSS + 0.849 g/L Na_2HPO_4	7.78–7.97	*Steel–WSS–Na_2HPO_4*
WSS + 1.014 g/L MBT	7.33–7.52	*Steel–WSS–MBT*
WSS + 0.849 g/L Na_2HPO_4 + 1.014 g/L MBT	7.38–7.82	*Steel–WSS–Na_2HPO_4–MBT*
NaCl 3.5 wt %	7.71–7.72	*Steel–NaCl–blank*
NaCl 3.5 wt % + 0.849 g/L Na_2HPO_4	7.78–7.97	*Steel–NaCl–Na_2HPO_4*
NaCl 3.5 wt % + 1.014 g/L MBT	7.33–7.52	*Steel–NaCl–MBT*
NaCl 3.5 wt % + 0.849 g/L Na_2HPO_4 + 1.014 g/L MBT	7.38–7.82	*Steel–NaCl–Na_2HPO_4–MBT*

The corrosion inhibition efficiency, $\eta(\%)$, was estimated by Equation (1):

$$\eta(\%) = \frac{R_{ct(inhibitor)} - R_{ct(bareMetal)}}{R_{(inhibitor)}} \times 100 \tag{1}$$

where $R_{ct(inhibitor)}$ and $R_{ct(bareMetal)}$ represent the charge transfer resistances in the presence and in the absence of an inhibitor, respectively [41,42].

The synergistic parameter (S_i) was calculated from Equation (2) given by Aramaki and Hackerman [43]:

$$S_i = \frac{1 - \eta_{1+2}}{1 - \eta^*_{1+2}}, \tag{2}$$

where η_1 is the inhibition efficiency of the Na_2HPO_4, η_2 is the inhibition efficiency of the MBT, and $\eta_{1+2} = (\eta_1 + \eta_2) - (\eta_1 \times \eta_2)$, η^*_{1+2} is the measured inhibition efficiency of the MBT in combination with Na_2HPO_4.

3. Results and Discussion

3.1. Morphology Evaluation

The visual pictures of mild steel immersed in WSS for 96 h, in the presence of inhibitors or not, are demonstrated in Figure 1a–d. Onto the slightly rinsed corroded surface of Figure 1a, ferrous yellowish, orange, and brownish red species create a hardly homogenous layer, as expected from the literature for the mild steel surface in the absence of inhibitors (*Steel–WSS–blank*). Susceptible to a series of local corrosion phenomena as well as to the creation of local galvanic cells and pits, the final layer emerges in dependence with the local composition and physicochemical conditions of the system [44]. In depth, it is claimed that reddish hues stem from hematite (α-Fe_2O_3) in a hexagonal crystal system, in which the face-sharing octahedral keeps the Fe centers at the shortest distance (0.29 nm), whereas the yellowish ones are caused by the edge- or/and corner-sharing octahedral of lepidocrocite [γ-FeO(OH)], maghemite (γ-Fe_2O_3), goethite [α-FeO(OH)] and other compounds, with Fe center distances from 0.30 to 0.35 nm. In the presence of each inhibitor separately, the mild steel surface is either converted by the phosphate invasion into the crystal lattice, in case of phosphate solution in *Steel–WSS–Na_2HPO_4* (Figure 1b), or covered by the organic three-dimensional organic layer precipitated, in the case of MBT in *Steel–WSS–MBT* (Figure 1c). In mere phosphate solutions, the lack of reddish corrosion products and the development of a thinner and yellower final layer, as shown in Figure 1b (*Steel–WSS–Na_2HPO_4*) and Figure 1f (*Steel–NaCl–Na_2HPO_4*), are characteristic of the corrosion process due to the hindrance of hematite formation, as detailed below. Since the earlier stages and according to the literature, the inner layer (2D layer) is expected to consist of gray and blackish magnetite (Fe_3O_4) and maybe wustite (FeO). Supplementary to the electrochemical and spectroscopic measurement analyses that are discussed in detail below, the visual image of Figure 1d confirms the synergistic action of the inhibitors (*Steel–WSS–Na_2HPO_4–MBT*), overcoming the defects and enhancing the advantages of the preceded separate use of MBT and Na_2HPO_4. After the exposure tests, a compact, resistant, uniform, protective layer was created, appearing to have a light-colored saline coating in the outer and a metallic gray or blackish inner layer, which is the magnetite layer. At a first glance, the lack of brown, reddish, and yellowish hues in the corrosion products incline us toward the conception that the main and more time-consuming corrosion processes were restricted successfully, resulting in a resistant and adherent protective layer on mild steel, which was almost intact after 96 h of subjection in a slightly saline environment and even after drying.

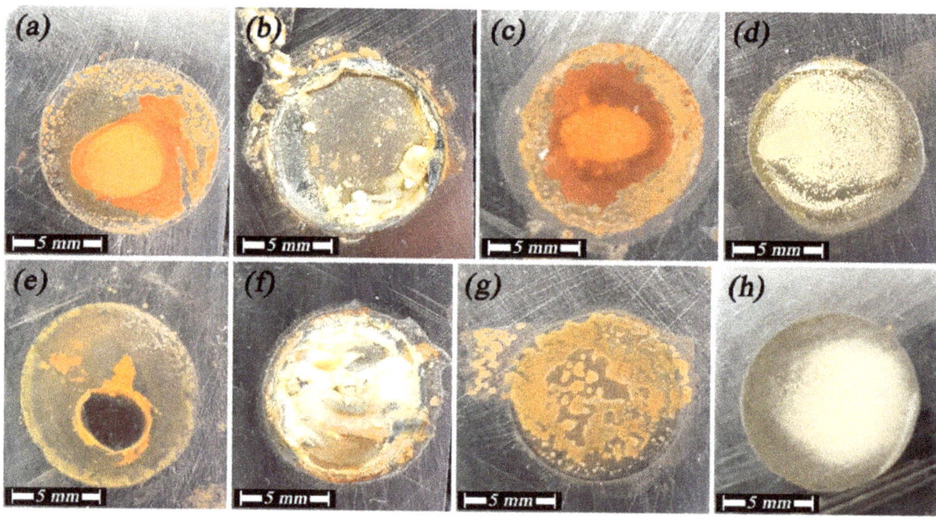

Figure 1. Visual images of mild steel after submission for 96 h: in WSS (**a**) *Steel–WSS–blank*, (**b**) *Steel–WSS–Na_2HPO_4*, (**c**) *Steel–WSS–MBT*, (**d**) *Steel–WSS–Na_2HPO_4–MBT*; in 3.5 wt % NaCl (**e**) *Steel–NaCl–blank*, (**f**) *Steel–NaCl–Na_2HPO_4*, (**g**) *Steel–NaCl–MBT*, (**h**) *Steel–NaCl–Na_2HPO_4–MBT*.

The mild steel visual pictures after exposure to 3.5 wt % NaCl for 96 h, in the presence or absence of inhibitors, are illustrated in Figure 1e–h. It can be seen that the surface of mild steel without the influence of inhibitors has lots of corrosion signs, as it is the bluish gray stain, indicating that uniform and local corrosion phenomena took place, with an accelerated rate regarding to the blank system (*Steel–NaCl–blank*, Figure 1e). On the other hand, corrosion products are also observed on the surface of both *Steel–NaCl–Na_2HPO_4* and *Steel–NaCl–MBT* (Figure 1f,g), reassuring that the chosen inhibitors are appropriate for the present system. Moreover, the surface of the sample with the presence of both inhibitors seems free from corrosion products in the outer area and a more compact, dark-colored adherent layer onto the metallic surface was developed, with brown-gray color, which was probably brown due to the organic tail of MBT and blackish/gray due to magnetite and/or wustite compounds, as will be discussed below (*Steel–NaCl–Na_2HPO_4–MBT*, Figure 1h) [45].

The SEM images of mild steel prior to exposure to electrolytes (*Steel–blank*) are illustrated in Figure 2. It can be seen that there are holes, dents, and scratches on the surface either due to the fabrication of the specimens or due to the cleaning process. Taking into account the EDS analysis, the iron value concentration was measured to be approximately 92.9 wt %, while the corresponding oxygen value was estimated very low, roughly 0.1 wt % indicating the absence of iron oxide products (Table 3).

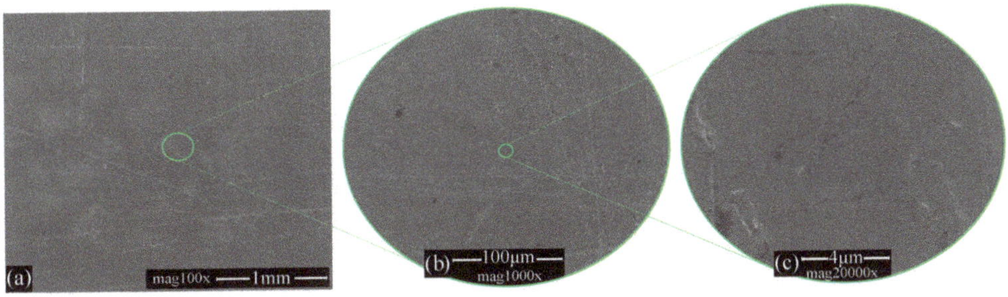

Figure 2. SEM images of mild steel prior exposure to electrolytes (*Steel–blank*).

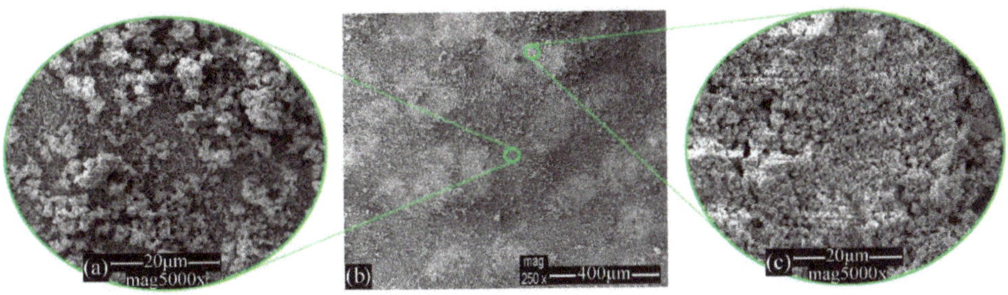

Figure 3. SEM images of mild steel immersed in WSS for 96 h in the absence of inhibitors.

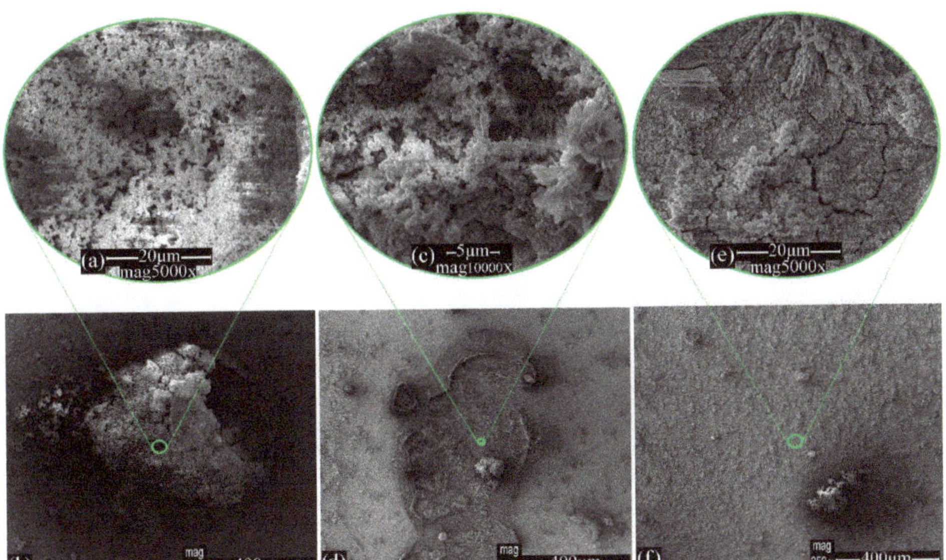

Figure 4. SEM images of mild steel immersed in WSS for 96 h in the presence of Na_2HPO_4.

Table 3. Tabulated values of wt % element concentration of mild steel panels prior and after subjection in either WSS, or 3.5 wt % NaCl for 96 h in the presence of inhibitors or not.

Sample	Spectrum at	wt % Element Concentration						
		O	Fe	C	Si	P	S	Na
Steel–blank	Figure 2a	0.1	92.9	3.7	2.0	-	-	-
Steel–WSS–blank	Figure 3b	52.3	44.0	1.5	2.3	-	-	-
Steel–WSS–Na_2HPO_4	Figure 4b	39.9	39.5	2.6	0.2	14.2	-	2.4
	Figure 4f	16.9	72.4	3.0	1.7	2.7	-	2.3
Steel–WSS–MBT	Figure 5e	44.6	46.1	4.2	1.2	-	3.6	-
	Figure 5f	44.4	46.9	4.3	1.5	-	2.9	-
Steel–WSS–Na_2HPO_4–MBT	Figure 6c	31.2	40.0	7.8	0.4	11.3	3.6	2.6
Steel–NaCl–Blank	Figure 7e	61.3	38.7	-	-	-	-	-
Steel–NaCl–Na_2HPO_4	Figure 8b	39.3	39.0	3.0	0.3	14.9	-	-
	Figure 8d	35.5	49.9	3.3	0.1	6.5	-	3.0
Steel–NaCl–MBT	Figure 9b	23.0	67.4	4.7	1.5	-	2.2	-
Steel–NaCl–Na_2HPO_4–MBT	Figure 10b	47.1	26.8	4.5	-	16.0	5.2	0.4

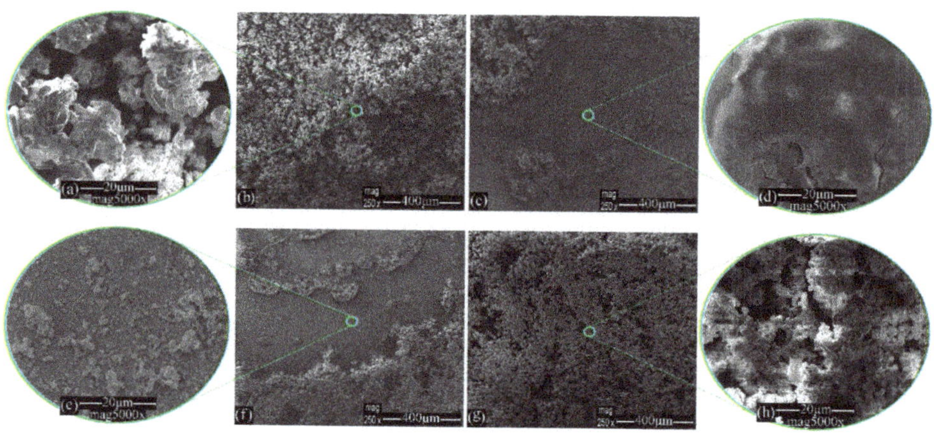

Figure 5. SEM images of mild steel immersed in WSS for 96 h in the presence of MBT.

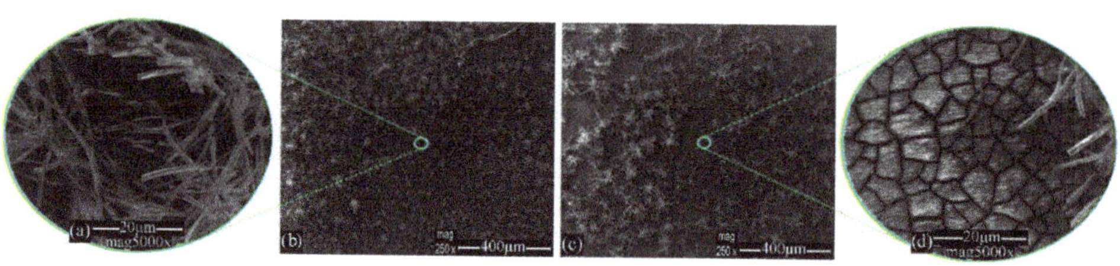

Figure 6. SEM images of mild steel immersed in WSS for 96 h in the presence of Na_2HPO_4 and 2-mecraptobenzothiazole (MBT).

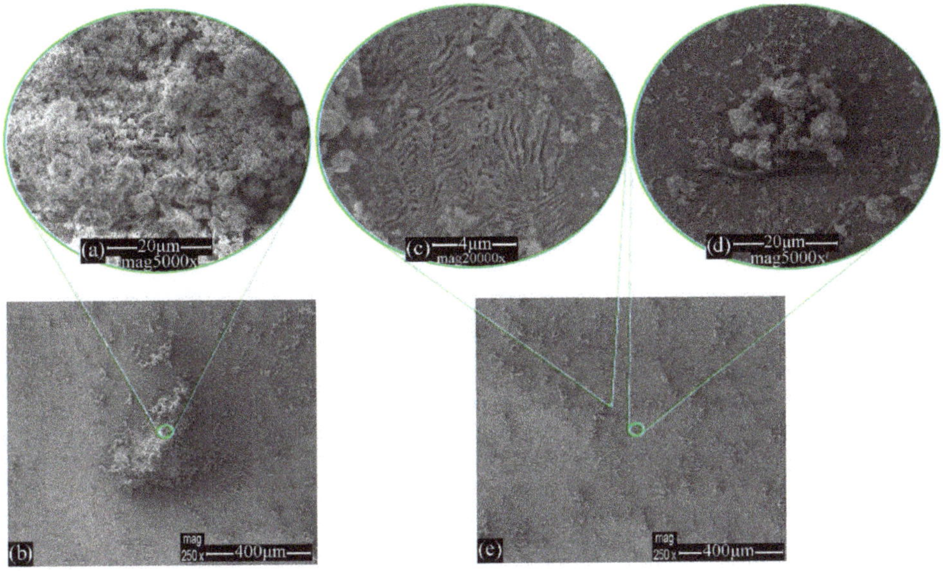

Figure 7. SEM images of mild steel immersed in 3.5 wt % NaCl for 96 h in the absence of inhibitors.

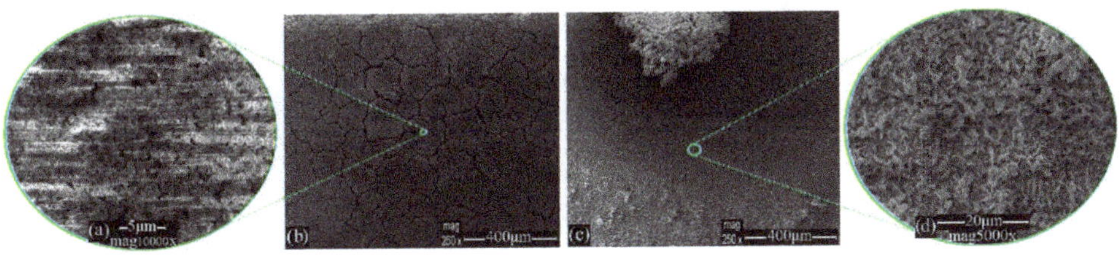

Figure 8. SEM images of mild steel immersed in 3.5 wt % NaCl for 96 h in the presence of Na_2HPO_4.

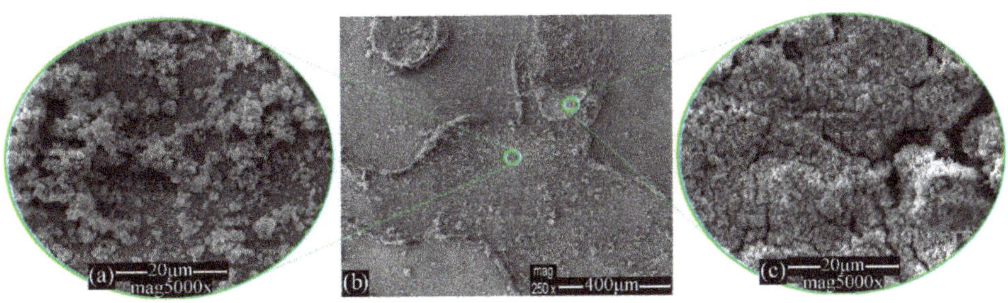

Figure 9. SEM images of mild steel immersed in 3.5 wt % NaCl for 96 h in the presence of MBT.

In general, the surface characterization of mild steel corrosion products remains a quite demanding issue, as denoted by previous works [46]. Various image analyzing techniques were deployed in order to assess the mild steel surface relating to its corrosion products structure and behavior after immersion in the studied inhibitor solutions.

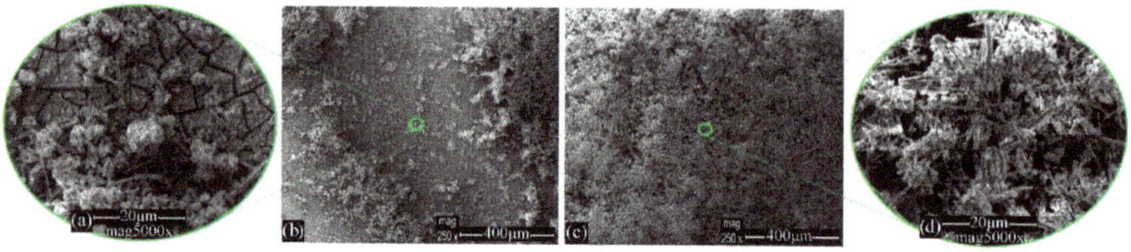

Figure 10. SEM images of mild steel immersed in 3.5 wt % NaCl for 96 h in the presence of Na_2HPO_4 + MBT.

In order to extract as much information as possible, it is a prerequisite to adopt a circumstantial model for the mild steel surface corrosion system during subjection in aquatic solutions, such as those proposed in the literature [7,47]. It is commonly accepted that the initial mixture of corrosion products consists of a primary viscous layer on a mild steel surface, to undergo afterwards gradation and separation into two distinguishable layers (an electrochemical double layer), an adherent rust layer, and a loosely adherent anodic oxide film [3,48]. At an early stage, amorphous masses of oxides with a spongy appearance are created, to transform primarily to lepidocrocite, part of which is further transformed in goethite. The studied phases have been generated either from the direct precipitation from a ferrous or a ferric solution, or by transformation of another iron oxide, at room temperature [3,49]. At a first glance at preceding investigations, Evan's model [50] regarding phase transition during wetting and drying conditions could contribute to illustrating the expended phase composition of a mild steel surface, after the prolonged exposure tests [51]. Consequently, during exposure, the reduction of lepidocrocite to a hydrated Fe^{2+}- intermediate takes place (phenomenon of olation) due to metal anodic dissolution, especially near the steel surface, where magnetite appears. However, oxidizing reactions can result in the emersion of lepidocrocite and maghemite phases [1,50,51] when oxygen circulation is evitable by drying [7].

The SEM images of mild steel immersed in WSS (*Steel–WSS–blank*) for 96 h, in the absence of inhibitors, are illustrated in Figure 3. Characteristic laminar morphologies of lepidocrocite were captured on the reference sample surface, exhibiting a large spectrum of more or less known crystalline formations, as it is denoted in the literature. Near the mild steel surface broken glass type [3], flowery and worm nest shapes dominate, in parallel with the gradual growth of miscellaneous laminar volumes of lepidocrocite formed radially outward from the metal surface. This particular formation is also described by many as cotton balls [3] and defined as crystalline phases due to the aggregation of flat grains of lepidocrocite grown on top of amorphous γ-$Fe_2O_3 \cdot H_2O$ [46,48]. The EDS analysis of Figure 3b discloses enhanced oxygen value concentration compared to that of the *Steel–blank* sample due to the creation of the aforementioned iron oxides (Table 3).

The surface SEM images of *Steel–WSS–*Na_2HPO_4 after subjection in WSS for 96 h are depicted in Figure 4. It can be seen that characteristic tassel-like structures, mainly consisting of [Fe^{II}-OH] compounds, with substituted hydroxyls by phosphate anions ($H_2PO_4^-$, HPO_4^{2-}, PO_4^{3-}) [52], reveal the phosphates incorporation, thus making possible the identification of iron phosphates, appearing in amorphous or crystal phases. The latter are often observed as filamentous agglomerates [53] (Figure 4a,b). However, phosphorous detection through microscopy techniques of EDS analysis of the surface has been considered harsh, as phosphate anions get incorporated in steel crystal lattice [26]. Furthermore, ordinary structures of lepidocrocite spongy structure (and probably parts of maghemite, a phase that usually adopts the shape of its precursor [54]) and smaller aggregations of magnetite (as more globular, smooth) seem to be present, while a compact continuous inner layer (with cracks from drying process) has been obviously generated, which is indicative of steel surface passivation by phosphates (Figure 4c,d). In addition, large platy crystals of nanocrystalline phases are captured, reassuring the creation of the expected thin adherent passive layer of iron phosphates, maghemite, and lepidocrocite in crystalline and amorphous formations [26,53] (Figure 4e,f). It should be taken into account that the special influence of phosphate anion is participating in various chemical paths. Although adsorbed phosphate is assumed to act as a precursor for hematite creation and often for goethite nucleation and growth, at different concentrations and especially at locally more basic areas, phosphate anions increase the negative electronic charge of ferrihydrite (present at earlier stages) between particles. Therefore, the formation of hematite retards or is limited [44]. Since the samples have undergone primary aging, in conjunction with the spectroscopic measurement results, it is considered that a part of the magnetite layer has decomposed in a binary solid-state transformation. At the same time, a structural conversion of magnetite, oxidizing partially to form maghemite and a compositional frontal transformation to hematite take place [55]. The choice of the physicochemical path that will be followed depends primarily on the anion concentration [56,57]. The EDS analysis of Figure 4b,f reveals elevated oxygen value concentrations compare to that of the sample *Steel–blank* due to the formation of the aforementioned iron oxides as well as the presence of phosphorus that is assigned to the adsorbed phosphates (Table 3).

Figure 5 pictures the surface SEM images of mild steel submitted to WSS for 96 h in the presence of MBT (*Steel–WSS–MBT*). It can be seen that hexagonal double-plates are illustrated, revealing the presence of magnetite [58], the precipitation of Fe^{2+} and Fe^{3+} ions from the solution, and the creation of green rust particles (general type $[Fe^{2+}{}_{(1-x)}Fe^{3+}{}_{x}(OH)_2]^{x+}$) (Figure 5a,b). Positively charged ions from corroded iron forms sediments of double-hydroxide layers, consisting of coordinated Fe^{2+} and Fe^{3+} octahedra with hydroxylate groups between them [48,59]. Green rust acts as an ion exchange surface withholding other species of metal ions and decreasing locally the negative charge between the particles of the corrosion products, conferring more stability to the final layer [54,59]. Moreover, unattached geode type formations are present, enhancing the assumption of magnetite creation with smaller crystals of lepidocrocite inside the cavity, as the former is mostly detected near the metallic surface as a blackish oxide [58], while the latter usually constitutes the biggest part of the total corrosion products above (Figure 5c,d). Additionally, the creation of primarily small amounts of akaganeite [β-FeO(OH)] is considered, whereas the precise identification of these phases was complicated using the SEM technique [46,47]. Taking into account the capture of cracks (Figure 5e,f), it may be remarked that a quite thick and compact inner layer was generated from the inhibitor (MBT) adsorption [29], in parallel with the conventional corrosion products of the mild steel, such as crystals of lepidocrocite and other aforementioned bulk structures of goethite, maghemite and lepidocrocite. The EDS analysis of Figure 5e,f discloses elevated oxygen value concentrations compared to that of the *Steel–blank* sample due to the formation of the aforementioned iron oxides as well as the presence of sulfur that is ascribed to the adsorbed organic MBT (Table 3).

The surface SEM images of mild steel immersed in WSS for 96 h in the presence of MBT together with Na_2HPO_4 (*Steel–WSS–Na_2HPO_4–MBT*) are demonstrated in Figure 6. Bulky corrosion products as tubular rods are predicting the creation of α and β phases of iron oxides, as well as more fine-grained phases of the expended corroded phases, as the interactivity of the mild steel with the inhibitors lessened the rate and the extent of the corrosion process (Figure 6a,b). The successful generation of a layer with cracks is probably due to the drying process, implying the conjunction of the organic inhibitor molecule (MBT) to a mild steel surface, as well as the formation of tertiary phosphates [26,60] (Figure 6c,d). The EDS analysis of Figure 6c reveals elevated oxygen value concentrations compared to that of the sample *Steel–blank* due to the creation of the aforementioned iron oxides along with the presence of sulfur and phosphorus elements that are attributed to the adsorbed organic MBT and inorganic phosphates, respectively (Table 3).

The surface SEM images of mild steel subjected to 3.5 wt % NaCl (*Steel–NaCl–blank*) for 96 h are illustrated in Figure 7. It can be seen that there is a surface distinctly damaged by the saline solution, with the characteristic appearance of an etched layer near the metal surface, as chloride anions accelerate corrosion phenomena (Figure 7a,b). In several areas, bigger bulk crystalline tubular or rod-like forms of goethite (often visually yellowish to reddish rust) or hematite (visually metallic gray rust) [54,61] and distinguishable cloud-like [3] lepidocrocite crystals are depicted (Figure 7c,e).

As commented in preceding studies, it seems that small amounts of goethite were formed after a significant exposure time [3] by lepidocrocite solid transformation due to the presence of chloride ions as well as by the transformation of green rust ($Fe^{II}{}_xFe_y{}^{III}(OH)_{3x+2y-z}(A^-)_z$ [54]) during the drying process (Figure 7d,e). The chloride ions, as aggressive anions, are considered [49] to accelerate the corrosion process, penetrating the protective layer and hence attacking the fresh unharmed metal surface beneath the layer, triggering pitting corrosion. Therefore, chloride anions may not be detected onto the corrosion product layer neither as ferrous ($FeCl_2$) nor as ferric ($FeCl_3$) salts, which are initially formed [47,62], since these salts are meant to be significantly soluble in water. The EDS analysis of Figure 7e reveals an elevated oxygen value concentration compared to that of the *Steel–blank* sample due to the formation of the previously mentioned iron oxides (Table 3).

The surface SEM images of *Steel–NaCl–Na$_2$HPO$_4$* after exposure in 3.5 wt % NaCl for 96 h are depicted in Figure 8. It seems that globular axially generated spongy agglomerations have constructed a quite homogenous voluminous layer, currently hindering the exposure of the inner layer on the metal surface. As mentioned before, these agglomerations are attributed to lepidocrocite, maghemite, and precursors of green rust (as magnetite) created in NaCl solution [15] and are not yet removed, due to the prolonged exposure process [3,63] (Figure 8a,b). Furthermore, miscellaneous laminas accreting in thorny three-dimensional structures [3] resemble hematite and goethite (sharper, more acicular) α phases with small aggregates of more globular lepidocrocite (or/and maghemite), which are precipitated during the submission together with tertiary phosphates [46,60,64] (Figure 8c,d). As mentioned before, cracks with diameters of a few micrometers are considered to be due to the drying stage (Figure 8b). The EDS analysis of Figure 8b,d reveals elevated oxygen value concentrations compare to that of the sample *Steel–blank* due to the formation of the aforementioned iron oxides as well as the presence of phosphorus, which is assigned to the adsorbed phosphates (Table 3).

The surface SEM images of mild steel immersed in 3.5 wt % NaCl for 96 h in the presence of MBT (*Steel–NaCl–MBT*) are demonstrated in Figure 9. Characteristic laminar and eroded rosettes of green rust, lepidocrocite, and hematite consist of the three-dimensional structures (Figure 9a), while the ruptured inner layer (Figure 9c) indicates the inhibitor (MBT) interaction with the steel surface [29,64,65]. The EDS analysis of Figure 9b reveals elevated oxygen value concentrations compared to that of the *Steel–blank* sample due to the creation of the previously mentioned iron oxides along with the presence of sulfur that is assigned to the adsorbed organic MBT (Table 3).

The surface SEM images of mild steel submitted to NaCl for 96 h in the presence of MBT together with Na_2HPO_4 (*Steel–NaCl–Na$_2$HPO$_4$–MBT*) are demonstrated in Figure 10. The accelerated corrosion test validates the aforementioned protective layer creation, in parallel with the emergence of both less and more fine-grained phases (tubular rods and laminar lepidocrocite and maghemite structures), attributing to the enhanced corroding environment in saline solution and to the prolonged submission process as well as to the formation of tertiary phosphates [26,60] (Figure 10). The EDS analysis of Figure 10b reveals elevated oxygen value concentrations compared to that of the *Steel–blank* sample due to the formation of the aforementioned iron oxides along with the presence of phosphorus and sulfur elements that are attributed to the adsorption of inorganic phosphates and organic MBT, respectively (Table 3).

3.2. *Raman Spectroscopy Analysis*

The Raman spectra of the mild steel panels after subjection to WSS for 96 h, in the presence of inhibitors or not, are depicted in Figure 11. The corresponding classification of the Raman spectra peaks is in Table S1. Regarding the *Steel–WSS–blank*, it can be mentioned that the region between 200 and 450 cm^{-1} of the Raman graph belongs to α, β, and γ-crystalline iron oxides and oxyhydroxides, accompanied with amorphous products, but is not considered as appropriate for the precise identification of the rust compounds (Figure 11a). Quite guardedly, it could be claimed that the pair of peaks near 225 and 290 cm^{-1} is assigned to the presence of hematite, the peak near 298 cm^{-1} is assigned to the presence of goethite, and that around 663 cm^{-1} is assigned to the presence of magnetite [66]. The pair of peaks with the shoulders at about 1320 and 1630 cm^{-1} is attributed to ferrihydrites, which is the main iron oxyhydroxide rust compound of the present system [7].

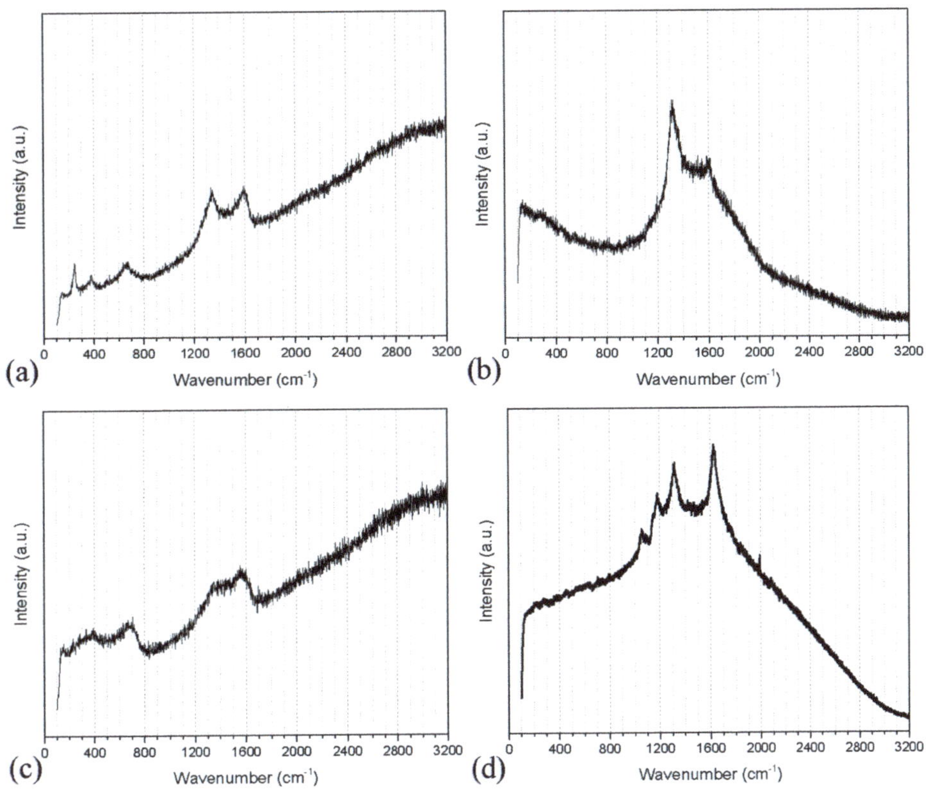

Figure 11. Raman spectra of the mild steel panels after subjection in WSS for 96 h: (**a**) *Steel–WSS–blank*, (**b**) *Steel–WSS–*Na_2HPO_4, (**c**) *Steel–WSS–MBT*, (**d**) *Steel–WSS–*Na_2HPO_4*–MBT*.

Taking into account the Raman spectrum of *Steel–WSS–*Na_2HPO_4 (Figure 11b), it may be remarked that except for the initial noisy fluctuations from iron oxyhydroxides and oxides [63,67] or either from Fe–O lattice vibrations [68], the most intense peak at about 1310 cm^{-1} arrives from iron phosphate species and may be addressed to PO_2 asymmetric stretch [69], whereas the less intense (accompanied with a shoulder) near 1600 cm^{-1} could be attributed to bending vibrations of water molecules [68] attached to the more hygroscopic phases.

Considering the *Steel–WSS–MBT*, it can be mentioned that the organic inhibitor adsorption on [Fe–O–OH] species mitigated their characteristic Raman shifts, assuming that the elevated experimental line in the region 200–400 cm^{-1} implies their presence under MBT (Figure 11c). The two most discrete peaks were captured near 1380 cm^{-1} and 1590 cm^{-1} with a neck between them, which was a shifted pattern that was commented previously to emerge due to ferrihydrites. Furthermore, it could be claimed that the pair of peaks at 220 and 290 cm^{-1} with the detection of the primarily lower peak near 700 cm^{-1} is related to pyrite characteristic Raman shift, meaning the identification of the Fe–S band [70]. Enhancing this option, it should be considered that after drying at specific areas, the organic voluminous MBT molecule would be detected, hiding the rest, more crystalline structures underneath, whereas there would be sites where the organic layer would have been partially detached.

In view of the *Steel–WSS–Na_2HPO_4–MBT* Raman spectrum, it can be said that despite the fact that the first peak at 1050 cm^{-1} seems to be attributed to bending vibrations of C–H bands of the MBT aromatic ring [65], its emergence in combination with the shoulder up to the next peak around 1185 cm^{-1} and the smaller peak at 1316 cm^{-1} reveal the identification of iron phosphate crystalline or amorphous compounds [69] (Figure 11d). The latter peak at 1628 cm^{-1} is believed to correspond to the thione mode of the MBT molecule, which is indicative of being perpendicular to the metal surface orientation of the aromatic ring [65].

The Raman spectra of the mild steel panels after exposure to 3.5 wt % NaCl for 96 h, in the presence of inhibitors or not, are illustrated in Figure 12. Regarding the *Steel–NaCl–blank*, it can be mentioned that the generated thick three-dimensional amorphous corrosion products and the most dense and stable generated passive layer on the mild steel measured surface are considered to hinder the detection of the Raman shifts of more crystalline phases beneath (Figure 12a). Hence, the initial expected peaks of iron oxides and oxyhydroxides [67] are distorted due to the voluminous amorphous compounds. An intense peak at 1319 cm^{-1} [63] is attributed to hematite (α-Fe_2O_3) and the less intense peak at 1593 cm^{-1} [46] is attributed to (2-line) ferrihydrite ($Fe_5HO_8{\cdot}4H_2O$), which makes its appearance between 1380 cm^{-1} and 1600 cm^{-1} [7].

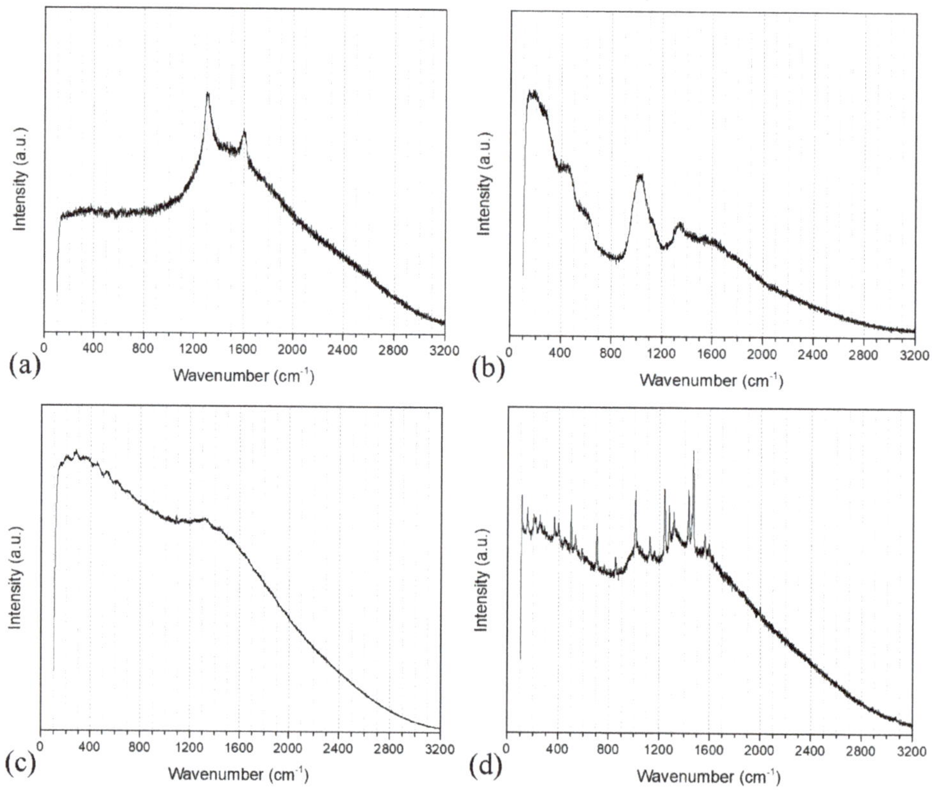

Figure 12. Raman spectra of the mild steel panels after subjection in 3.5 wt % NaCl for 96 h: (**a**) *Steel–NaCl–blank*, (**b**) *Steel–NaCl–Na_2HPO_4*, (**c**) *Steel–NaCl–MBT*, (**d**) *Steel–NaCl–Na_2HPO_4–MBT*.

Taking into account the Raman spectrum of *Steel–NaCl–Na_2HPO_4* (Figure 12b), it may be remarked that the absence of a peak near 218–221 cm^{-1}, which is attributed in the literature to the symmetric stretch of interlayer chloride ions in green rust [63,68] in combination with the detection of iron phosphate salts vibrations, at 430–450 cm^{-1} for $Fe_4(P_2O_7)_3$ and at 1011 cm^{-1} for a symmetric $PO_3{}^{2-}$ symmetric stretch of $Fe_3(PO_4)_2A$ (A for anion) [69], reassures mild steel surface phosphating and consequently the increase of crystallinity in corrosion products after exposure and drying processes.

Considering the *Steel–NaCl–MBT* (Figure 12c), it can be noticed that although low wavenumbers of Raman shift are connected with iron oxides as commented, the differentiation of mild steel spectrum enhances the surmise of MBT physical and chemical adsorption, which is in accordance with the absence of other characteristic peaks of MBT, as at 717 cm^{-1}, 1013 cm^{-1}, 1564 cm^{-1}, and 1593 cm^{-1}. The strongest peak near 278 cm^{-1} in conjunction with noisy, less intense peaks at 350 cm^{-1}, 470–480 cm^{-1}, 530–540 cm^{-1}, and 610 cm^{-1} reveal the generation of the Fe–S bond [71]. A less intense but relatively obvious peak obtained near 1320 cm^{-1} is connected to the N = C–S ring stretching mode, with its intensity considered in the literature as proportionate to the inhibitor chemisorption progress [71]. The peak at 1395 cm^{-1} is considered to stem from stretching vibrations of the C–C band in the aromatic ring of MBT [65].

In view of the *Steel–NaCl–Na_2HPO_4–MBT* (Figure 12d), it can be said that a more clarifying Raman spectrum with more than 15 Raman shift peaks was gained under accelerating corrosion conditions, as there are in the saline solution. The smaller peaks near 300 cm^{-1} (shoulder) and 400 cm^{-1} in addition to those near 500 cm^{-1} and 530 cm^{-1} are assumed to be caused by Fe–S generated by the chemisorption of MBT on the surface of the sample [71]. The peak that arises near 700 cm^{-1} may be attributed to P–O–P symmetric stretch vibrations, taking into account the simultaneous appearance of a smaller peak near 850 cm^{-1} generated by $[HPO_4]^{-2}$ anion symmetric stretch [54,63]. The next pair of Raman shifts near 1000 cm^{-1} (shoulder) and 1120 cm^{-1} comes to identify iron phosphate crystalline or amorphous phases [69] in conjunction with the combination of two other shifts near 1250 cm^{-1} and 1270 cm^{-1} and a smaller shift at 1310 cm^{-1}, the latter of which may have been gained from PO_2 asymmetric stretch. Moreover, the pair of 1430 cm^{-1} and 1480 cm^{-1} peaks in combination with those at 1580 cm^{-1} and 1590 cm^{-1} were noted to be in accordance with experimental Raman shift values for the adsorbed MBT molecule, especially when it is measured in its thione mode [65,72].

3.3. X-ray Diffraction Analysis

The XRD patterns of the mild steel panels after subjection in WSS for 96 h in the presence or absence of the evaluated corrosion inhibitors are depicted in Figure 13. The collected XRD spectra reveal that the exposure of the mild steel panels to WSS + Na_2HPO_4 leads to the detection of XRD diffraction peaks corresponding to [65-4899 Fe iron], [30-0662 $Fe_3(PO_4)_2$*$8H_2O$ vivianite, syn], [01-074-5846 $Fe_{0.96}P_{0.04}$ iron phosphide], goethite [α-FeO(OH)], and lepidocrocite [γ-FeO(OH)] [73] (Figure 13b). Moreover, the same diffraction peaks are detected via the exposure of mild steel panels to WSS in the presence of Na_2HPO_4 and MBT (Figure 13d). Furthermore, the exposure of mild steel panels to WSS in the absence or presence of MBT leads to the verification of XRD diffraction peaks that are attributed to [65-4899 Fe Iron], goethite [α-FeO(OH)], and lepidocrocite [γ-FeO(OH)] (Figure 13a,c).

The mild steel specimen surfaces after subjection in 3.5 wt % NaCl for 96 h were also analyzed by XRD to determine the crystal structure (Figure 14). XRD diffraction peaks corresponding to [65-4899 Fe iron], [30-0662 $Fe_3(PO_4)_2$*$8H_2O$ vivianite, syn], [01-074-5846 $Fe_{0.96}P_{0.04}$ iron phosphide], goethite [α-FeO(OH)], and lepidocrocite [γ-FeO(OH)] are ascertained after the exposure of mild steel panels to 3.5 wt % NaCl in the presence of either Na_2HPO_4 (Figure 14b) or Na_2HPO_4 along with MBT (Figure 14d) [73]. Finally, the submission of the mild steel panels in 3.5 wt % NaCl in the absence or presence of MBT leads to the verification of XRD diffraction peaks that are attributed to [65-4899 Fe iron], goethite [α-FeO(OH)], and lepidocrocite [γ-FeO(OH)] (Figure 14a,c). According to the literature [74], vivianite is a common and complicated breakdown product of the interaction of phosphate, iron, and H_2O.

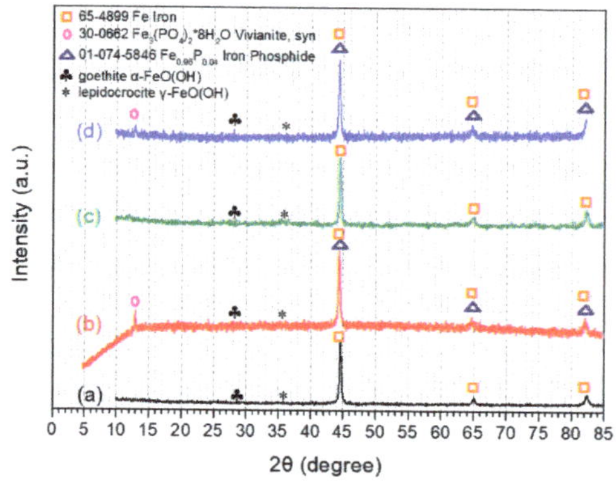

Figure 13. XRD spectra of the mild steel panels after submission in WSS for 96 h: (**a**) *Steel–WSS–blank*, (**b**) *Steel–WSS–*Na_2HPO_4, (**c**) *Steel–WSS–MBT*, (**d**) *Steel–WSS–*Na_2HPO_4*–MBT*.

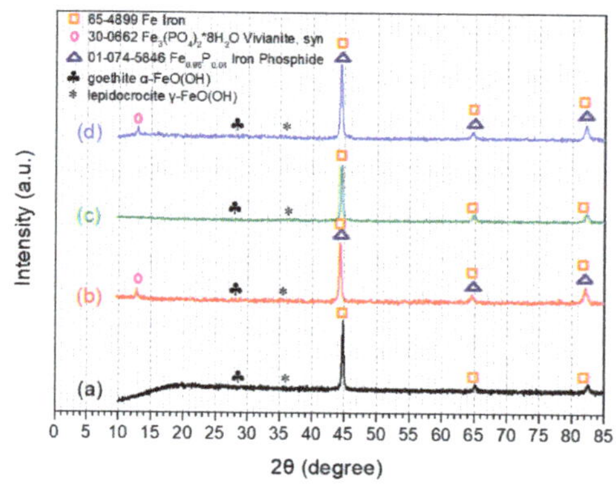

Figure 14. XRD spectra of the mild steel panels after submission in 3.5 wt % NaCl for 96 h: (**a**) *Steel–NaCl–blank*, (**b**) *Steel–NaCl–*Na_2HPO_4, (**c**) *Steel–NaCl–MBT*, (**d**) *Steel–NaCl–*Na_2HPO_4*–MBT*.

3.4. Electrochemical Studies

All the EIS, PP, and LPR electrochemical characterizations were performed in a pH range from 7.33 to 7.97 (Table 2). It should be taken into consideration that all the electrochemical measurements were conducted in triplicate, and a corresponding representative curve is illustrated for each sample.

In general, the presence of a compound into an electrolyte solution can shift the OCP either into positive values if the compound reacts as an anodic inhibitor retarding the oxidation process, or into negative values if the compound inhibits the cathodic reactions. The observation of the OCP values in respect of the time can provide information related to the stability of the electrolytic system in order for linear sweep voltammetry characterizations such as PP to be conducted.

Taking into account the evolution of OCP versus time in the case of mild steel panels after subjection in WSS for 23 h at room temperature (Figure 15), it is observed that the system including the mild steel together with the phosphates (*Steel–WSS–*Na_2HPO_4) shifts the OCP to more positive values after the fourth hour compared to the one of the *Steel–WSS–blank*. This indication clearly denotes that Na_2HPO_4 acts as an anodic inhibitor for mild steel, which is a result that is also confirmed by the literature [75]. On the other hand, for the system *Steel–WSS–MBT*, the OCP shifts to negative values, indicating that the presence of MBT inhibits the cathodic reactions [76]. Finally, the presence of both Na_2HPO_4 and MBT into the WSS (*Steel–WSS–*Na_2HPO_4*–MBT*) results in OCP values more negative than those of *Steel–WSS–blank* and more positive than those of *Steel–WSS–MBT*. This behavior can be assigned to the simultaneous inhibition of the cathodic reactions as well as to the formation of a protective layer that inhibits the anodic reactions. Moreover, it can be mentioned that during the beginning of exposure (0–2 h), the corrosion potential (OCP) is lower than in isolated inhibitor systems due to passivation, but as the exposure time elapses, the OCP increases up to an almost stable moderate value for many hours. This behavior is characteristic of the accelerated formation of metastable pits at first and the gradual passivation of all of them, up to the restoration of the passive layer and the return of the system to the initial OCP, which is a process known as re-passivation [77].

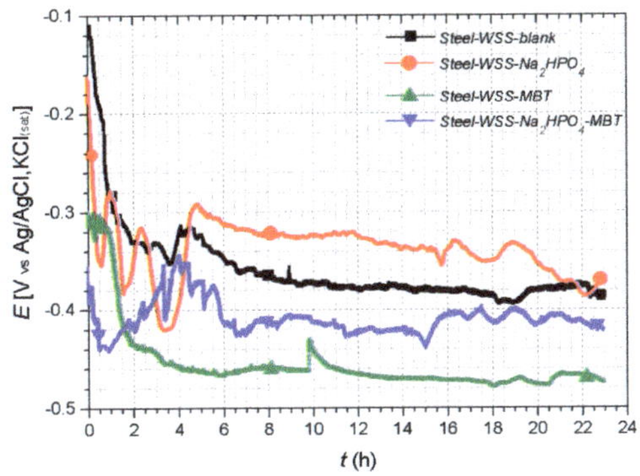

Figure 15. The evolution of open circuit potential (OCP) versus time in case of the mild steel panels after submission in WSS for 23 h: (■ **filled square**) *Steel–WSS–blank*, (● **filled circle**) *Steel–WSS–Na_2HPO_4*, (▲ **filled triangleD**) *Steel–WSS–MBT*, (▼ **filled triangleU**) *Steel–WSS–Na_2HPO_4–MBT*.

Considering the evolution of OCP versus time in the case of the mild steel panels after subjection in 3.5 wt % NaCl for 23 h at room temperature (Figure 16), it can be noticed that the presence of Na_2HPO_4 and MBT in the electrolytic solutions increases the OCP compared to the *Steel–NaCl–blank*. These facts denote that there is an inhibition to the anodic reactions resulting in the corrosion protection of the mild steel. The increase in potential is assigned to the creation of protective films onto the metal surface, making it more resistant to corrosion than the panel subjected to the solution without inhibitor additives. Moreover, it may be remarked that MBT acts as a mixed inhibitor for mild steel in NaCl solutions [78].

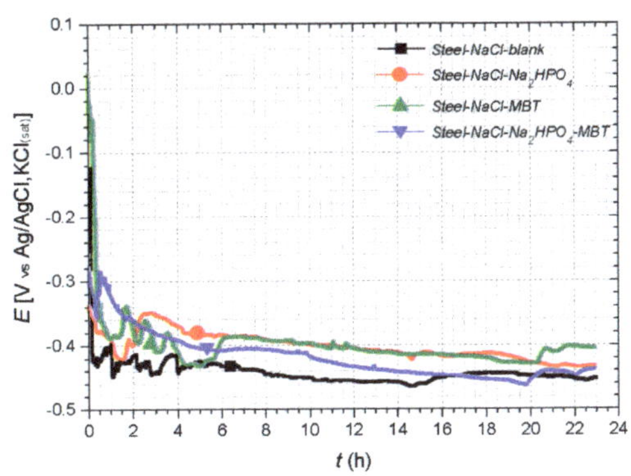

Figure 16. The evolution of OCP versus time in case of the mild steel panels after submission in 3.5 wt % NaCl for 23 h: (■ **filled square**) *Steel–NaCl–blank*, (● **filled circle**) *Steel–NaCl–Na_2HPO_4*, (▲ **filled triangleD**) *Steel–NaCl–MBT*, (▼ **filled triangleU**) *Steel–NaCl–Na_2HPO_4–MBT*.

The polarization curves recorded for mild steel panels in the presence or absence of inhibitors after subjection in WSS for 24 h are demonstrated in Figure 17. Considering the cathodic branches of the diagram (Figure 17a), it can be seen that the presence of MBT into the solution (*Steel–WSS–MBT*) results to the appearance of lower cathodic current densities compared to the *Steel–WSS–blank*, indicating a possible reduction of the cathodic reactions. On the other hand, according to the same diagram, there is no evidence that the other two systems (*Steel–WSS–*Na_2HPO_4, *Steel–WSS–*Na_2HPO_4*–MBT*) affect the cathodic reactions. Taking into account the anodic branches of the polarization curves (Figure 17b), it may be remarked that the addition of Na_2HPO_4 into the commensurate solution retards the anodic reactions effectively, since the corresponding anodic current densities are decreased compared to those of the *Steel–WSS–blank*. On the contrary, there is no indication that the *Steel–WSS–MBT* and *Steel–WSS–*Na_2HPO_4*–MBT* systems influence the anodic reactions.

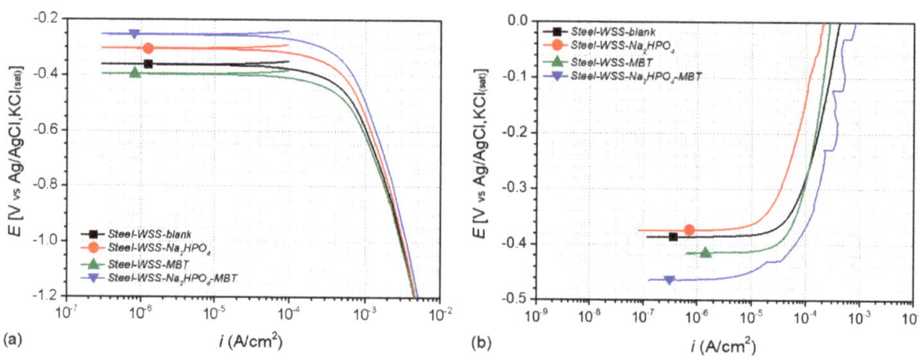

Figure 17. The polarization curves (**a**) cathodic, (**b**) anodic, recorded for mild steel panels after submission in WSS for 24 h: (■ **filled square**) *Steel–WSS–blank*, (● **filled circle**) *Steel–WSS–*Na_2HPO_4, (▲ **filled triangleD**) *Steel–WSS–MBT*, (▼ **filled triangleU**) *Steel–WSS–*Na_2HPO_4*–MBT*.

Furthermore, studying the cathodic branches of the polarization curves recorded for mild steel panels in the presence or absence of inhibitors after subjection in 3.5 wt % NaCl for 24 h, it is mentioned that the addition of Na_2HPO_4 either alone or in combination with MBT affected the cathodic reactions by lowering the corresponding cathodic current densities (Figure 18a). Therefore, there is an indication of a possible reduction of the cathodic reactions for both the *Steel–NaCl–*Na_2HPO_4, and *Steel–NaCl–*Na_2HPO_4*–MBT* systems. Considering the anodic branches, it could be documented that the introduction of the inhibitors suppresses the anodic reactions because of the reduction of the correlative anodic corrosion current densities (Figure 18b). This outcome can be attributed to the formation of corresponding protective layers.

Taking into account the Pourbaix diagram of iron [79], it may be remarked that in all the PP characterizations, reductions of both water and oxygen take place because the PP measurements were conducted in a pH range from 7.33 to 7.97 (Table 2) as well as the cathodic and anodic branches of polarization curves being recorded in a potential range between −1.6 V and 0 V versus Ag/AgCl,$KCl_{(sat)}$ electrode. Furthermore, in the aforementioned ranges of pH and potential, dissolution of iron exists but, be that as it may, the oxidation of iron does not involve the oxidation of water. In addition, it should be taken into consideration that the PP characterization technique was used only for qualitative interpretations, because the calculation of both the corrosion current (i_{corr}) and R_p values via the linear fitting of the polarization curves according to the Tafel method is not feasible, due to the existence of two reduction processes in the system: the reduction of water and the reduction of oxygen [80]. Moreover, the corrosion current density cannot be determined from the anodic branches, as there is no clear linear region. Finally, under the aforementioned experimental conditions, steel is also corroded via localized corrosion with the formation of pits [81,82]. Therefore, the Tafel method is not appropriate to be used for localized corrosion, such as pitting corrosion, as it only yields an average uniform corrosion rate [83].

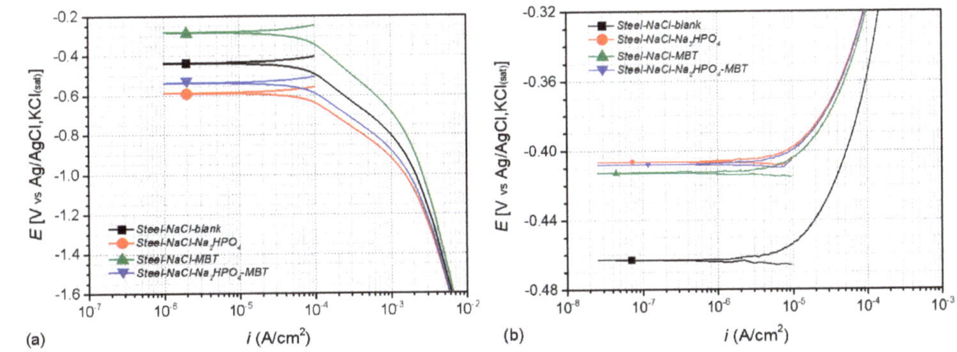

Figure 18. The polarization curves (**a**) cathodic and (**b**) anodic, recorded for mild steel panels after submission in 3.5 wt % NaCl for 24 h: (■ **filled square**) *Steel–NaCl–blank*, (● **filled circle**) *Steel–NaCl–Na$_2$HPO$_4$*, (▲ **filled triangleD**) *Steel–NaCl–MBT*, (▼ **filled triangleU**) *Steel–NaCl–Na$_2$HPO$_4$–MBT*.

Regarding the LPR measurements, it can be noted that all the mild steel panels were exposed to either WSS or 3.5 wt % NaCl for 96 h, in the presence of inhibitors or not. Bearing in mind the obtained R_p values (Table 4), it can be assumed that the presence of both *Na$_2$HPO$_4$* and *MBT* in the electrolytic solutions has as an outcome enhanced corrosion protection of mild steel, as the *Steel–WSS–Na$_2$HPO$_4$–MBT* and *Steel–NaCl–Na$_2$HPO$_4$–MBT* systems exhibit R_p values higher compared to either the systems without inhibitors (*Steel–WSS–blank* and *Steel–NaCl–blank*) or to those systems including only one inhibitor.

Table 4. Tabulated values of E_{OC} [V vs. Ag/AgCl,KCl$_{(sat)}$] obtained from the potentiodynamic polarization (PP) technique and of R_p estimated via the linear polarization resistance (LPR) method acquired for all mild steel panels after subjection in either WSS or 3.5 wt % NaCl in the presence of inhibitors or not.

Sample	E_{OC} (V) (Cathodic Branch) (24 h Exposure)	E_{OC} (V) (Anodic Branch) (24 h Exposure)	R_p (Kohm cm^2) (96 h Exposure)
Steel–WSS–blank	−0.361	−0.386	6.279
Steel–WSS–Na$_2$HPO$_4$	−0.303	−0.374	2.852
Steel–WSS–MBT	−0.394	−0.415	4.731
Steel–WSS–Na$_2$HPO$_4$–MBT	−0.253	−0.463	6.730
Steel–NaCl–blank	−0.435	−0.462	2.822
Steel–NaCl–Na$_2$HPO$_4$	−0.586	−0.406	1.800
Steel–NaCl–MBT	−0.282	−0.412	3.132
Steel–NaCl–Na$_2$HPO$_4$–MBT	−0.536	−0.408	4.054

The EIS plots of mild steel panels in the presence or absence of inhibitors after subjection in WSS are demonstrated in Figure 19. Taking into account the EIS Bode curves after 24 h of exposure (Figure 19a), it can be seen that the *Steel–WSS–Na$_2$HPO$_4$* and *Steel–WSS–Na$_2$HPO$_4$–MBT* systems illustrate higher EIS moduli compared to the one of the *Steel–WSS–blank*, indicating that the presence of the Na_2HPO_4 inhibitor alone or together with MBT into the corrosive environment increases the mild steel corrosion protection. On the contrary, the addition of an MBT compound into the WSS solution does not seem to improve the corrosion protection of the mild steel. Regarding the EIS Bode curves obtained after 96 h of exposure to WSS (Figure 19b), the same results are observed denoting that the introduction of the Na_2HPO_4 inhibitor alone or together with MBT in to the WSS solution protects the mild steel from corrosion.

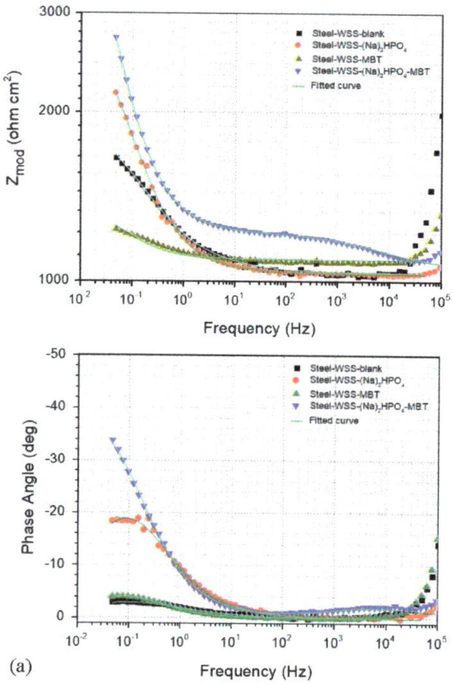

Figure 19. Electrochemical impedance spectroscopy (EIS) plots recorded for mild steel panels after submission in WSS for (**a**) 24 h and (**b**) 96 h: (**■ filled square**) *Steel–WSS–blank*, (**● filled circle**) *Steel–WSS–Na_2HPO_4*, (**▲ filled triangleD**) *Steel–WSS–MBT*, (**▼ filled triangleU**) *Steel–WSS–Na_2HPO_4–MBT*, (—) Fitted curve.

Considering the EIS plots of mild steel panels in the presence or absence of inhibitors after subjection in 3.5 wt % NaCl (Figure 20), it is clearly seen that after 24 h of exposure (Figure 20a), the system *Steel–NaCl–Na_2HPO_4–MBT* illustrates the highest EIS modulus in the low-frequency range, denoting that the presence of the Na_2HPO_4 inhibitor in accordance with MBT into the corrosive environment increases the mild steel corrosion protection. On the other hand, the insertion of either Na_2HPO_4 or MBT compounds into the 3.5 wt % NaCl electrolytic solution does not remarkably improve the corrosion protection of the mild steel. In view of the EIS Bode curves acquired after 96 h of exposure to 3.5 wt % NaCl (Figure 20b), it may be remarked that the addition of Na_2HPO_4 inhibitor alone or together with *MBT* into the electrolytic solution protects the mild steel from corrosion.

The EIS Bode plots for both the untreated mild steel and the systems including MBT (*Steel–WSS–MBT*, *Steel–NaCl–MBT*) after exposure to either WSS or 3.5 wt % NaCl for 24 h can be described with one time constant in the middle–low-frequency range that is attributed to the corrosion process (Figures 19 and 20). However, as corrosion progresses from 24 h to 96 h, the relative relaxation process moves to lower frequencies because of the increase of the corrosion active area and the corresponding double-layer capacitance enhancement.

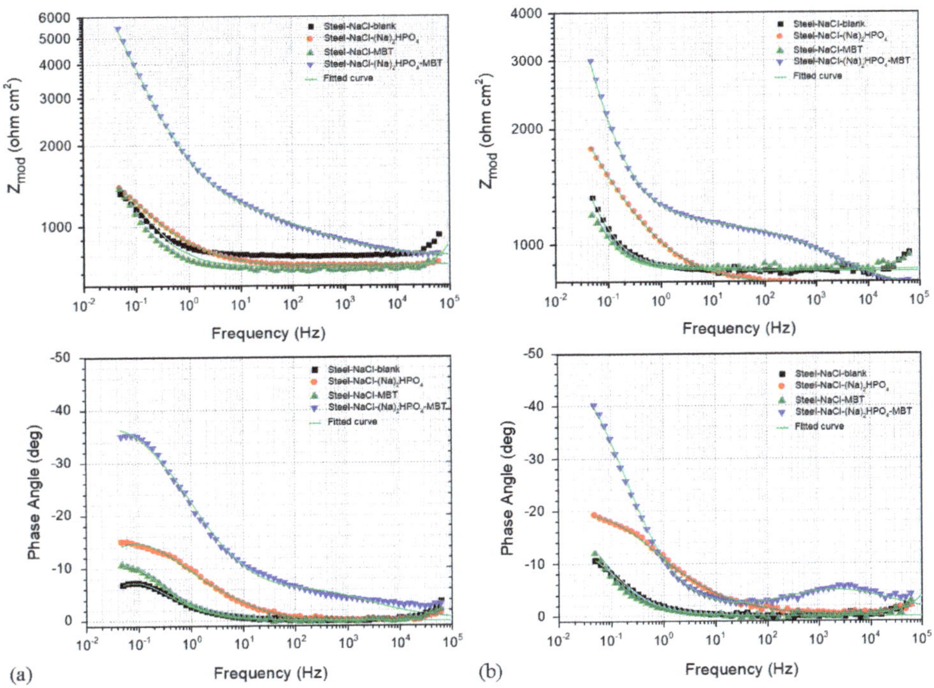

Figure 20. EIS plots recorded for mild steel panels after submission in 3.5 wt % NaCl for (**a**) 24 h and (**b**) 96 h: (■ **filled square**) *Steel–NaCl–blank*, (● **filled circle**) *Steel–NaCl–Na_2HPO_4*, (▲ **filled triangleD**) *Steel–NaCl–MBT*, (▼ **filled triangleU**) *Steel–NaCl–Na_2HPO_4–MBT*, (–) Fitted curve.

On the other hand, estimating the EIS Bode plots for the systems including both Na_2HPO_4–MBT or only Na_2HPO_4 in either WSS or 3.5 wt % NaCl for 24 h, it can be noted that two time constants are illustrated: one in the high–middle frequency range that is ascribed to a film formation between the mild steel surface and the phosphate compounds, and a second one in the middle–low-frequency range that is assigned to corrosion process (Figures 19 and 20). As the exposure time elapses from 24 h to 96 h, the time constant assigned to corrosion process is shifted to lower frequencies due to the enhancement of the corrosion area and the corresponding double-layer capacitance increment.

The interpretation of the obtained EIS results for the mild steel panels after subjection in the electrolytic solutions in the absence or presence of corrosion inhibitors was conducted by numerical fitting using the equivalent circuits depicted in Figure 21. According to these equivalent circuits, the pure capacitors have been replaced by constant phase elements (CPE). This adaptation is mandatory when the phase shift of a capacitor is different from −90° [84]. The impedance of an R-CPE parallel connection is given by:

$$Z_{R\text{-}CPE} = \frac{R}{1 + RY_0(j\omega)^n}, \tag{3}$$

where Y_0 is the admittance of the CPE and n is the CPE exponent. The n takes the value of 1 when it corresponds to a capacitor, it is between $0.5 < n < 1$ when a non-ideal capacitor behavior occurs, n equals 0.5 when the CPE corresponds to a Warburg impedance describing a diffusion process, and finally, n equals 0 when it represents a resistor [85]. Using the CPE together with the Cole–Cole approach [83,86], the capacitance can be calculated from the fittings by:

$$C = \sqrt[n]{\frac{RY_0}{R^n}}. \tag{4}$$

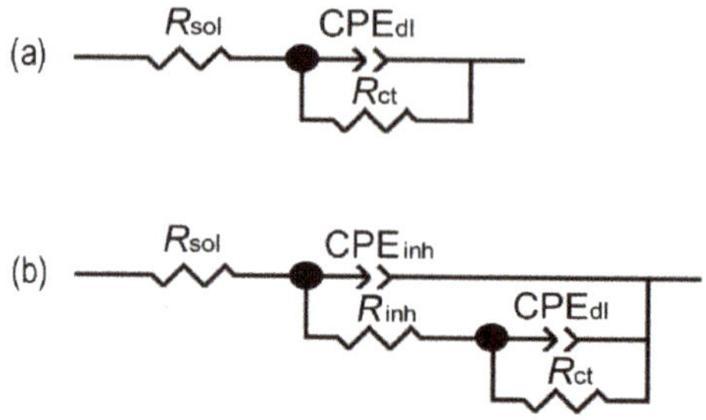

Figure 21. Electrochemical impedance spectroscopy (EIS) equivalent circuits used for numerical simulation of the EIS data obtained for the mild steel panels after submission in the electrolytic solutions in the absence or presence of corrosion inhibitors: (**a**) equivalent circuit with three elements, (**b**) equivalent circuit with five elements.

The equivalent circuit for both the untreated mild steel and the systems including MBT (*Steel–WSS–MBT*, *Steel–NaCl–MBT*) after exposure to either WSS or 3.5 wt % NaCl includes three components and corresponds to an EIS spectrum comprising of one time constant. In particular, it consists of the solution resistance (R_{sol}), the charge transfer resistance together with a double-layer CPE (R_{ct}-CPE_{dl}) that are attributed to the presence of corrosion active pits either at the mild steel surface or at the metal–inhibitor interface, as shown in Figure 21a.

Regarding the systems including both Na_2HPO_4–MBT or only Na_2HPO_4 in either WSS or 3.5 wt % NaCl (*Steel–WSS–*Na_2HPO_4, *Steel–WSS–*Na_2HPO_4*–MBT*, *Steel–NaCl–*Na_2HPO_4, *Steel–NaCl–*Na_2HPO_4*–MBT*), the relative equivalent circuit consists of five components and corresponds to an EIS spectra of two relaxation times. More specifically, it includes the resistance of the solution (R_{sol}), the pore resistance, and the CPE of the inhibitor layer (R_{inh}-CPE_{inh}) in parallel connection, which are attributed to the response of the electrolyte inside the pores of the layer, and finally, the charge transfer resistance and the double-layer CPE (R_{ct}-CPE_{dl}) that are due to the presence of the corrosion process.

Taking into account the aforementioned Equation (4), the C_{inh}, C_{int}, and C_{dl} parameters were calculated. In principal, the C_{dl} value is influenced by the roughness of the metal surface [38,39]. The determined values of the EIS fitting parameters for the WSS and 3.5 wt % NaCl electrolytic solutions are tabulated in Tables S2 and S3, respectively. However, taking into account the CPE_{inh}-P values [CPE_{inh}-P represents the *n* which is the CPE exponent in Equations (3) and (4)] calculated for the *Steel–WSS–*Na_2HPO_4*–MBT* (0.50334) and *Steel–WSS–*Na_2HPO_4 (0.50486) samples after exposure to WSS solution for 96 h, it can be mentioned that these elements cannot be associated to a non-ideal capacitor (Table S2) [84]. Similar assessments can be reported for the CPE_{inh}-P values determined for the *Steel–NaCl–*Na_2HPO_4*–MBT* sample after exposure to 3.5 wt % NaCl solution for 24 h (0.5009), 48 h (0.50201), 72 h (0.50471), and 96 h (0.50474), as well as for the *Steel–NaCl–*Na_2HPO_4 sample after exposure to 3.5 wt % NaCl solution for 96 h (0.50913) (Table S3). Nevertheless, it should be mentioned that the introduction of a Warburg impedance component did not improve the numerical fitting of the EIS results. Moreover, it should be taken into consideration that Tables S2 and S3 do not demonstrate the mean values of the EIS fitting parameters. Finally, considering the obtained R_p and R_{ct} values of Table 4 and Tables S2 and S3, it can be stated that both LPR and EIS methods confirm that the systems including both Na_2HPO_4 and *MBT* inhibitors exhibit the best anticorrosive behavior after 96 h of exposure to either WSS or 3.5 wt % NaCl electrolytic solutions.

The evolution of the mean values together with the corresponding error bars of the EIS fitting parameters R_{ct}, C_{dl} of the mild steel panels after their exposure to WSS solution for 96 h in the presence or absence of corrosion inhibitors is depicted in Figure 22. Concerning the determined values, it can be mentioned that the system including both the inhibitors Na_2HPO_4 and MBT exhibits the highest R_{ct} values for all the exposure time intervals. It is important to notice that high R_{ct} values indicate enhanced resistance to the corrosion process [87]. On the other hand, the lowest low C_{dl} values are observed for the system containing the inhibitors Na_2HPO_4, revealing a low corrosion reaction area [87]. Furthermore, small variations are discerned in the C_{dl} of this system, indicating that the corrosion reaction area does not undergo changes. Consequently, the inhibitor Na_2HPO_4 either alone or in combination with MBT provides the best protection to mild steel against corrosion in WSS solution (Table S2).

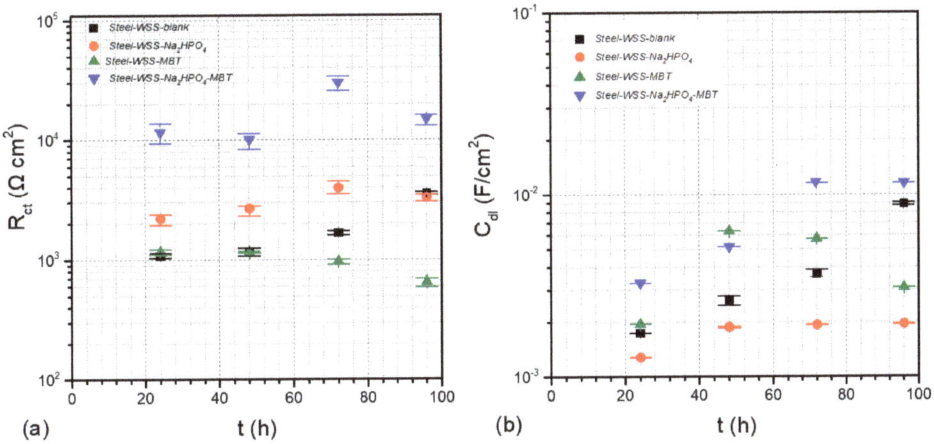

Figure 22. The evolution of the mean values of the EIS fitting parameters as a function of time for mild steel panels after submission in WSS: (**a**) R_{ct}, (**b**) C_{dl}.

Analyzing the evolution of the EIS fitting parameters R_{ct}, C_{dl} of the mild steel panels after their subjection in 3.5 wt % NaCl solution for 96 h in the presence or absence of corrosion inhibitors (Figure 23), it can be noticed that that the system including both the inhibitors Na_2HPO_4 and MBT reveals the highest R_{ct} values together with the lowest C_{dl} values for all the exposure time intervals. Therefore, the combination of both inhibitors Na_2HPO_4 and MBT performs the best protection to mild steel against corrosion in 3.5 wt % NaCl solution (Table S3). Furthermore, it appears that all the systems including inhibitors exhibit higher R_{ct} values compared to the *Steel–NaCl–blank*, indicating effective corrosion protection to mild steel.

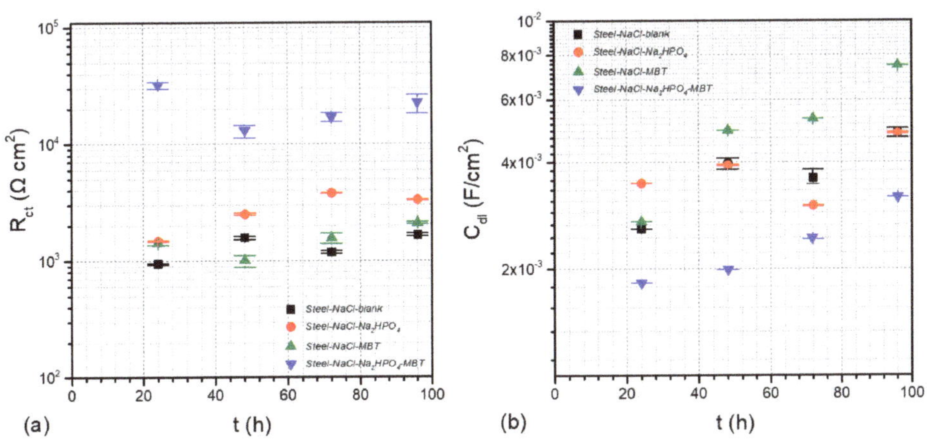

Figure 23. The evolution of the mean values of the EIS fitting parameters as a function of time for mild steel panels after submission in 3.5 wt % NaCl: (**a**) R_{ct}, (**b**) C_{dl}.

The tabulated values of $\eta(\%)$ derived from Equation (1) after 24, 48, 72, and 96 h of mild steel samples submission either in WWS solution or into 3.5 wt % NaCl solution, in the presence or absence of inhibitor, are illustrated in Tables S2 and S3, respectively. The results reveal that for all the time intervals, the highest values of $\eta(\%)$ appear for the systems including both inhibitors Na_2HPO_4 and MBT, denoting enhanced corrosion inhibition to mild steel. Furthermore, according to the obtained $\eta(\%)$ values, it may be remarked that the system containing the Na_2HPO_4 provides also improved protection to mild steel against corrosion. Finally, it should be noted that for the *Steel–WSS–MBT* sample, the $\eta(\%)$ values could be calculated for 48, 72, and 96 h since the corresponding R_{ct} values are lower than those of the *Steel–WSS–blank* sample. The same result was revealed for the *Steel–WSS–*Na_2HPO_4 sample after 96 h of subjection in WSS solution as well as for the *Steel–NaCl–MBT* after 48 h of exposure to 3.5 wt % NaCl solution. These outcomes clearly denote that the presence of the inhibitor MBT into the WWS solution cannot provide adequate corrosion protection to mild steel after 48 h.

3.5. Corrosion Inhibition Mechanism

Taking into account the aforementioned obtained results, the following corrosion inhibition mechanism can be claimed that is in accordance with the dry–wet–dry model, which is illustrated by the equal model of Evans (differential aeration corrosion cell), considering the four-stage model of Fontana and Greene [77].

3.5.1. The Main Pathway of Corrosion Process in an Fe–C/NaCl $_{(aq)}$ System

Taking into account the mild steel submission in the saline solution, a milestone for the initiation of corrosion could be considered the time when the first ferrous ions (Fe^{2+}) or ferric ions (Fe^{3+}) ions release (Reactions 5 and 6), defining the anodic site that emerges near the metallic surface. Subsequently, the local charge density increases, causing the appearance of the cathodic current, and taking into account that the pH of the electrolytic solutions was in a range from 7.33 to 7.97, the dissolved O_2 reacts with H_2O and becomes reduced to OH^- ions (Reaction 7). In the same way, as in the case of mild steel, the alloying elements release Mn^{2+}, Cu^{2+}, and other cations, originating from the dissolution of the alloy equally to Fe oxidation and maintaining the stoichiometric ratio they used to have in the alloy. The consequent excess either of Fe^{2+} or Fe^{3+} and OH^- of the main oxidation–reduction reactions (8, 9) results in the precipitation of iron hydroxides from the solution to the corroded surface.

$$Fe \rightarrow Fe^{2+} + 2e^-, \text{ anode} \quad (5)$$

$$Fe \rightarrow Fe^{3+} + 3e^-, \text{ anode} \quad (6)$$

$$O_2 + 2H_2O + 4e^- \rightarrow 4OH^-, \text{cathode} \quad (7)$$

$$Fe^{+2} + 2OH^- \rightarrow Fe(OH)_2(s), \quad (8)$$

$$Fe^{+3} + 3OH^- \rightarrow Fe(OH)_3(s), \quad (9)$$

The obtained $Fe(OH)_2(s)$ and $Fe(OH)_3(s)$ create a thick gel, either precipitating from the solution or created very close to the metallic surface where oxygen concentration is very low to negligible. This thick gel is characteristic of wet rapid corrosion [88], which represent an appropriate precursor for the first types of iron oxides and oxide–hydroxides. Moreover, the visualization of the primary redox processes arises as an initial thin adherent layer on the steel surface, which is empirically metallic gray or black in color. This inner layer consists mainly of Fe_3O_4 (magnetite) and γ-Fe_2O_3 (maghemite) and/or γ-FeO(OH) (lepidocrocite), insoluble Fe(III) solid phases, as was also commented in spectroscopic measurements. On the other hand, the bulk layer (of the overall electrochemical double layer) consists of (i) oxides from the alloying elements dissolution such as heavy metal oxides, i.e., Mn_2O_7, (ii) interstitial complex phases, i.e., the known ferrites and ferrates (mixtures chiefly of iron oxides and other metallic oxides) [89], and (iii) various intermediates stemming from olation and corrosion phenomena, local physicochemical conditions (pH, ions concentration etc.) and physical parameters such as the humidity or aeration level of the surface.

Consequently, charge separation takes place and signals the start of the second phase of the corrosion mechanism on a mild steel surface, according to the Evans model perspective. As mentioned, in anodic areas, $Fe(OH)_2$ and $Fe(OH)_3$ create a characteristic thick gel that restricts the diffusion of the products of reduction semi-reactions such as O_2 and OH^-, which are usually responsible for the rust layer deterioration at a later stage. Additionally, Cl^-, which originates in the $NaCl_{(aq)}$ solution and the soluble forming Fe–Cl salts ionization (reactions 10, 11), since they are more reactive than OH^-, they tend to attack the protective layer, reacting with the agile Fe^{2+} of the anodic dissolution, triggering a series of local corrosion phenomena, which would open the way even to reach and hazard the inner layer (reactions 12, 13). For example, in wet conditions, the chemical reactions (5) and (14) occur, indicatively:

$$Fe^{+2} + 2Cl^- \rightarrow FeCl_2, \tag{10}$$

$$Fe^{+3} + 3Cl^- \rightarrow FeCl_3, \tag{11}$$

$$FeCl_2 + 2H_2O\ ^- \rightarrow Fe(OH)_2(s) + 2HCl, \tag{12}$$

$$FeCl_3 + 3H_2O \rightarrow Fe(OH)_3(s) + 3HCl, \tag{13}$$

$$8FeOOH + Fe^{2+} + 2e^- \rightarrow 3Fe_3O_4 + 4H_2O, \tag{14}$$

After a relative stabilization of the primary olation processes and iron oxides and hydroxides interaction, the eventually generated [Fe–O–OH] thick gel, which is chemically attached to the inner layer of corrosion products, obtains a unique for each system constitution of crystalline iron oxide phases. This constitution depends on the Fe^{2+}, Fe^{3+}, and dissolved O_2, OH^- availability, the mixture of which is responsible for the emergence of a pioneering family of materials, the layered double hydroxides (LDH) [90,91], with green rust as the main representatives in literature [92,93]. Taking into account that the direct oxidation of iron or low-alloyed steel to the unstable ferric ions (Fe^{3+}) is restricted thermodynamically against the generation of ferrous ions (Fe^{2+}), it could be assumed that the concentration of the former would be found to be negligible, referring to the whole rust layer structure.

In particular, the identification of green rusts (LDH structures) in the obtained corrosion products denotes that ferric ion creation is favored. Moreover, as the forming Fe–Cl salts were expected to have been dissolved, and were considered totally soluble in the present aqueous system, the chlorides detection in the rust layer indicates that sulfate green rust structures could have captured chlorides [94–96]. After a relative stabilization of the double layer, with the inner layer to consist mainly of magnetite (Fe_3O_4), mass transfer, electromigration and more time-consuming phenomena take place, depending on the local physicochemical conditions and compounds concentrations. In the mere aqueous saline system of mild steel, it was expected that the Cl^- would attack and damage the magnetite passive film, triggering a series of local corrosion phenomena, as analyzed in the aforementioned paragraph. Hence, as referred also in spectroscopic measurements, Cl^- were not expected to be detected in solid phases, as they are very soluble in the current system [97].

Essentially, the generated Fe^{2+} and e^- from the mild steel dissolution in anode diffuse through the inner layer (or magnetite layer) and provoke the cathodic precipitation of magnetite, reducing the continuously forming oxyhydroxides according to reaction (9) and conferring a particular thickness to the bulk rust layer. In case of aeration or in the presence of a sufficient quantity of dissolved oxygen, for the wet condition, water and oxygen molecules pass through the FeO(OH) outer layer and oxidize magnetite, forming again stacking oxyhydroxides and interstitial species mainly of [Fe, O, OH], bonding with oxygen bridges and hydrogen bonds in a near-neutral environment (reaction (15)).

$$3Fe_3O_4 + 3/4O_2 + 9/2H_2O \rightarrow 9FeOOH, \tag{15}$$

A more acceptable approximation for the latter would be the formation of Fe·OH·OH, which is more stable [44]. A significant variable is the surface charge of the outer FeO(OH) layer, depending on the protonation of Fe·OH·OH for a wet near-neutral environment, and deprotonation of the Fe·OH·OH layer in more acidic conditions. In the conventional natural mild steel saline system, the more time-consuming mechanisms and solid-state transformations are considered to proceed evermore, due to the thermodynamic equilibriums of each particular molecular interaction (aging). Characteristic processes of that stage are the further dehydration and crystallization of the oxyhydroxides and of the $Fe(OH)_3$, as previously observed even in wet conditions, underlying the outer rust layer. Additionally, complicated structures of crystalline flowers, honeycombs, nests, rods, and other unique formations emerge, as described elaborately in our SEM analysis, which were generated axially with regard to the metallic surface, as part of a family of processes that is responsible for illustrating the rust layer constitution for the most of the mild steel corrosion investigations in the past.

3.5.2. Corrosion Inhibitors

According to the literature, two main corrosion inhibition mechanisms for mild steel protection have proved to be more reliable [77]. The first one is through chemisoption onto the mild steel surface, limiting the metal dissolution rate by forming an adherent passive film, as MBT does for lightly acidic and near-neutral aqueous systems [78]. The second one is by the formation of a new phase onto the metallic surface, creating a conversion coating, as during the phosphating process.

Mercapto functional azole compounds present enhanced inhibitive efficiency to mild steel. The coexistence of sulfur and nitrogen atoms in the heterocyclic ring of the inhibitor molecule provides increased binding capability for the inhibitor attachment [98,99]; the aromatic ring of the molecule is known for its contribution to increase electronic density [100], whilst the higher electronegativity of sulfur than that of nitrogen renders the thiol group an especially efficient electron donor [29,99]. Particularly, according to the works of Chen et al. [29] and Obot et al. [101], it can be claimed that of the two tautomers of MBT, thiol (MBT-thiol) or thione (MBT-thione) (Figure 24a), coexisting usually in ionic solutions and ambient temperatures, the second seems to be more stable, especially in solid state, as in rusts. The latter group of researchers mentioned by the means of chemical hardness measurements that MBT-thione has increased electronegativity compared to MBT-thiol, which confers to the former a denser electronic cloud around the functional mercapto group of the molecule, and hence enhanced inhibition ability. Considering that electronegativity is proportionate to the chemisorption of the inhibitor and therefore to the level of its inhibition ability [100], they found that the heterocyclic ring could bond to the metallic surface through the endocyclic nitrogen, as the most electronegative of the three more possible donors. Furthermore, and at the same time, the heterocyclic ring could also bond with the exocyclic sulfur, chelating in that manner the iron cations and attaching, with its planar aromatic ring parallel to the metallic surface.

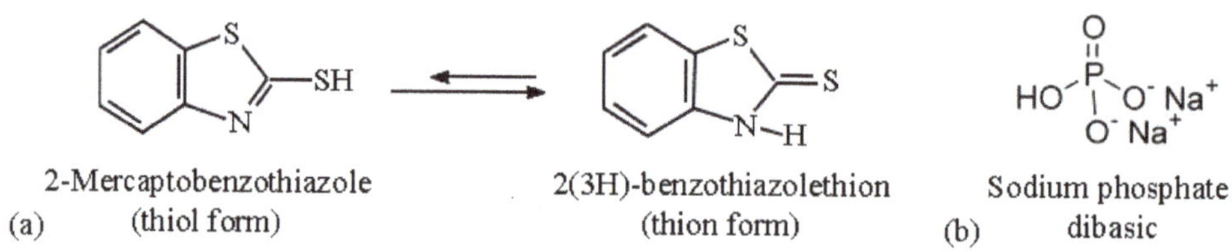

Figure 24. Schematic representation of (**a**) 2-mercaptobenzothiazole (thiol form) and 2(3H)-benzothaizolethion (thion form); (**b**) sodium phosphate dibasic [Na_2HPO_4].

In our study, the most electronegative positions would be the two atoms of sulfur. In spite of the electrostatic interaction preference, due to steric hindrance, the more stable coordination geometry of the MBT-thione with mild steel was commented to realize by means of the exocyclic sulfur, with the aromatic plane perpendicular to the metallic surface. The two tautomeric types of MBT usually coexist in circumneutral aquatic and ionic solutions [102], but be that as it may, a supplementary clue in favor of the thione mode, as predominant in the rust layer, emerges from a characteristic Raman shift in the *Steel–WSS–*Na_2HPO_4*–MBT* sample.

Taking into account that MBT-thione inhibition efficiency upon time depends on the rate of the attachment of the exocyclic sulfur to mild steel, two stages of the inhibitor adsorption were distinguished [29]. The first stage is meant to be a physisorption process of the organic hydrolyzed molecule to the solid metallic surface, leading to the creation of a primary monomolecular layer. Physical binding of the inhibitor needs relatively low quantities of exchangeable energy, takes place to a limited extent as a front reaction, and is the prominent and necessary process for the inhibitor 's attaching and hence to initiate the adsorption of the inhibitor [99]. The second stage, chemisorption, is also quickly initiated, as Fe–S chemical bonding is considered to be prompt and more powerful. The two adsorption processes were identified as similar systems to MBT-thione and iron by Obot et al. in 2018. However, the duration of each stage and even of the whole adsorption process and the enhanced stability and uniformity of the final layer imply that a more complicated process had actualized. It was found that MBT has the tendency to form hydrophobic bonds in neutral aquatic electrolytic solutions [103], followed by a self-assembly ordering with characteristic duration [104]. In case of parallel plane geometry to the metallic surface, hydrophobic-originating complexes create locally hydrophobic complexes as metal islands that combine afterwards to a uniform protective layer. In perpendicular orientation, similar mechanisms with hydrophobic intermolecular connections were suggested [65].

Despite the appearing complexity of the adsorption process of MBT, its kinetics were already specified and recorded in the literature, as also the self-assembly process through which the inhibitor lays the protective layer [29,65]. At a second glance, a precise orientation of the planar organic inhibitor could be similar to that proposed in the study of Venkataramanan et al. on the Ag surface [72]. Due to the increased local electronic density and the negative δ contribution on the thiol edge, in conjunction with a commonly observed attitude of S to alter the planarity of MBT-thione [100,105] and bend the S–C bond [103], it seems as the most probable scenario that the inhibitor attaches onto a mild steel surface, with the exocyclic sulfur–carbon bond perpendicular to metal and the remainder planar molecule angled [65]. In that way, the imminent rearrangement process, which initiates the self-assembly of the protective layer, would lack steric hindrance, as each molecule-plane could be simultaneously strongly chemically bonded onto the metal through an exocyclic sulfur atom. On the other hand, the slightly negatively charged remainder planes would be free to rotate, providing a relevant ionic permeability to the protective layer and diminishing the possibility for the layer rupture.

Phosphorus consists of a significant alloying element of low-alloyed and carbon steels against atmospheric corrosion [106] and Cl^- attack [107], and it is considered to improve machinability, fatigue, and hardenability. Phosphating through immersion baths, as conventionally realized in industry, is considered to be complete after several days, mainly due to recrystallization and solid-state transformation processes and in order to enhance surface coherence with metallic coatings, as zinc phosphating [53] and even with organic protective coatings. There has been a variety of chiefly organic substances proposed as accelerators of phosphating, such as certain imines, pyridines, and other oxidizing substances [108]. Phosphate anions have been considered to confer satisfying corrosion resistance, stabilizing both crystalline and amorphous phases in the corrosion product layer [109]. Although inorganic phosphorus-containing compounds were broadly utilized against the corrosion and wear phenomena of mild steel, and its crystallization kinetics and pathways were intricately investigated in ambient conditions [110], there are impenetrable aspects to explore up to date, from the interatomic to hypercrystallic level. Not only does phosphorus bind strongly to common non-metals such as oxygen and nitrogen, it also does so with metals and minerals consisting of oligomeric or near-polymeric complexes, such as ferrates and polyoxometallates. In phosphate tetrahedra, the phosphorus is strongly bonded to oxygen atoms, and their bond length could found to be even fourfold inside the same solid (bond–stretch isomerism) [59,110]. In particular, inorganic phosphate tetrahedra (mainly PO_4^{3-}, HPO_4^{2-}) behave as relatively strong Lewis bases, maintain their bonding in aquatic near-neutral solutions, and precipitate forming insoluble salts with the locally available cations (Figure 24b). In ambient conditions, they coordinate in hypercrystallic structures and minerals through tetrahedra or octahedra (or connected trioctahedra by others [111], linking by corners or edge sharing that form chains, clusters, and sheets [112–114].

In the aquatic system of our study, equally to the increase in local electronegativity and valence of the cations, phosphates tend to form the most thermodynamically stable hypercrystallic structures, with an ascending affinity to cyclic or/and branched complexes, with the latter to be achieved either through electrostatic attraction to small cations, as Cl^-, SO_4^{2-}, CO_3^{2-} or with ligand exchange, as with solid crystalline phases and metastable iron aqua ions [indicatively $FeOH^{2+}$, $Fe(OH)_4^-$, $Fe(OH)^{2+}$, and $Fe(H_2O)_6^{2+}$], which were formed through the hydration, olation, or oxolation of ferrous and ferric species [109,113,115,116]. It is established that transition metals can exchange a mere electron, whereas non-metal reactive species, as dioxygen and radical species, are restricted by Pauli's principle, reacting in a slower way [109]. Specifically, between 5 and 9 pH values, the instantly forming ferric salts hydrolyze immediately, while ferrous salts, in the absence of oxidizing substances, react with water, resulting in aqua ions generation, as $Fe(H_2O)_6^{2+}$.

Generally, ferric complexes tend to condense from very acidic (near pH 1, room temperature) up to near neutral solutions, conferring interstitial oligomeric or/and polymeric Fe(III) compounds that are named as ferrihydrites [117], while ferrous complexes condense only above pH 6 and in oxygen deprivation onto the hardly accessible metal surface, where they are hydrolyzed, to precipitate as $Fe(OH)_2$. A chain of reactions occur in the meanwhile, as the olation of the dimeric $[Fe_2(OH)_8(H_2O)_8]^{2+}$ complex to planar tetrameric $[Fe_4(OH)_8(H_2O)_8]^0$, and the following rapid nucleation of the mentioned tetrahedral and/or octahedral interstitial formations, simultaneously at many sites of each emerging plane, initiating a process of crystalline epitaxial growth for the generation of the layered double hydroxides. The ferrous phases, either in solution or in solid state, have expressed a propensity to oxidation, from which mixed ferric-ferrous phases are considered to be formed, known as green rust, magnetite, goethite and lepidocrocite. In the literature, two physicochemical pathways for phosphorus bonding are distinguished. A fraction of it is rapidly adsorbed and triggers the conversion process of mild steel surface, whereas most of the phosphorus is gradually diffusing into the crystalline iron oxide and hydroxide compounds of the inner layer, as in case of goethite, either into the pores or into the crystal lattice [118].

It was reported that at pH 7, the transformation of $Fe(OH)_2$ to lepidocrocite is a characteristic reaction, in conjunction with a noted elevation of the potential [109], whereas the hydrolysis and then the aggregation of ferrihydrites often results in the crystallization of goethite and hematite, which is a process chiefly depending on the oxolation of metastable [Fe,OH,O] species [107,117,119]. Phosphate anions are considered to have a unique coordination effect in the corrosion process by executing selective adsorption in different phases [21,107]. Ascending to the microscale, phosphates seem to play a binary role in ferrous corrosion products, either by being adsorbed into iron oxides and oxyhydroxides, as into goethite and lepidocrocite crystal lattices, with the creation of bidentate surface complexes by replacing two hydroxyls—their main ion competitor in aquatic solutions [118,120–123]—or by binding on mild steel surface molecules and forming phosphate and iron phosphate crystalline structures, creating new active sites for further phosphorus adsorption [122]. After all, it could be claimed that the dominant structures of the double-layer compounds are those of iron oxides and hydroxides, whereas into the inner layer of mild steel, hybrid crystalline and amorphous combinations of $[Fe,[O_2],H_2O,HPO_4^{2-},PO_4^{3-}]$ compounds are commonly solidified, providing a unique composition and behavior for each layer.

The investigation of ambient aquatic circumneutral systems of corroding mild steel revealed that a variety of crystalline iron–phosphate products were to be detected, interstitial of the different stoichiometry and oxidation states of iron, during the successive alteration of local conditions, from vivianite (ferrous phosphate), $Fe_3(PO_4)_2{\cdot}8H_2O$, to strengite (ferric), $FePO_4{\cdot}8H_2O$ (reactions 16,

17) [74,124,125]. The final constitution of the converted coating is meant to emerge by the balancing of the competitive processes of ionic precipitation, solid-state transformations, and crystallization, in conjunction with the local reagent concentrations and aeration level. In accordance with the experimental results, we assume that phosphate anions converted successfully onto the steel surface, participating in green rust stabilization, preventing GRI(Cl^-) oxidation [107] and fine-grained crystalline well-ordered structures in the final rust layer.

$$H_3PO_4 \rightarrow H_2PO_4^- + H^+ \rightarrow HPO_4^{2-} + 2H^+ \rightarrow PO_4^{3-} + 3H^+ \quad (16)$$

$$3Fe^{2+} + 2PO_4^{3-} + 8H_2O \rightarrow Fe_3(PO_4)_2{\cdot}8H_2O(s) \quad (17)$$

3.5.3. The Synergistic Effect

Table 5 presents the calculated values of the synergistic parameter (S_i) using Equation (2). In general, the synergistic effect can be described as the trend, whereas the combination of two or more compounds results in greater action than of those of the individual compounds. Regarding the corrosion inhibition, the synergistic effect refers to the effective interaction of two or more substances in respect of the protection of a material against corrosion.

Table 5. Tabulated values of the synergistic parameter (S_i) calculated using Equation (2).

	Time of Exposure (h)	Synergistic Parameter (S_i)
Steel–WSS–Na$_2$HPO$_4$–MBT	24	4.76
Steel–NaCl–Na$_2$HPO$_4$–MBT	24	22.6
Steel–WSS–Na$_2$HPO$_4$–MBT	96	1.99
Steel–NaCl–Na$_2$HPO$_4$–MBT	96	4.22

The synergistic effect can be considered as an effective method to enhance the inhibitive properties of a compound corrosive media. A value of (S_i) >1 indicates the existence of inhibition synergism between the two compounds, whereas (S_i) = 1 means that no interaction between the two compounds exists; when (S_i) < 1, it denotes toward an antagonistic effect. The synergistic effect of the corrosion inhibition behavior of MBT and Na_2HPO_4 in a molar ratio of 1:1 revealed that the admixture performed effectively with an inhibition efficiency above 90%, R_{ct} up to 6.7 kohm cm^2 and with a synergistic parameter above 22. The synergistic effect of the corrosion inhibition behavior of MBT and Na_2HPO_4 is increased in the presence of NaCl due to the Cl^- interaction. In synergism with the organic inhibitor, phosphate was capable of converting appropriately the corroding surface and in parallel to enhancing the local positive electronic cloud, above the inner layer, encountering the negative abundance of the organic voluminous inhibitor, accelerating and boosting in that way the corrosion inhibitive action of both substances. By the means of electrochemical, spectroscopic, microscopic, and image analyzing techniques of characterization, it is excluded that the acceleration and enhancement of the highly efficient Na_2HPO_4 inhibition was achieved, in combination with the mitigation of the disadvantages of MBT, especially in near-neutral aquatic solutions of mild and low-alloyed steel. The synergistic effect resulted in the formation of more compact and more durable film, which significantly retards the corrosion process.

Surface hardening was achieved as well as a considerable increase in the coherence of corrosion product layer, by developing microstructures with ameliorated resistance against the wear effects of mild steel at the micro and nanoscale, originating from fine-grained phases with powerful bonding not only inside the various chemical species but also between them [124]. Achieving the creation of α-crystals, as goethite and hematite, and the prolonged protection of the converted magnetite layer, it seems reasonable that we have the most thermodynamically stable phases due to the face-centered cubic (fcc)and hexagonal close-packed (hcp)systems, which are the crystalline structures with the highest density, and due to their connectivity provided by phosphorus [126]. The unique chemical connection of the phosphated mild steel surface, by means of the electronegative, non-polar –SH edge of the organic planar part of the layer, is able to rotate, conferring to the bonding flexibility, as it facilitates ions mobility, which confers the most powerful chemical bonding to the metal (Figure 25). The complex interacts with the outer iron oxides and hydroxides, thus maintaining relatively electrical surface neutrality. Moreover, it was intervened not only in interatomic but also in intermolecular interactions, with the final layer to have significantly increased resistance to corrosive factors, encountering both mass transfer and electrochemical attacking phenomena. The rust layer's outer surface is considered to have a chemical composition consisting of green rusts, FeO(OH), and Fe_xO compounds, binding with hydrogen bonds and oxygen bridges; therefore, the point of zero charge of them fluctuates, especially for the α-crystals, between iron oxides and oxyhydroxides, from 7.5 to 9.38 at 22 °C. That confers to the layer to have a dynamic response to the increase of the pH locally, for a specific quantity i.e., of a basic compound and to function as an ion trap for the metallic surface [111,124,127].

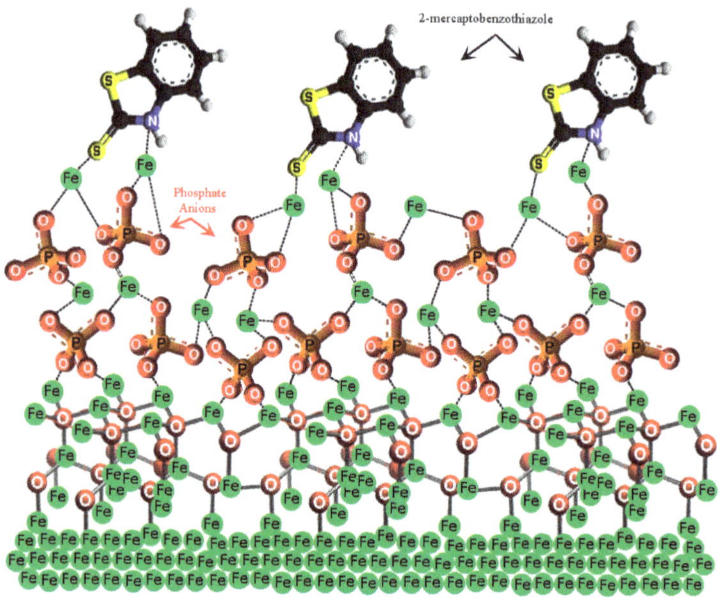

Figure 25. Schematic representation of the phosphate anions and MBT connections onto the mild steel surface, resulting in the formation of the protective film.

4. Conclusions

The electrochemical characterizations demonstrated that MBT and Na_2HPO_4 can be considered as corrosion inhibitors of steel. The initial mixture of corrosion products constitutes a primary viscous layer on the mild steel surface, which afterwards undergoes gradation and is separated into two distinguishable layers: an adherent rust layer and a loosely adherent anodic oxide film. At an early stage, amorphous masses of oxides with spongy appearance are created that transform chiefly to lepidocrocite, part of which is further transformed in goethite. During submission, the reduction of lepidocrocite to hydrated Fe^{2+} intermediate takes place due to metal anodic dissolution, especially near the steel surface, where magnetite appears. The chloride ions are considered to accelerate the corrosion process as aggressive anions, penetrating the protective layer and hence attacking the fresh unharmed metal surface beneath the layer, triggering pitting corrosion.

Adsorbed phosphate is assumed to act as a precursor for hematite creation and often for goethite nucleation and growth; at different concentrations and especially at locally more basic areas, phosphate anions increase the negative electronic charge of ferrihydrite (present at earlier stages) between particles, and therefore, the formation of hematite retards or is limited. The interaction of ions or neutral molecules at the electrical double layer changes its properties and structures. The water molecules pre-adsorbed at the metal surface in contact with the aqueous solution are involved in the successive adsorption processes. A thin multimolecular adsorption layer via the self-assembly process was formed (phosphonate layer formation) due to the intermolecular interaction between phosphonate groups. The inhibition mechanism is anodic type, hindering the active iron dissolution to a large extent.

Since MBT contains polar groups with an atom of nitrogen and sulfur, the inhibiting properties were determined by the electron density at the reaction center. With an increase in the electron density at the reaction center, the chemisorption bonds between the MBT and the metal are strengthened. Therefore, MBT is adsorbed on the surface of mild steel, formatting stable produced self-assembled monolayers of MBT compounds on the surface of mild steel, protecting from aggressive ions. The inhibitive action of MBT takes place through the adsorption of its molecules onto the mild steel surface via the lone pair of electrons of the N and S atoms. The presence of both aforementioned inhibitors into the corrosive environment exhibited the highest impedance modulus and R_{ct} value as the exposure time elapsed. The synergistic effect of the corrosion inhibition behavior of MBT and Na_2HPO_4 in a molar ratio of 1:1 revealed that the admixture performed effectively with inhibition efficiency above 90% and with a synergistic parameter above 22. The synergistic effect of the corrosion inhibition behavior of MBT and Na_2HPO_4 is increased in the presence of NaCl due to the Cl^- interaction. Considering the electrochemical, spectroscopy, and morphology characterizations, the corrosion protection mechanisms of steel can be attributed to the protective layers formed onto the metal surface due to the presence of the inhibitors, which prevent chloride's penetration.

Supplementary Materials: The following are available online at http://www.mdpi.com/2076-3417/10/1/290/s1, Table S1. Classification of the Raman spectra peaks. Table S2. Fitting parameters according to equivalent circuits of Figure 21 obtained for the mild steel immersed in WSS solution. Table S3. Fitting parameters according to equivalent circuits of Figure 21 obtained for the mild steel immersed 3.5 wt % NaCl solution.

Author Contributions: I.A.K., P.S., and C.A.C. conceived and designed the experiments. P.S. and I.A.K. conducted the experiments. I.A.K., P.S., and E.K.K. performed the characterization and evaluation of the data. I.A.K., P.S., and C.A.C. discussed the data and wrote the paper. All authors have read and agreed to the published version of the manuscript.

Funding: This research received no external funding.

Conflicts of Interest: The authors declare no conflict of interest.

References

1. Evans, U.R.; Taylor, C.A.J. Mechanism of atmospheric rusting. *Corros. Sci.* **1972**, *12*, 227–246. [CrossRef]
2. Smith William, F.; Hashemi, J.; Wang, S.-H. *Foundations of Materials Science and Engineering*, 4th ed.; McGraw Hill: New York, NY, USA, 2006.
3. Raman, A.; Nasrazadani, S.; Sharma, L. Morphology of rust phases formed on weathering steels in various laboratory corrosion tests. *Metallography* **1989**, *22*, 79–96. [CrossRef]
4. Chen, Y.Y.; Tzeng, H.J.; Wei, L.I.; Wang, L.H.; Oung, J.C.; Shih, H.C. Corrosion resistance and mechanical properties of low-alloy steels under atmospheric conditions. *Corros. Sci.* **2005**, *47*, 1001–1021. [CrossRef]
5. Malaibari, Z.; Kahraman, R.; Saricimen, H.; Quddus, A. Investigation of atmospheric corrosion of mild steel after treatment by several inhibitor solutions. *Corros. Eng. Sci. Technol.* **2013**, *42*, 112–118. [CrossRef]
6. De la Fuente, D.; Díaz, I.; Simancas, J.; Chico, B.; Morcillo, M. Long-term atmospheric corrosion of mild steel. *Corros. Sci.* **2011**, *53*, 604–617. [CrossRef]
7. Allam, I.M.; Arlow, J.S.; Saricimen, H. Initial stages of atmospheric corrosion of steel in the Arabian Gulf. *Corros. Sci.* **1991**, *32*, 417–432. [CrossRef]
8. Kamimura, T.; Hara, S.; Miyuki, H.; Yamashita, M.; Uchida, H. Composition and protective ability of rust layer formed on weathering steel exposed to various environments. *Corros. Sci.* **2006**, *48*, 2799–2812. [CrossRef]

9. Morcillo, M.; Chico, B.; Díaz, I.; Cano, H.; de la Fuente, D. Atmospheric corrosion data of weathering steels. A review. *Corros. Sci.* **2013**, *77*, 6–24. [CrossRef]
10. Kolotyrkin, Y.M. Third International Congress on Metallic Corrosion. *Br. Corros. J.* **2013**, *1*, 89. [CrossRef]
11. Siddique, M.; Anwar-ul-Islam, M.; Butt, N.M.; Hussain, N.; Rehman, S.; Arshed, M. Mössbauer study of corrosion of mild steel induced by acid rain. *J. Radioanal. Nucl. Chem.* **1999**, *241*, 239–240. [CrossRef]
12. Migneault, S.; Koubaa, A.; Perré, P.; Riedl, B. Effects of wood fiber surface chemistry on strength of wood–plastic composites. *Appl. Surf. Sci.* **2015**, *343*, 11–18. [CrossRef]
13. McCafferty, E. *Surface Chemistry of Aqueous Corrosion Processes*; Springer: New York, NY, USA, 2015. [CrossRef]
14. Sha, W. *Steels: From Materials Science to Structural Engineering*; Springer: London, UK, 2013. [CrossRef]
15. Refait, P.; Génin, J.M.R. The transformation of chloride-containing green rust one into sulphated green rust two by oxidation in mixed Cl^- and SO_4^{2-} aqueous media. *Corros. Sci.* **1994**, *36*, 55–65. [CrossRef]
16. Parkins, R.N. The intergranular corrosion and stress corrosion cracking of mild steel in clarke's solution. *Corros. Sci.* **1994**, *36*, 2097–2110. [CrossRef]
17. Mobin, M.; Malik, A.U.; Andijani, I.N. The effect of heavy metal ions on the localized corrosion behavior of steels. *Desalination* **2007**, *217*, 233–241. [CrossRef]
18. Brinis, H.; Samar, M.E.H. A method for inhibiting scale formation and corrosion in a cooling water system. *Desalin. Water Treat.* **2013**, *52*, 2609–2619. [CrossRef]
19. Song, F.M. Predicting the mechanisms and crack growth rates of pipelines undergoing stress corrosion cracking at high pH. *Corros. Sci.* **2009**, *51*, 2657–2674. [CrossRef]
20. Eliyan, F.F.; Kish, J.R.; Alfantazi, A. Corrosion of New-Generation Steel in Outer Oil Pipeline Environments. *J. Mater. Eng. Perform.* **2016**, *26*, 214–220. [CrossRef]
21. Reeves, N.J.; Mann, S. Influence of inorganic and organic additives on the tailored synthesis of iron oxides. *J. Chem. Soc. Faraday Trans.* **1991**, *87*, 3875. [CrossRef]
22. Bockris, J.O.M.; Conway, B.E. Hydrogen Overpotential and the Partial Inhibition of the Corrosion of Iron. *J. Phys. Colloid Chem.* **1949**, *53*, 527–539. [CrossRef]
23. El Ibrahimi, B.; Jmiai, A.; Bazzi, L.; El Issami, S. Amino acids and their derivatives as corrosion inhibitors for metals and alloys. *Arab. J. Chem.* **2017**. [CrossRef]
24. McCafferty, E. *Introduction to Corrosion Science*; Springer Science: Washington, DC, USA, 2010. [CrossRef]
25. Sanyal, B. Organic compounds as corrosion inhibitors in different environments—A review. *Prog. Org. Coat.* **1981**, *9*, 165–236. [CrossRef]
26. Refaey, S.A.M.; Abd El-Rehim, S.S.; Taha, F.; Saleh, M.B.; Ahmed, R.A. Inhibition of chloride localized corrosion of mild steel by PO_4^{3-}, CrO_4^{2-}, MoO_4^{2-}, and NO^{2-} anions. *Appl. Surf. Sci.* **2000**, *158*, 190–196. [CrossRef]
27. Marczewska-Boczkowska, K.; Kosmulski, M. 2-Mercaptobenzothiazole as a Corrosion Inhibitor in Low Temperature Ionic Liquids. In *Trends in Colloid and Interface Science XXIV*; Springer: Berlin/Heidelberg, Germany, 2011. [CrossRef]
28. Goudarzi, N.; Farahani, H. Investigation on 2-mercaptobenzothiazole behavior as corrosion inhibitor for 316-stainless steel in acidic media. *Anti Corros. Methods Mater.* **2013**, *61*, 20–26. [CrossRef]
29. Feng, Y.; Chen, S.; Zhang, H.; Li, P.; Wu, L.; Guo, W. Characterization of iron surface modified by 2-mercaptobenzothiazole self-assembled monolayers. *Appl. Surf. Sci.* **2006**, *253*, 2812–2819. [CrossRef]
30. Finšgar, M.; Jackson, J. Application of corrosion inhibitors for steels in acidic media for the oil and gas industry: A review. *Corros. Sci.* **2014**, *86*, 17–41. [CrossRef]
31. Gunasekaran, G.; Natarajan, R.; Muralidharan, V.S.; Palaniswamy, N.; Appa Rao, B.V. Inhibition by phosphonic acids—An overview. *Anti Corros. Methods Mater.* **1997**, *44*, 248–259. [CrossRef]
32. *Multicylinder Test Sequences for Evaluating Automotive Engine Oils: Sequence IID*; Sponsored by ASTM Committee B.01 on Automotive Lubricants and D-2 on Petroleum Products and Lubricants; American Society for Testing and Materials: West Conshohocken, PA, USA, 1993.
33. Heit, A.H.; Calvin, C. Corrusion Inhibition Compositions. U.S. Patent C 3, 291 C, 11 September 1962.
34. ASTM A568/A568M-09. *Standard Specification for Sheet, Carbon, Structural, and High-Strength, Low-Alloy, Hot-Rolled and Cold-Rolled, General Requirements for*; ASTM International: West Conshohocken, PA, USA, 2009. [CrossRef]
35. ASTM D6386-99. *Standard Practice for Preparation of Zinc (Hot-Dip Galvanized) Coated Iron and Steel Product and Hardware Surfaces for Painting*; ASTM International: West Conshohocken, PA, USA, 1999. [CrossRef]

36. Matsushima, I. Carbon Steel—Atmospheric Corrosion. In *Corrosion Handbook*; Revie, R.W., Ed.; John Wiley & Sons, Inc.: Hoboken, NJ, USA; Etobicoke, ON, Canada, 2008; pp. 515–528.
37. Kelly, R.G.; Scully, J.R.; Shoesmith, D.W.; Buchheit, R.G. The Polarization Resistance Method for Determination of Instantaneous Corrosion Rates. In *Electrochemical Techniques in Corrosion Science and Engineering*; Schweitzer, P.A., Ed.; Marcel Dekker, Inc.: New York, NY, USA, 2002; pp. 125–150.
38. Luo, X.; Ci, C.; Li, J.; Lin, K.; Du, S.; Zhang, H.; Li, X.; Cheng, Y.F.; Zang, J.; Liu, Y. 4-aminoazobenzene modified natural glucomannan as a green eco-friendly inhibitor for the mild steel in 0.5 M HCl solution. *Corros. Sci.* **2019**, *151*, 132–142. [CrossRef]
39. Fernandes, C.M.; Ferreira Fagundes, T.D.S.; Escarpini dos Santos, N.; Shewry de, M.; Rocha, T.; Garrett, R.; Borges, R.M.; Muricy, G.; Valverde, A.L.; Ponzio, E.A. Ircinia strobilina crude extract as corrosion inhibitor for mild steel in acid medium. *Electrochim. Acta* **2019**, *312*, 137–148. [CrossRef]
40. ASTM G44-99. *Standard Practice for Exposure of Metals and Alloys by Alternate Immersion in Neutral 3.5% Sodium Chloride Solution*; ASTM International: West Conshohocken, PA, USA, 2013. [CrossRef]
41. Migahed, M.A.; Al-Sabagh, A.M.; Khamis, E.A.; Zaki, E.G. Quantum chemical calculations, synthesis and corrosion inhibition efficiency of ethoxylated-[2-(2-{2-[2-(2-benzenesulfonylamino-ethylamino)-ethylamino] -ethylamino}-ethylamino)-ethyl]-4-alkyl-benzenesulfonamide on API X65 steel surface under H2S environment. *J. Mol. Liq.* **2015**, *212*, 360–371. [CrossRef]
42. Zhu, Y.; Free, M.L.; Yi, G. Electrochemical measurement, modeling, and prediction of corrosion inhibition efficiency of ternary mixtures of homologous surfactants in salt solution. *Corros. Sci.* **2015**, *98*, 417–429. [CrossRef]
43. Aramaki, K.; Hackerman, N. Inhibition Mechanism of Medium-Sized Polymethyleneimine. *J. Electrochem. Soc.* **1969**, *116*, 568. [CrossRef]
44. Tamura, H. The role of rusts in corrosion and corrosion protection of iron and steel. *Corros. Sci.* **2008**, *50*, 1872–1883. [CrossRef]
45. Revie, R.W. (Ed.) *Uhlig's Corrosion Handbook*; John Wiley & Sons: Hoboken, NJ, USA, 2011. [CrossRef]
46. Alcántara, J.; Chico, B.; Simancas, J.; Díaz, I.; de la Fuente, D.; Morcillo, M. An attempt to classify the morphologies presented by different rust phases formed during the exposure of carbon steel to marine atmospheres. *Mater. Charact.* **2016**, *118*, 65–78. [CrossRef]
47. Ma, Y.; Li, Y.; Wang, F. The effect of β-FeOOH on the corrosion behavior of low carbon steel exposed in tropic marine environment. *Mater. Chem. Phys.* **2008**, *112*, 844–852. [CrossRef]
48. Misawa, T.; Hashimoto, K.; Shimodaira, S. The mechanism of formation of iron oxide and oxyhydroxides in aqueous solutions at room temperature. *Corros. Sci.* **1974**, *14*, 131–149. [CrossRef]
49. De la Fuente, D.; Díaz, I.; Alcántara, J.; Chico, B.; Simancas, J.; Llorente, I.; García-Delgado, A.; Jiménez, J.A.; Adeva, P.; Morcillo, M. Corrosion mechanisms of mild steel in chloride-rich atmospheres. *Mater. Corros.* **2016**, *67*, 227–238. [CrossRef]
50. Evans, U.R. Mechanism of rusting. *Corros. Sci.* **1969**, *9*, 813–821. [CrossRef]
51. Stratmann, M.; Bohnenkamp, K.; Engell, H.J. An electrochemical study of phase-transitions in rust layers. *Corros. Sci.* **1983**, *23*, 969–985. [CrossRef]
52. Mathew, V.; Kim, S.; Kang, J.; Gim, J.; Song, J.; Baboo, J.P.; Park, W.; Ahn, D.; Han, J.; Gu, L.; et al. Amorphous iron phosphate: Potential host for various charge carrier ions. *NPG Asia Mater.* **2014**, *6*, e138. [CrossRef]
53. Tiwari, A.; Hihara, L.; Rawlins, J. (Eds.) *Intelligent Coatings for Corrosion Control*; Butterworth-Heinemann: Oxford, UK, 2015. [CrossRef]
54. Cornell, R.M.; Schwertmann, U. *The Iron Oxides: Structure, Properties, Reactions, Occurrences and Uses*, 2nd ed.; Wiley-VCH Verlag GmbH & Co. KGaA: Weinheim, Germany, 2003.
55. Farzanian, K.; Pimenta Teixeira, K.; Perdigão Rocha, I.; De Sa Carneiro, L.; Ghahremaninezhad, A. The mechanical strength, degree of hydration, and electrical resistivity of cement pastes modified with superabsorbent polymers. *Constr. Build. Mater.* **2016**, *109*, 156–165. [CrossRef]
56. Gálvez, N. Effect of Phosphate on the Crystallization of Hematite, Goethite, and Lepidocrocite from Ferrihydrite. *Clays Clay Miner.* **1999**, *47*, 304–311. [CrossRef]
57. Schwertmann, U. Effect of pH on the Formation of Goethite and Hematite from Ferrihydrite. *Clays Clay Miner.* **1983**, *31*, 277–284. [CrossRef]
58. Antunes, R.A.; Costa, I.; Faria, D.L.A.D. Characterization of corrosion products formed on steels in the first months of atmospheric exposure. *Mater. Res.* **2003**, *6*, 403–408. [CrossRef]

59. Labinger, J.A. Bond-stretch isomerism: A case study of a quiet controversy. *C. R. Chim.* **2002**, *5*, 235–244. [CrossRef]
60. Tamilselvi, M.; Kamaraj, P.; Arthanareeswari, M.; Devikala, S.; Selvi, J.A. Development of nano SiO_2 incorporated nano zinc phosphate coatings on mild steel. *Appl. Surf. Sci.* **2015**, *332*, 12–21. [CrossRef]
61. Dwivedi, D.; Lepková, K.; Becker, T. Carbon steel corrosion: A review of key surface properties and characterization methods. *RSC Adv.* **2017**, *7*, 4580–4610. [CrossRef]
62. De la Fuente, D.; Alcántara, J.; Chico, B.; Díaz, I.; Jiménez, J.A.; Morcillo, M. Characterisation of rust surfaces formed on mild steel exposed to marine atmospheres using XRD and SEM/Micro-Raman techniques. *Corros. Sci.* **2016**, *110*, 253–264. [CrossRef]
63. Simard, S.; Odziemkowski, M.; Irish, D.E.; Brossard, L.; Ménard, H. In situ micro-Raman spectroscopy to investigate pitting corrosion product of 1024 mild steel in phosphate and bicarbonate solutions containing chloride and sulfate ions. *J. Appl. Electrochem.* **2001**, *31*, 913–920. [CrossRef]
64. Dhaiveegan, P.; Elangovan, N.; Nishimura, T.; Rajendran, N. Weathering Steel in Industrial-Marine-Urban Environment: Field Study. *Mater. Trans.* **2016**, *57*, 148–155. [CrossRef]
65. Sandhyarani, N.; Skanth, G.; Berchmans, S.; Yegnaraman, V.V.; Pradeep, T. A Combined Surface-Enhanced Raman-X-ray Photoelectron Spectroscopic Study of 2-mercaptobenzothiazole Monolayers on Polycrystalline Au and Ag Films. *J. Colloid Interface Sci.* **1999**, *209*, 154–161. [CrossRef]
66. Socrates, G. *Infrared and Raman Characteristic Group Frequencies*, 3rd ed.; John Wiley & Sons Ltd.: Chichester, UK, 2001; pp. 283–327.
67. De Faria, D.L.A.; Venâncio Silva, S.; de Oliveira, M.T. Raman microspectroscopy of some iron oxides and oxyhydroxides. *J. Raman Spectrosc.* **1997**, *28*, 873–878. [CrossRef]
68. Kolitsch, U.; Bernhardt, H.-J.; Lengauer, C.L.; Blass, G.; Tillmanns, E. Allanpringite, $Fe_3(PO_4)2(OH)_35H_2O$, a new ferric iron phosphate from Germany, and its close relation to wavellite. *Eur. J. Mineral.* **2006**, *18*, 793–801. [CrossRef]
69. Zhang, L.; Brow, R.K. A Raman Study of Iron-Phosphate Crystalline Compounds and Glasses. *J. Am. Ceram. Soc.* **2011**, *94*, 3123–3130. [CrossRef]
70. Liu, Q.; Chen, M.; Zheng, K.; Yang, Y.; Feng, X.; Li, H. In Situ Electrochemical Investigation of Pyrite Assisted Leaching of Chalcopyrite. *J. Electrochem. Soc.* **2018**, *165*, H813–H819. [CrossRef]
71. Weber, I.; Böttger, U.; Pavlov, S.G.; Hübers, H.W.; Hiesinger, H.; Jessberger, E.K. Laser alteration on iron sulfides under various environmental conditions. *J. Raman Spectrosc.* **2017**, *48*, 1509–1517. [CrossRef]
72. Venkataramanan, M.; Skanth, G.; Bandyopadhyay, K.; Vijayamohanan, K.; Pradeep, T. Self-assembled Monolayers of Two Aromatic Disulfides and a Diselenide on Polycrystalline Silver Films: An Investigation by SERS and XPS. *J. Colloid Interface Sci.* **1999**, *212*, 553–561. [CrossRef]
73. Ali Asaad, M.; Sarbini, N.N.; Sulaiman, A.; Ismail, M.; Huseien, G.F.; Abdul Majid, Z.; Bothi Raja, P. Improved corrosion resistance of mild steel against acid activation: Impact of novel Elaeis guineensis and silver nanoparticles. *J. Ind. Eng. Chem.* **2018**, *63*, 139–148. [CrossRef]
74. McGowan, G.; Prangnell, J. The significance of vivianite in archaeological settings. *Geoarchaeology* **2006**, *21*, 93–111. [CrossRef]
75. Morozov, Y.; Calado, L.M.; Shakoor, R.A.; Raj, R.; Kahraman, R.; Taryba, M.G.; Montemor, M.F. Epoxy coatings modified with a new cerium phosphate inhibitor for smart corrosion protection of steel. *Corros. Sci.* **2019**, *159*, 108128. [CrossRef]
76. Singh, I.; Banerjee, T. Effect of 2-mercaptobenzothiazole on aciddissolution and hydrogen absorption of steel. *Corros. Sci.* **1972**, *12*, 503–515. [CrossRef]
77. Kelly, R.G.; Scully, J.R.; Shoesmith, D.W.; Buchheit, R.G. Passivity and localised corrosion. In *Electrochemical Techniques in Corrosion Science and Engineering*; Schweitzer, P.A., Ed.; Marcel Dekker Inc.: New York, NY, USA, 2002; pp. 55–124. [CrossRef]
78. Jafari, H.; Akbarzade, K.; Danaee, I. Corrosion inhibition of carbon steel immersed in a 1M HCl solution using benzothiazole derivatives. *Arab. J. Chem.* **2019**. [CrossRef]
79. Kasai, N.; Utatsu, K.; Park, S.; Kitsukawa, S.; Sekine, K. Correlation between corrosion rate and AE signal in an acidic environment for mild steel. *Corros. Sci.* **2009**, *51*, 1679–1684. [CrossRef]
80. Kelly, R.G. Electrochemical thermodynamics and kinetics of relevance to corrosion. In *Electrochemical Techniques in Corrosion Science and Engineering*; Schweitzer, P.A., Ed.; Marcel Dekker, Inc.: New York, NY, USA, 2002; pp. 9–54.

81. Matsushima, I. Carbon Steel—Corrosion in Fresh Waters. In *Corrosion Handbook*; Revie, R.W., Ed.; John Wiley & Sons, Inc.: Hoboken, NJ, USA; Etobicoke, ON, Canada, 2008; pp. 529–544.
82. Matsushima, I. Carbon Steel—Corrosion by Seawater. In *Corrosion Handbook*; Revie, R.W., Ed.; John Wiley & Sons, Inc.: Hoboken, NJ, USA; Etobicoke, ON, Canada, 2008; pp. 545–553.
83. Shifler, D.A.; Aylor, D.M. Testing in environments-seawater. In *Corrosion Tests and Standards Application and Interpretation*; Baboian, R., Ed.; ASTM International: Baltimore, MD, USA, 2005; pp. 362–379.
84. Hsu, C.H.; Mansfeld, F. Technical Note:Concerning the Conversion of the Constant Phase Element Parameter Y0into a Capacitance. *Corrosion* **2001**, *57*, 747–748. [CrossRef]
85. Kong, G.; Lingyan, L.; Lu, J.; Che, C.; Zhong, Z. Corrosion behavior of lanthanum-based conversion coating modified with citric acid on hot dip galvanized steel in aerated 1M NaCl solution. *Corros. Sci.* **2011**, *53*, 1621–1626. [CrossRef]
86. Cole, K.S.; Cole, R.H. Dispersion and Absorption in Dielectrics II. Direct Current Characteristics. *J. Chem. Phys.* **1942**, *10*, 98–105. [CrossRef]
87. Shen, S.; Zuo, Y.; Zhao, X. The effects of 8-hydroxyquinoline on corrosion performance of a Mg-rich coating on AZ91D magnesium alloy. *Corros. Sci.* **2013**, *76*, 275–283. [CrossRef]
88. Kaesche, H. *Corrosion of Metals: Physicochemical Principles and Current Problems*; Springer: Berlin, Germany, 2003. [CrossRef]
89. Sharma, V.K. *Ferrites and Ferrates: Chemistry and Applications in Sustainable Energy and Environmental Remediation*; American Chemical Society: Washington, DC, USA, 2016; pp. 489–491.
90. Evans, D.G.; Slade, R.C.T. Structural Aspects of Layered Double Hydroxides. In *Layered Double Hydroxides*; Springer: Berlin/Heidelberg, Germany, 2005. [CrossRef]
91. K. Karaxi, E.; Kartsonakis, I.A.; Charitidis, C.A. Chloride ion entrapment by calcined layered double hydroxides. *Int. J. Struct. Integr.* **2016**, *7*, 788–806. [CrossRef]
92. Trolard, F.; Bourrié, G. Chapter 5 Geochemistry of Green Rusts and Fougerite. *Adv. Agron.* **2008**, *99*, 227–288. [CrossRef]
93. Davesne, E.; Dideriksen, K.; Christiansen, B.C.; Sonne, M.; Ayala-Luis, K.B.; Koch, C.B.; Hansen, H.C.B.; Stipp, S.L.S. Free energy of formation for green rust sodium sulphate (NaFeII6FeIII3(OH)18(SO4)2(s)). *Geochim. Cosmochim. Acta* **2010**, *74*, 6451–6467. [CrossRef]
94. Duan, X.; Evans, D.G. *Layered Double Hydroxides*; Springer: Berlin/Heidelberg, Germany, 2006; Volume 119. [CrossRef]
95. Li, Y.-H.; Tsui, T.-F. The solubility of CO_2 in water and sea water. *J. Geophys. Res.* **1971**, *76*, 4203–4207. [CrossRef]
96. Refait, P.; Drissi, S.H.; Pytkiewicz, J.; Génin, J.M.R. The anionic species competition in iron aqueous corrosion: Role of various green rust compounds. *Corros. Sci.* **1997**, *39*, 1699–1710. [CrossRef]
97. Génin, J.M.R.; Refait, P.; Olowe, A.A.; Abdelmoula, M.; Fall, I.; Drissi, S.H. Identification of Green Rust Compounds in the Aqueous Corrosion Processes of Steels; the Case of Microbially Induced Corrosion and Use of 78 K CEMS. *Hyperfine Interact.* **1998**, *112*, 47–51. [CrossRef]
98. Amar, H.; Tounsi, A.; Makayssi, A.; Derja, A.; Benzakour, J.; Outzourhit, A. Corrosion inhibition of Armco iron by 2-mercaptobenzimidazole in sodium chloride 3% media. *Corros. Sci.* **2007**, *49*, 2936–2945. [CrossRef]
99. Mahdavian, M.; Ashhari, S. Corrosion inhibition performance of 2-mercaptobenzimidazole and 2-mercaptobenzoxazole compounds for protection of mild steel in hydrochloric acid solution. *Electrochim. Acta* **2010**, *55*, 1720–1724. [CrossRef]
100. Gece, G.; Bilgiç, S. Quantum chemical study of some cyclic nitrogen compounds as corrosion inhibitors of steel in NaCl media. *Corros. Sci.* **2009**, *51*, 1876–1878. [CrossRef]
101. Albrakaty, R.H.; Wazzan, N.A.; Obot, I.B. Theoretical Study of the Mechanism of Corrosion Inhibition of Carbon Steel in Acidic Solution by 2-aminobenzothaizole and 2- Mercatobenzothiazole. *Int. J. Electrochem. Sci.* **2018**, 3535–3554. [CrossRef]
102. Wang, G.; Harrison, A.; Li, X.; Whittaker, G.; Shi, J.; Wang, X.; Yang, H.; Cao, P.; Zhang, Z. Study of the adsorption of benzimidazole and 2-mercaptobenzothiazole on an iron surface by confocal micro-Raman spectroscopy. *J. Raman Spectrosc.* **2004**, *35*, 1016–1022. [CrossRef]

103. Gholami, M.; Danaee, I.; Maddahy, M.H.; RashvandAvei, M. Correlated ab Initio and Electroanalytical Study on Inhibition Behavior of 2-Mercaptobenzothiazole and Its Thiole–Thione Tautomerism Effect for the Corrosion of Steel (API 5L X52) in Sulphuric Acid Solution. *Ind. Eng. Chem. Res.* **2013**, *52*, 14875–14889. [CrossRef]
104. Wang, C.T.; Chen, S.-H.; Qi, C.-S. Protection of copper corrosion by carbazole and N-vinylcarbazole self-assembled films in NaCl solution. *J. Appl. Electrochem.* **2003**, *33*, 179–186. [CrossRef]
105. Şahin, M.; Bilgiç, S.; Yılmaz, H. The inhibition effects of some cyclic nitrogen compounds on the corrosion of the steel in NaCl mediums. *Appl. Surf. Sci.* **2002**, *195*, 1–7. [CrossRef]
106. Zhou, G.P.; Liu, Z.Y.; Qiu, Y.Q.; Wang, G.D. The improvement of weathering resistance by increasing P contents in cast strips of low carbon steels. *Mater. Des.* **2009**, *30*, 4342–4347. [CrossRef]
107. Sahoo, G.; Fujieda, S.; Shinoda, K.; Suzuki, S. Influence of phosphate species on green rust I transformation and local structure and morphology of γ-FeOOH. *Corros. Sci.* **2011**, *53*, 2446–2452. [CrossRef]
108. Sankara, T.S.N. Surface pretreatment by phosphate conversion coatings—A review. *Rev. Adv. Mater. Sci.* **2005**, *9*, 130–177.
109. Nespolo, M. Iron Metabolism. From Molecular Mechanisms to Clinical Consequences. *Acta Crystallogr. Sect. D Struct. Biol.* **2017**, *73*, 952–954. [CrossRef]
110. Skinner, H.C.W. Phosphates Geochemical, Geobiological and Materials Importance.: Edited by Matthew, J.; Kohn, John Rakovan and John, M. Hughes. Reviews in Mineralogy and Geochemistry, Vol. 48 (2002). Mineralogical Society of America, 1015 Eighteenth Street, Washington, D.C. 20036, U.S.A. US $40, $30 to MSA, GS, CMS members (ISBN 0 939950 60 X). *Can. Mineral.* **2003**, *41*, 1071–1073. [CrossRef]
111. Guilbaud, R.; White, M.L.; Poulton, S.W. Surface charge and growth of sulphate and carbonate green rust in aqueous media. *Geochim. Cosmochim. Acta* **2013**, *108*, 141–153. [CrossRef]
112. Huttner, G.; Strittmatter, J.; Sandhöfner, S. Phosphorus Tripodal Ligands. In *Comprehensive Coordination Chemistry II*; Elsevier Science: Amsterdam, The Netherlands, 2003; pp. 297–322. [CrossRef]
113. O'Keeffe, M. The Bond Valence Method in Crystal Chemistry. In *Modern Perspectives in Inorganic Crystal Chemistry*; Springer Science & Business Media: Berlin/Heidelberg, Germany, 1992. [CrossRef]
114. Jolivet, J.-P.; Tronc, E.; Chanéac, C. Iron oxides: From molecular clusters to solid. A nice example of chemical versatility. *C. R. Geosci.* **2006**, *338*, 488–497. [CrossRef]
115. Twigg, M.V.; Burgess, J. Iron. In *Comprehensive Coordination Chemistry II*; Elsevier Science: Amsterdam, The Netherlands, 2003; pp. 403–553. [CrossRef]
116. Downing, J.H.; Smith, M.B. Phosphorus Ligands. In *Comprehensive Coordination Chemistry II*; Elsevier Science: Amsterdam, The Netherlands, 2003; pp. 253–296. [CrossRef]
117. Das, S.; Hendry, M.J.; Essilfie-Dughan, J. Transformation of two-line ferrihydrite to goethite and hematite as a function of pH and temperature. *Environ. Sci. Technol.* **2011**, *45*, 268–275. [CrossRef]
118. Chitrakar, R.; Tezuka, S.; Sonoda, A.; Sakane, K.; Ooi, K.; Hirotsu, T. Phosphate adsorption on synthetic goethite and akaganeite. *J. Colloid Interface Sci.* **2006**, *298*, 602–608. [CrossRef]
119. Flynn, C.M. Hydrolysis of inorganic iron (III) salts. *Chem. Rev.* **1984**, *84*, 31–41. [CrossRef]
120. Navarro, J.M.; Botella, M.A.; Cerdá, A.; Martinez, V. Phosphorus uptake and translocation in salt-stressed melon plants. *J. Plant Physiol.* **2001**, *158*, 375–381. [CrossRef]
121. Hongshao, Z.; Stanforth, R. Competitive adsorption of phosphate and arsenate on goethite. *Environ. Sci. Technol.* **2001**, *35*, 4753–4757. [CrossRef] [PubMed]
122. Ler, A.; Stanforth, R. Evidence for surface precipitation of phosphate on goethite. *Environ. Sci. Technol.* **2003**, *37*, 2694–2700. [CrossRef] [PubMed]
123. Génin, J.-M.R.; Bourrié, G.; Trolard, F.; Abdelmoula, M.; Jaffrezic, A.; Refait, P.; Maitre, V.; Humbert, B.; Herbillon, A. Thermodynamic Equilibria in Aqueous Suspensions of Synthetic and Natural Fe(II)–Fe(III) Green Rusts: Occurrences of the Mineral in Hydromorphic Soils. *Environ. Sci. Technol.* **1998**, *32*, 1058–1068. [CrossRef]
124. Teodorovich, V.I.; Beznosikov, B.O. Morphological and functional study of the thrombocytes in essential polycythemia and thrombocytopenia (Werlhof's disease). *Lab. Anal. Clin. Bacteriol. Inmunol. Parasitol. Hematol. Anat. Patol. Quim. Clin.* **1961**, *39*, 77–83. [CrossRef]
125. Nriagu, J.O. Stability of vivianite and ion-pair formation in the system $Fe_3(PO_4)_2$-$H_3PO_4H_3PO_4$-H_2O. *Geochim. Cosmochim. Acta* **1972**, *36*, 459–470. [CrossRef]

126. Mishra, M.; Chun, D.-M. α-Fe_2O_3 as a photocatalytic material: A review. *Appl. Catal. A Gen.* **2015**, *498*, 126–141. [CrossRef]
127. McCafferty, E. Acid-Base Properties of Surface Oxide Films. In *Surface Chemistry of Aqueous Corrosion Processes*; Springer: Berlin/Heidelberg, Germany, 2015; pp. 1–54. [CrossRef]

applied sciences

MDPI

Article

Glycerin-Grafted Starch as Corrosion Inhibitor of C-Mn Steel in 1 M HCl Solution

Sihem Lahrour [1], Abderrahim Benmoussat [2,*], Brahim Bouras [1], Asma Mansri [1], Lahcene Tannouga [1] and Stefania Marzorati [3]

1 Research Laboratory of the Organic Electrolytes and Polyelectrolytes Application (LAEPO), Department of Chemistry Tlemcen, Faculty of Sciences, University of Tlemcen, Tlemcen 13000, Algeria; lahrour91sihem@gmail.com (S.L.); bourasbrahim_m8@yahoo.fr (B.B.); asma.mansri@yahoo.fr (A.M.); l14_ten@yahoo.fr (L.T.)

2 LAEPO Research Laboratory, Matter Sciences Department, Institute of Sciences and Technology, CUTAM University Center of Tamanrasset, Tamanrasset 11000, Algeria

3 Department of Environmental Science and Policy, Università degli Studi di Milano, 7-20122 Milano, Italy; stefania.marzorati@unimi.it

* Correspondence: abbenmoussa@gmail.com

Received: 29 September 2019; Accepted: 29 October 2019; Published: 3 November 2019

Featured Application: In the last decade, many research works were focused on the use of "green" corrosion inhibitors, lacking in the adverse health issues connected to the organic compounds used in the past. The use of biomaterials as corrosion inhibitors takes considerable attention due to their inherent stability, availability, cost effectiveness, and environmentally-friendly nature.

Abstract: C-Mn steels, commonly employed in structural applications, are often exposed to near-neutral aerated environments and hence subjected to general corrosion. In broader contexts, for example during pickling, acidizing treatments, or acid-releasing processes, where steel comes in contact with more aggressive solutions, the use of corrosion inhibitors is a supplementary strategy to cathodic protection and/or coating. This work focuses on the C-Mn steel corrosion protection in the presence of HCl, either as process fluid or by product. In order to avoid the toxicological issues related to conventional synthetic products, a bio-copolymer containing glycerin-grafted starch, synthesized by modification of maize starch, was studied as a "green" corrosion inhibitor by the weight loss method and electrochemical techniques (open circuit potential, potentiodynamic polarization and electrochemical impedance spectroscopy). Corrosion-related parameters, such as inhibitor concentration and temperature, were varied and optimized to characterize the corrosion process. Results showed that inhibition efficiency increases with increasing bio-copolymer concentration, reaching a maximum of 94%at the concentration of 300 mg L^{-1}. The kinetic and thermodynamic parameters were determined and discussed. The obtained values of corrosion potential and corrosion current density, E_{corr} and i_{corr}, obtained by potentiodynamic polarization, are in agreement with the weight loss method. The corrosion current densities decrease when the concentration of the inhibitor increases.

Keywords: C-Mnsteel; corrosion inhibitors; bio-copolymer; starch; glycerin

1. Introduction

Hypoeutectoid C-Mn steels are typically employed in many structural applications and, being commonly exposed to near-neutral aerated environment, undergo general corrosion in the form of rust. Cathodic protection and/or coating, together with the use of adequate corrosion allowance, enables the phenomenon to be controlled, thus reducing the risk of failure [1]. The use of C-Mn steel in a broader context, where the steel comes in contact with more aggressive solutions (for example, during pickling or oil well acidizing treatments as well as acid-releasing processes [2,3], might need diverse technologies, such as addition of inhibitor products.

In this work, we propose to investigate a method to control and minimize the steel corrosion in the presence of HCl by adding a corrosion inhibitor of bio-polymeric nature. The test solutions are meant to mimic the aggressiveness of the environment due to in-situ generation of harsh byproducts (i.e., during plastic or biomass combustion) or due to an environment that is aggressive itself in the exercise conditions (i.e., pickling or acidizing treatments).

The corrosion inhibition strategy is well-known to help in reducing the economic input of corrosion damage [4,5]. Moreover, through a proper selection of the nature of the used inhibitor compounds, the related toxicological aspects can be controlled. Efficient inhibitors are compounds whose activity is related to the presence in their structure of hetero atoms like nitrogen, oxygen, or sulfur [6]. It was observed that adsorption depends mainly on the presence of lone pair electrons, electron-donating groups and π-orbital character of the molecule [7]. However, even if many synthetic organic compounds are effective inhibitors, regarding their cost and toxic nature, researches were compelled [8,9]. The concern is not only related to the environmental poisoning, needing effective procedures of inhibitor removal before flowing the used solution out of the industrial plant, but also to the contact of the protected metal, for example, with food, beverages, and medicals, during its lifecycle. In the last decade, many research works were focused on the use of "green" corrosion inhibitors, lacking in the adverse health issues connected to the organic compounds used in the past [8,9]. The use of biomaterials as corrosion inhibitors takes considerable attention due to their inherent stability, availability, cost effectiveness, and environmentally-friendly nature [10,11].

The protection of a steel structure against corrosion has already been studied by means of biopolymers as corrosion inhibitors [12,13]. Bio-copolymers derived from starch, already used in the literature as a corrosion inhibitor, motivated our research [14,15].

The present work reports on the establishment of different possibilities of grafting glycerin on starch. We looked at the different grafting sites on amylopectin that can receive glycerin molecules. By the conditions, this was possible by eliminating amylose chains. The obtained product was used as corrosion inhibitor of C-Mn steel in 1 M HCl solution in the temperature range 25–50 °C. The corrosion inhibition was investigated by weight loss measurements, potentiodynamic polarization, and electrochemical impedance spectroscopy (EIS). Appropriate corrosion parameters such as corrosion current density (i_{corr}), polarization resistance (R_p), and parameters derived from EIS were derived to characterize the corrosion process. The kinetic and thermodynamic parameters were determined and discussed.

2. Materials and Methods

2.1. Reagents

The used reagents were maize starch (St), glycerin (Gly), hydrochloric acid 37%, and sodium hydroxide. They were supplied by the Aldrich Company. The only solvent used was bi-distilled water.

2.2. C-Mn Steel

Tests were conducted on C-Mn steel samples obtained and treated according to the procedure described in our previous work [4].

Atomic composition was compared to the one quoted in the material test certificates [16]. Elemental composition of the cutting samples (Table 1) was determined by emission spectroscopy analysis type "Spectro RP 212". Composition and microstructure can vary significantly, resulting in substantial differences in corrosion performances in a corrosion regime. Chemical composition of steel, displayed in Table 1, was in compliance with the API norm (American Petroleum Institute).

Table 1. Chemical composition of C-Mn steel.

Element	C	Si	P	Mn	S	Al	Mo
Mass %	0.129	0.290	0.016	1.590	0.018	0.024	0.008
Element	**Cr**	**Cu**	**Pb**	**Zn**	**Ni**	**V**	**Fe**
Mass %	0.015	0.024	0.0016	0.003	0.007	0.004	98.06

We observed a low carbon composition (0.129%), providing high chemical resistance, high concentration of manganese (1.590%), and low sulfur and phosphorus content. This composition can lead to the formation of manganese sulfide (MnS) inclusions, which are not desirable in the microstructure, being able to start corrosion pitting [7,17].

The surfaces of steel samples were observed under an optical microscope and SEM (Figure 1). The metallographic images revealed a fine microstructure of ferrito-perlitictype, with ferritic prevalencein the presence of clusters of pearlite in the grain boundaries withsome inclusion fields.

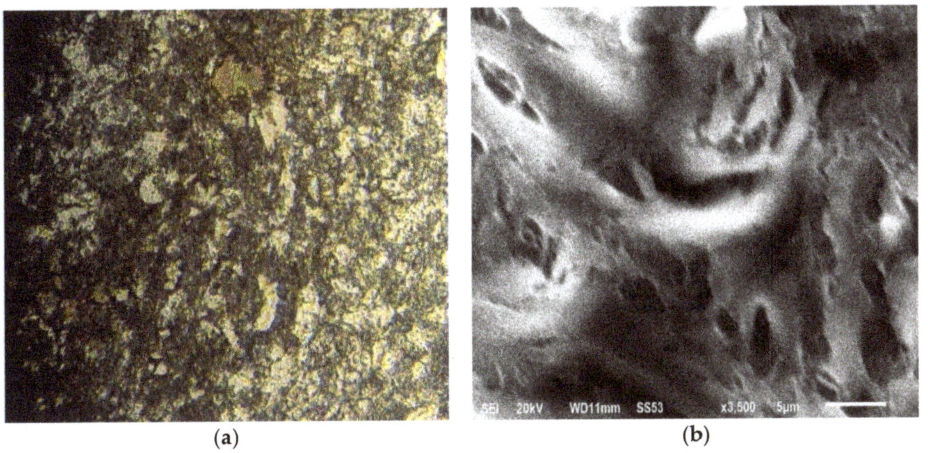

(a) (b)

Figure 1. Micrographs of ×60 steel. (**a**) Optical microscope observation; (**b**) SEM imaging (scale bar = 5 μm).

The refinement of ferritic size was obtained by different hardening and precipitation mechanisms based on a dislocation movement that increases elasticity limit and steel resistance. The SEM image (Figure 1b) confirms the fine microstructure of steel.

2.3. Test Solution

The chosen solution for corrosion tests was 1 M HCl solution prepared from analytical grade 37% (Aldrich) by dilution with bi-distilled water.

2.4. Synthesis of Bio-Copolymer (St63Gly37)

Anew glycerin-starch bio-copolymer (abbreviated from now on by (St63Gly37) was synthesized by modification of maize starch and used as corrosion inhibitor. Glycerin grafting on the starch takes place in three stages. The first and second steps consisted of the preparation of aqueous solutions of 50% glycerin (A) and 50% starch (B), respectively. In the third step, the glycerin (A) and starch (B) solutions were poured into a 200 mL Erlenmeyer flask, to which 3 mL of 1 M HCl were added under stirring and heating to evaporate all the water. Heating and stirring were continued for the sole purpose of evaporating all the water contained in the solution. The obtained mixture was neutralized by adding 3 mL of 1 M NaOH. The molecular structures of the involved species are displayed in Figure 2.

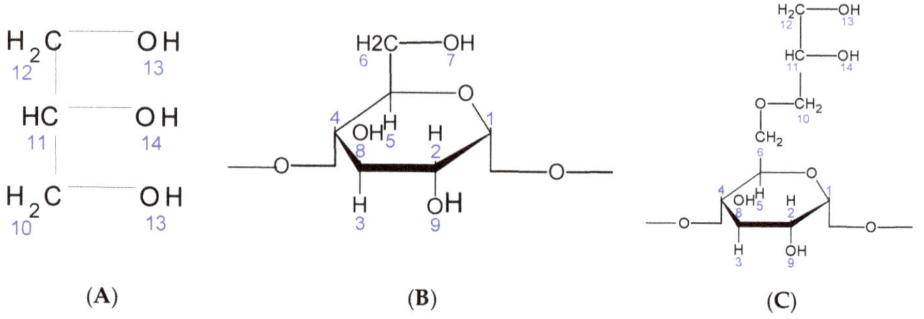

Figure 2. (**A**)Chemical structure of glycerin; (**B**) chemical structure of native starch; (**C**) chemical structure of St63Gly37.

2.5. ^1H NMR Characterization of St63Gly37

The ^1H NMR spectrum of the starch grafted with glycerin St63Gly37 was recorded on a Brucker spectrometer of 400 MHz and dissolved in DMSO-d_6 as solvent. The interpretation of the ^1H NMR results of the product St63Gly37 was carried out by comparison with the peaks of the native starch found in the literature [18,19].

2.6. FT-IR Characterization of St63Gly37 Bio-Copolymer

FT-IR spectra were recorded on a Fourier transform infrared (FT-IR) spectrometer Agilent Technologies Cary 600 Series with a resolution of 4 cm^{-1} in the LAEPO laboratory, University of Tlemcen. They were employed to observe the characteristic transmittance bands in St63Gly37. The FT-IR spectra were recorded over the frequency range between 500 and 4000 cm^{-1}.

2.7. Corrosion Inhibitor Tests

The weight loss (2 cm^2 apparent surface area) and electrochemical tests (1 cm^2 exposed surface to the corrosive solution) were carried out according to procedures described in previous works [4,13]. St63Gly37 testing was carried out in 1 M HCl solution, varying the concentration of the inhibitor between 5 and 300 mg L^{-1}. The temperature range was 25–50 °C.

The inhibition efficiency (E_w %) was calculated using the following Equation (1):

$$E_w = (1 - \frac{W_{corr}}{W^{\circ}_{corr}}) \times 100, \quad (1)$$

where W_{corr} and $W^{\circ}{}_{corr}$ are the corrosion rates of steel samples in the absence and presence of St63Gly37, respectively.

A potentiostat (Amel 549) and linear sweep generator (Amel 567) were used to record the current–voltage curves. The scan rate was 1 V min^{-1}. The reference electrode was a saturated calomel electrode (SCE); the counter electrode was the platinum electrode. The working electrode was polarized at 800 mV for 10 min before recording the cathodic curves. For the anodic curves, the potential of the electrode was swept from its open circuit value after 30 min. A Voltalab PGZ-100 electrochemical system was used for the determination of the electrochemical impedance spectroscopy (EIS) at E_{corr} after immersion in solution. After determination of the steady-state current at a given potential, sine wave voltage (10 mV) peak to peak, at frequencies between 100 kHz and 10 mHz, was superimposed on the rest potential. The measurements performed at rest potentials after 30 min of exposure were automatically controlled by computer programs. EIS diagrams are detailed in the Nyquist representations.

3. Results

3.1. St63Gly37 Characterization

The grafting of the glycerin may take place preferentially on the 7-position of the starch. It could also occur on positions 8 or 9 of the starch (Figure 2). On the other hand, glycerin can be grafted via its OH-13 or OH-14 functions. For the characterization of our products, we used ^{1}H NMR and Fourier transform infrared (FT-IR) spectroscopy techniques.

3.1.1. ^{1}H NMR Characterization

The ^{1}H NMR spectrum of the starch grafted with glycerin St63Gly37 was recorded on a Brucker spectrometer of 400 MHz and dissolved in DMSO as solvent. Results are displayed in Figure 3.

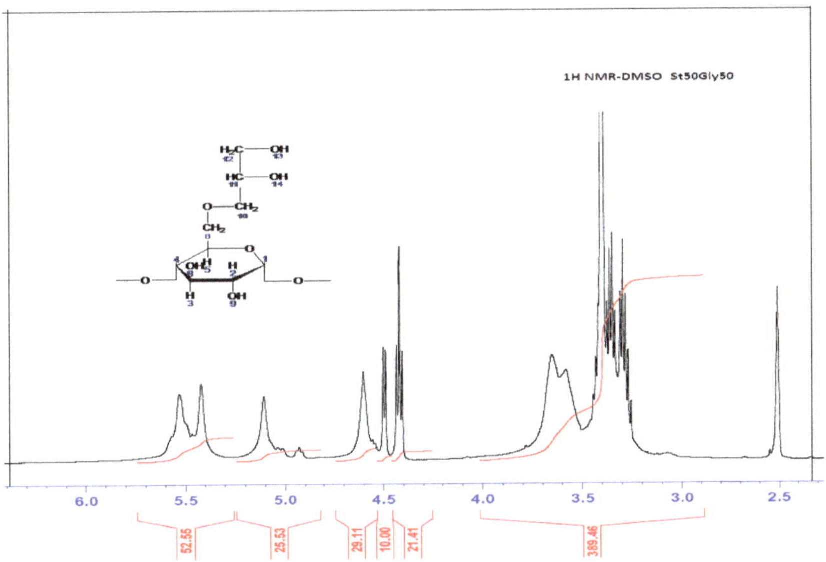

Figure 3. ^{1}H NMR Spectrum of St63Gly37.

3.1.2. FT-IR Characterization

The FT-IR technique is largely used for the characterization of starch as a natural polymer [20]. Examination of the FT-IR spectrum confirms the grafting of glycerin on the starch, by the CH_2 bands and the characteristic OH bands of glycerin. Figure 4 shows the FT-IR spectra of the native starch in comparison with the St63Gly37 copolymer.

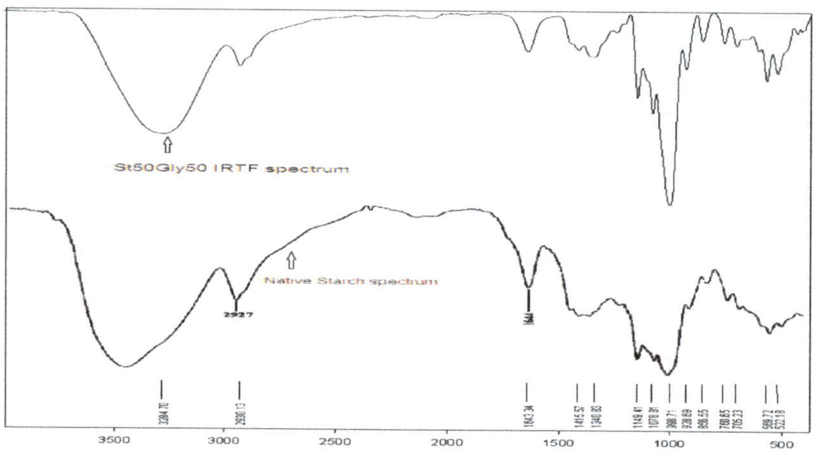

Figure 4. FT-IR spectra of the St63Gly37 copolymer (top) and native starch (bottom).

3.2. Weight Loss Measurements

3.2.1. Effect of Inhibitor Concentration

The results of C-Mn steel in HCl solution with different concentrations of St63Gly37 at 25 °C, using weight loss measurements, are reported in Table 2.

Table 2. Inhibition efficiency (E_w%) of C-Mn steel in 1 M HCl solution at different concentrations of St63Gly37, measured by weight loss at 25 °C.

Inhibitor Concentration (mg L^{-1})	W_{corr} (mgcm^{-2}h^{-1})	E_w (%)
0	0.661	-
5	0.471	29
10	0.404	39
50	0.330	50
100	0.225	66
200	0.113	83
300	0.038	94

E_w%—Inhibition efficiency, W_{corr}—corrosion rates.

3.2.2. Effect of Temperature

In order to study the effect of temperature on corrosion inhibition of C-Mn steel in HCl solution after two hours at different bio-copolymer concentrations, weight loss studies were carried out in a temperature range from 25 to 50 °C.

The variation of corrosion rate (W_{corr}) and inhibition efficiency E_w (%) with the temperature for different concentrations of St63Gly37 bio-copolymer are displayed in Figures 5 and 6, respectively.

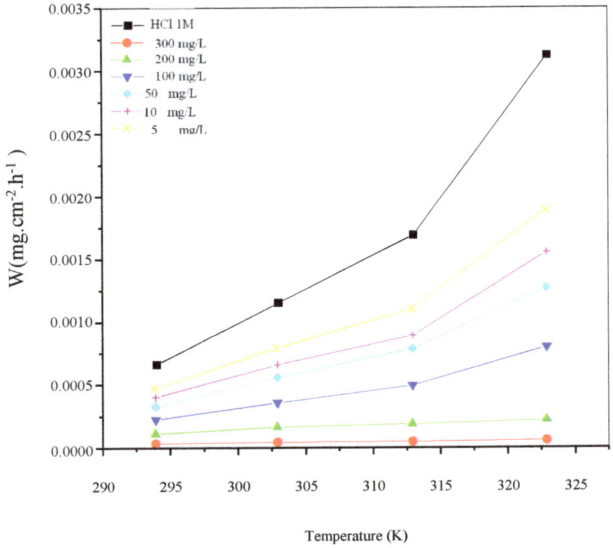

Figure 5. Variation of corrosion rate (W_{corr}) as a function of temperature for different concentrations of St63Gly37 bio-copolymer.

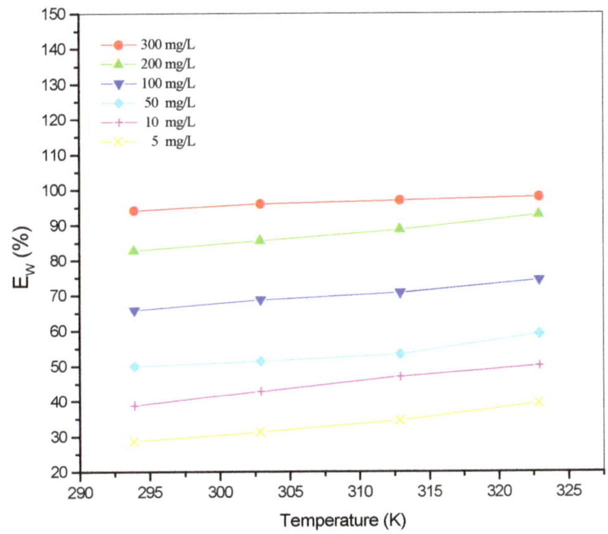

Figure 6. Variation of inhibition efficiency E_w (%) as a function of temperature for different concentrations of St63Gly37 bio-copolymer.

3.3. Thermodynamic and Kinetic Parameters

Thermodynamic and kinetic parameters, such as activation energy E_a, enthalpy, and entropy of adsorption of St63Gly37 on steel, were calculated building the Arrhenius plot. Figure 7 presents the Arrhenius plots of corrosion rate logarithm vs. 1000/T related to blank solution and bio-copolymer solution.

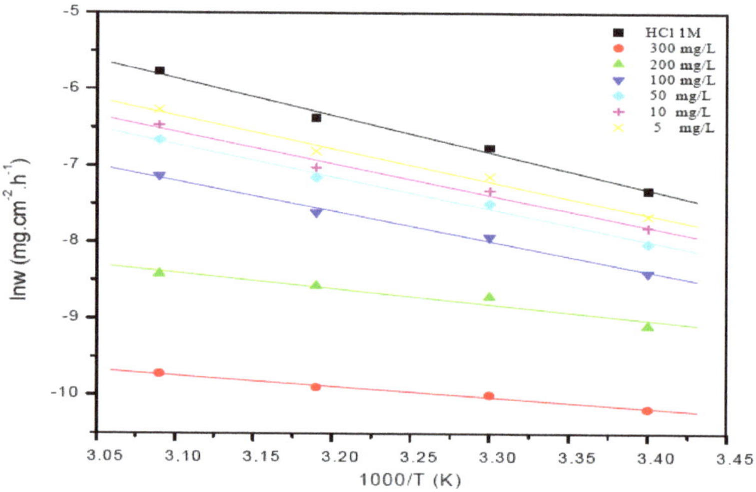

Figure 7. Arrhenius plots of the corrosion rate for both the blank solution and the solution of bio-copolymer.

The activation parameters for the corrosion process can be regarded as an Arrhenius-type process according to the following Equations (2) and (3):

$$\ln(W_{corr}) = \frac{-E_a}{RT} + A. \quad (2)$$

$$\ln(W'_{corr}) = \frac{-E'_a}{RT} + A. \quad (3)$$

E_a and E'_a are the apparent activation energies with and without the bio-copolymer, respectively. T is the absolute temperature, A is a constant, and R is the universal gas constant. W_{corr} and W'_{corr} are the steel corrosion rates in the absence and presence of the bio copolymer inhibitor, respectively.

3.4. Adsorption Isotherms

Figure 8 shows the results of adsorption isotherms. Linear plots were obtained in the studied temperature range.

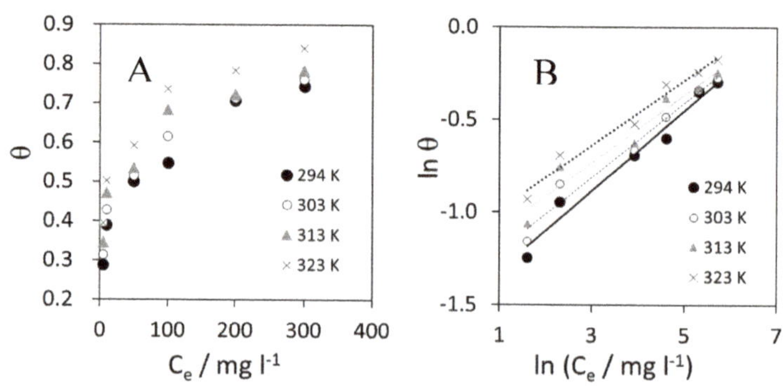

Figure 8. (**A**) Freundlich adsorption plot at different temperatures in the studied range; and (**B**) linearization of Freundlich isotherm.

The equilibrium constant of adsorption K is related to the standard free energy ΔG_{ads} [21]. ΔG_{ads} values at different temperatures can be calculated by Equation (4):

$$K = \frac{1}{55.5}\exp\left(\frac{-\Delta G_{ads}}{RT}\right), \quad (4)$$

where 55.5 represents the concentration of water in solution expressed in mol L^{-1}.

3.5. Electrochemical Tests

3.5.1. Polarizations

Current–potential plots resulting from cathodic and anodic polarization curves of steel in 1 M HCl in the presence of the studied bio-copolymer at various concentrations were recorded. The Tafel plots, recorded with a cathodic-to-anodic polarization of the system are shown in Figure 9.

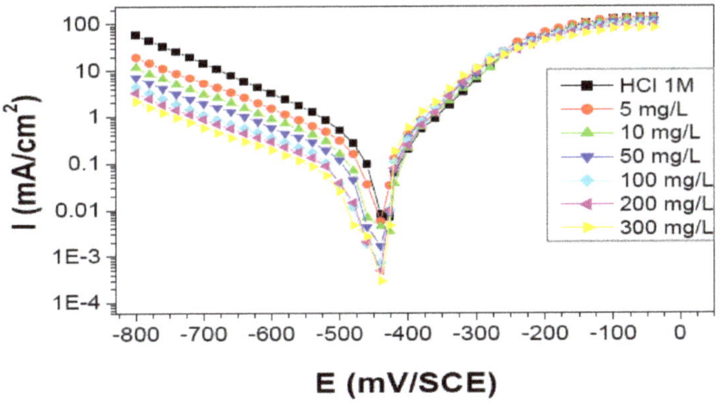

Figure 9. Cathodic and anodic polarization curves of C-Mn steel in 1 M HCl at different concentrations of St63Gly37. Scan rate 1 V min^{-1}.

The electrochemical parameters and the inhibition efficiencies (E_I), determined by the following Equation (5), are presented in Table 3:

$$E_I\% = \left(1 - \frac{I_{corr}}{I^{\circ}_{corr}}\right) \cdot 100, \quad (5)$$

where i_{corr} and i°_{corr} are the corrosion current density values with and without St63Gly37 bio-copolymer inhibitor, respectively, determined by extrapolation of the cathodicbranch of the Tafel plot.

Table 3. Polarization parameters (E_{corr} and I_{corr}) values and inhibition efficiencies of C-Mn steel corrosion in 1 M HCl at different concentrations of St63Gly37 bio-copolymer at 298 K.

Inhibitor Concentration (mg L^{-1})	E_{corr} (mV vs. SCE)	β_c ($Vdec^{-1}$)	i_{corr} (μAcm^{-2})	E_I (%)
0	−439	0.155	304	-
5	−440	0.180	223	27
10	−444	0.185	192	37
50	−455	0.190	157	48
100	−458	0.191	113	63
200	−459	0.187	56	82
300	−460	0.190	27	91

E_{corr}—corrosion potential, i_{corr}—corrosion current density, E_I—inhibition efficiencies, β_c—Tafel slope constant.

3.5.2. Electrochemical Impedance Spectroscopy

A study of C-Mn steel corrosion behavior in 1 M HCl solution with and without St63Gly37 at 298 K was studied by EIS after an immersion time of 30 min. The purpose was to compare and complete the results obtained by the previous weight loss and polarization methods [22]. Nyquist diagrams obtained in the presence of various concentrations of bio-copolymer are shown in Figure 10. The deduced impedance parameters, as charge transfer resistance R_t (Ω cm^2), double-layer capacitance C_{dl} (μFcm^{-2}), and inhibition efficiency (E_{Rt}%), are shown in Table 4.

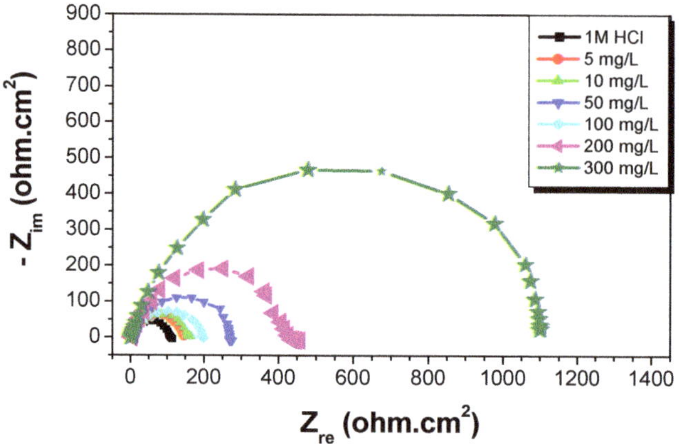

Figure 10. Nyquist plots for C-Mn steel in 1 M HCl at different concentrations of St63Gly37 bio-copolymer (298 K).

Table 4. Impedance parameters for corrosion of C-Mn steel in 1 M HCl at different concentrations of St63Gly37 at 298 K.

St63Gly37 (mg L^{-1})	R_t (Ωcm^2)	f_{max} (Hz)	C_{dl} (μFcm^{-2})	E_{Rt} (%)
0	110	16.32	88.7	-
5	146	12.88	84.6	24
10	167	12.81	74.4	34
50	200	11.39	69.9	45
100	275	9.08	63.8	60
200	458	6.52	53.3	76
300	1100	4.57	31.6	90

R_t—charge transfer resistance, f_{max}—maximum frequency, C_{dl}—double-layer capacitance, E_{Rt}—inhibition efficiency from the charge transfer resistance.

The charge transfer resistance values were qualitatively estimated from the difference in impedance at lower and higher frequencies [23]. The double-layer capacitance was obtained at the frequency f_m, at which the imaginary component of the impedance is maximal ($-Z_{i,max}$) according the following Equation (6):

$$C_{dl} = \frac{1}{2\pi f_m \cdot Rt}. \quad (6)$$

The inhibition efficiency from the charge transfer resistance was calculated by the following Equation (7):

$$E_{Rt}(\%) = \frac{R'_t - Rt}{R'_t} \times 100, \quad (7)$$

where R'_t and R_t are the charge transfer resistances with and without the St63Gly37 bio-copolymer as a corrosion inhibitor, respectively.

4. Discussions

4.1. St63Gly37 Synthetisis

The ^{1}H NMR spectrum interpretation of the product St63Gly37 is carried out by comparison with the peaks of the native starch found in the literature [20]. The spectrum of native starch is shown in Table 5. The glycerin has two chemical shifts at 4.3 ppm corresponding to the proton of the CH, and at 4.4 ppm corresponding to the protons of CH_2.

Table 5. Chemical shifts of native starch [20].

Chemical Shifts δ (ppm)	Attributions
3.6	H 2,3,4,5,6
4.6	H_7
5.1	H_1
5.6	H_8, H_9

Chemical shifts of native starch and St63Gly37 bio-copolymer are comparable. The starch and glycerin signals are also found together, thus proving the success of the grafting reaction. The glycerin OH signals can be confused with those of the starch OHs (H_8, H_9).

After concluding that glycerin is well grafted on starch, the question that still remains is how this grafting takes place. Glycerin grafting may occur preferentially on the 7-position of the starch. It could also occur on positions 8 or 9 of the starch (see Figure 2). On the other hand, glycerin can be grafted via its OH-13 or OH-14 functions. Indeed, the possible reacting OH hydrogens are: H_7, H_8 and H_9. For the positions H_8 and H_9, they are still visible on the ^{1}H NMR spectrum displayed in Figure 3 and Table 3. We can; therefore, conclude that H_7 reacted to give modified patterns according to the structure shown in Figure 2. The presence of the peaks at chemical shifts of 4.4 ppm (triplet) and 4.5 ppm (doublet) correspond, respectively, to the CH and CH_2 of the structure.

The ^{1}H NMR spectrum also enables the calculation of relative amounts of starch and glycerin in St63Gly37. The copolymer contains 63% starch and 37% glycerin. The chemical shifts are represented in the following Table 6.

Table 6. Chemical shifts δ (ppm) of St63Gly37.

Chemical Shifts δ (ppm)	Attributions
3.2–3.7	H 2,3,4,5,6
4.4	H 11
4.5	H 10
4.6	H 13, 12
5.1	H 1
5.4–5.6	H 8,9,14

The FT-IR technique is largely used for the characterization of starch as a natural polymer [20]. Examination of the FT-IR spectrum confirms the grafting of glycerin on the starch, by the CH_2 bands and the characteristic OH bands of glycerin, as reported in Table 7. Figure 4 shows the FT-IR spectra of the native starch in comparison with the St63Gly37 copolymer. It is visible that the C–O–C bond bands change frequency because of the effect of the proportion of glycerin in the copolymer. The two spectra are almost of the same nature. The assignment of the different vibration bands of the native starch have been previously reported [20]. The comparison of the two FT-IR spectra shows a slight displacement of the elongation bands.

Table 7. Attribution of the different vibration bands of St63Gly37.

Wavenumber: υ (cm^{-1})	Attribution	Nature
below 600	Vibration of the polysaccharide backbone	
998.71–1000.10	Antisymmetric C–O–C bond	Elongation
1078.59	Antisymmetric C–O–C bond	Elongation
1149.84	Antisymmetric C–O–C bond	Elongation
1362.59–1359.95	–OH in the plan	Deformation
1416.76–1415.06	C–H bond	Deformation
1641.98–1644.62	H–O–H vibration of adsorbed water	Deformation
2929.18–2931.80	C–H and CH2 of a polysaccharide	Elongation
3274.32–3284.70	OH associated	Elongation

4.2. EffectOfinhibitor Concentration

The results of C-Mn steel in HCl solution with different concentrations of St63Gly37 at 25 °C, using weight loss measurements, are reported in Table 8.

Table 8. Data from the literature correlated to the type, concentration, and maximum inhibition efficiency ($E_{w,\max}$) obtained at 25 °C in HCl for Carbon-steel.

Type of Inhibitor	Inhibitor Concentration ($mg\ L^{-1}$)	$E_{w,\max}$ (%)	Ref.
Glycerin-grafted starch	300	94	This work
Ochrosiaoppositifolia extract	25	94	[24]
Olive leavesextract	900	91	[25]
2-amino-4-methylpentanoicacid	7200	87.46	[26]
L-tryptophan	2000	90.8	[27]
Gallicacid	1000	59.64	[28]
3,5-bis(n-aminophenyl)-4-amino-1,2,4-triazole	300	99	[29]
Vanillin	1520	86.1	[30]

Results showed that the inhibition efficiency calculated according to Equation (1) increases with increasing inhibitor concentrations. The highest concentration of inhibitor, equal to 300 mg L^{-1}, corres ponded to a maximum efficiency of 94%. For comparison, in Table 8, some data from the literature are reported. The cited papers deal with corrosion inhibition in a HCl environment by some investigated bio-based or synthetic compounds. A comparison of reported data demonstrates how the inhibition efficiency varies depending on the species involved in the inhibition mechanism. Moreover, it is also noticeable that the quantity needed to achieve a satisfying corrosion rate control strongly depends on the selected compounds. In a few cases, reported in Table 8, the used quantities are even excessive, losing the reasonable meaning of "corrosion inhibitors" as "compounds added in low amount to the aggressive environment". The concentration of 300 ppm used in this work was not increased more for the purpose of keeping the inhibitor amount in a logical range.

4.3. Effect of Temperature

The variation of corrosion rate (W_{corr}) and inhibition efficiency E_w (%) in the temperature range 25–50 °C for different concentrations of St63Gly37 bio-copolymer, obtained by weight loss studies, are displayed in Figures 5 and 6.

Results showed that the inhibition efficiency increases with increasing temperature. As expected, corrosion process and inhibition efficiency are significantly dependent on the temperature [31–33]. Optimum temperature was found equal to 323 K, with a maximum efficiency of 98.07% with inhibitor concentration of 300 mg L^{-1}.

In the absence of corrosion inhibitor (blank solution), corrosion rate increases with increasing temperature, but when St63Gly37 bio-copolymeris added, the dissolution of C-Mn steel is widely retarded. These results indicate that the corrosion inhibition mechanism might be more complex than a simple physisorption process on the steel surface. The values of inhibition efficiency, obtained using the weight loss method in the experimented temperature range, show that higher temperatures might favor the inhibitor sorption onto the steel surface. This might be explained in terms of chemisorption of polymer on the steel surface. In fact, in case of chemisorption, the extent of adsorption increases with rise in temperature, as reported in a previous work [34].

4.4. Thermodynamic and Kinetic Parameters

Results showed that the corrosion process for C-Mn steel increases more rapidly according to the temperature in the absence of inhibitor, rather than in its presence. This result confirms that the inhibitor acts as an efficient corrosion inhibitor in the range of temperatures studied.

Enthalpy and entropy of the corrosion process may be evaluated from the effect of temperature by an alternative formulation of transition state, as displayed in the following Equation (8) [35].

$$W = \frac{RT}{Nh}\exp\left(\frac{\Delta S^{\circ}{}_{a}}{R}\right)\exp\left(-\frac{\Delta H^{\circ}{}_{a}}{RT}\right), \quad (8)$$

where h is Plank's constant, N is Avogadro number, and $\Delta S^{\circ}{}_{a}$ and $\Delta H^{\circ}{}_{a}$ are the entropy and enthalpy of activation, respectively. Table 9 presents the calculated values of E_a, $\Delta S^{\circ}{}_{a}$, and $\Delta H^{\circ}{}_{a}$ in inhibited and uninhibited corrosive solutions. It is observed that the activation energy value is higher in the presence of the bio-copolymer inhibitor than in the uninhibited solution. The obtained activation energy value of the corrosion process in the inhibitor's presence, compared to its absence, can be attributed to its sorption. It is the result of electrostatic attraction between charged metal surface and charged species in solution and/or chemical interaction between polymer and metal.

Table 9. Calculated parameters at different concentrations of the bio-copolymer.

Inhibitor Concentration (mg L^{-1})	E_a (kJmol^{-1})	ΔH_a (kJmol^{-1})	ΔS_a (Jmol^{-1})
0	40.9	38.5	−36.7
5	36.6	34.1	−37.1
10	35.7	33.3	−37.2
50	35.3	32.9	−37.4
100	26.5	24.0	−37.5
200	27.5	25.0	−38.0
300	28.8	26.3	−38.1

(E_a—activation energy, $\Delta S^{\circ}{}_{a}$ and $\Delta H^{\circ}{}_{a}$—the entropy and enthalpy of activation).

The values of $\Delta H^{\circ}{}_{a}$ are reported in Table 9. The positive sign of the reflects the endothermic nature of the steel dissolution process and values vary in the same way with inhibitor concentration and acid solutions [36].

On the other hand, values of are more positive in the uninhibited solutions and decrease by increasing the inhibitor concentration. Large and negative values of entropies imply that the activated complex in the rate-determining step represents an association rather than a dissociation step, meaning that a decrease in disordering takes place on going from reactants to the activated complex. A similar observation has been reported in the literature [37].

One can notice that E_a and $\Delta H^{\circ}{}_{a}$ values vary in the same way (Figure 11). This result allows for the verification of the known thermodynamic relationship between the E_a and $\Delta H^{\circ}{}_{a}$, as shown in Equation (9).

$$\Delta H^{\circ}{}_{a} = E_a - RT \quad (9)$$

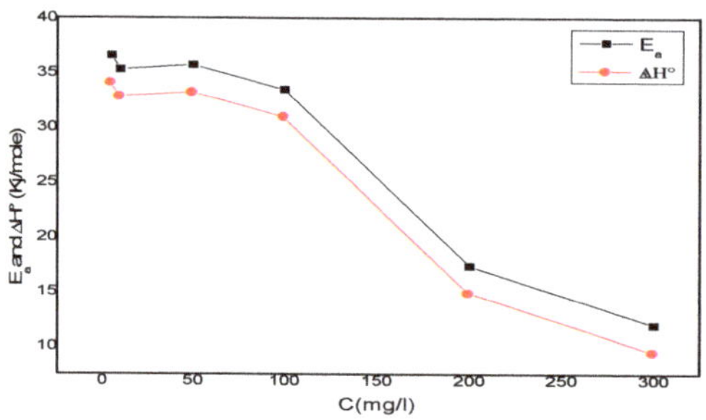

Figure 11. E_a and ΔH° variation depending on the corrosion inhibitor concentration.

4.5. Adsorption Isotherm

As reported above, data indicate that a complex interaction mechanism takes place between the steel and the inhibitor, and a chemisorption might occur. Given this awareness, the analysis, reported in this paragraph, follows anyway the literature approach of applying isotherm adsorption models to verify the most experimental and realistic data. Adsorption isotherms are in fact used to understand the mechanism of metal–inhibitor interaction. The most frequently used isotherms are Langmuir, Frumkin, Temkin, and Parson [38,39]. The type of adsorption isotherm provides information about the interaction among both the adsorbed molecules themselves and their interactions with the metal surface. The most widely used isotherm employs the Langmuir model, whose primary assumptions are: (i) Adsorbate molecules attach to the active sites of the adsorbent surface, (ii) the Langmuir equation assumes that adsorption is monolayer, and (iii) all the sites on the solid surface are equal in size and shape and have equal affinity for adsorbate molecules [40]. However, the last two conditions are hard to fulfill in the corrosion studies, and this is the main weak point in terms of using the Langmuir model, as already evidenced by the literature [8]. Many practical cases, in fact, cannot be described the Langmuir model. The more complex adsorption models take into consideration factors such as the surface heterogeneity and the presence of areas having different adsorption energy or interactions between the adsorbed molecules.

The Freundlich isotherm is empirical and very widely used to describe the adsorption characteristics when the energy of adsorption on a homogeneous surface is independent of surface coverage. Its linearized form is described by the following Equation (10):

$$\ln\,\theta = \ln\,K + \frac{1}{n}\ln\,C_e \tag{10}$$

where θ is the surface coverage (calculated as E_w/100), C_e is the adsorbate concentration in solution at equilibrium (in mg L^{-1}), Kis the equilibrium constant, and $1/n$ is a measure of intensity of adsorption. If the plotln θ vs. $\ln C_e$ displays a linear trend, it means that the adsorption process can be described by the Freundlich model [41].

If ΔG_{ads} is lessthan −10.40 kJmol^{-1}, it can be inferred that the inhibitor interacts on the C-Mn steel surface by electrostatic effect.

The negative values of $\Delta G^\circ{}_{ads}$, displayed in Table 8, confirmed the spontaneity of the process and stability of the adsorbed layer on the steel surface [42,43]. The obtained values of $\Delta G^\circ{}_{ads}$ show the dependence of $\Delta G^\circ{}_{ads}$ on temperature (Table 10), indicating a strong interaction between the bio-polymer molecules and the metal surface.

Thermodynamically, $\Delta G^\circ{}_{ads}$ is related to the standard enthalpy and entropy of the adsorption process, $\Delta H^\circ{}_{ads}$ and $\Delta S^\circ{}_{ads}$, respectively, via the Equation (11):

$$\Delta G^\circ{}_{ads} = \Delta H^\circ{}_{ads} - T\Delta S^\circ{}_{ads} \tag{11}$$

The positive value of $\Delta H°$ suggests an endothermic nature of the metal dissolution process in the presence of the inhibitor. The positive value of $\Delta S°$ indicates that the adsorption process is accompanied by an increase in entropy, which is the driving force for the adsorption of the inhibitor onto the metal surface [31].

Table 10. Variation of the thermodynamic parameters according to the Langmuir isotherm at the studied temperatures.

T (K)	K_{ads}	$\Delta G°_{ads}$ (kJmol^{-1})	$\Delta H°_{ads}$ (kJ mol^{-1})	$\Delta S°_{ads}$ (Jmol^{-1}K^{-1})
294	0.223	−6.15		
303	0.244	−6.56	9.22	52.2
313	0.278	−7.12		
323	0.312	−7.65		

K_{ads}—adsorption parmeter, $\Delta G°_{ads}$—adsorption enthalpy, $\Delta H°_{ads}$—enthalpy, $\Delta S°_{ads}$—entropy of the adsorption process.

4.6. ElectrochemicalResults

4.6.1. Polarizations

The values reported in Table 3 showed results in agreement with what was obtained by the weight loss method: Corrosion current densities decrease when the concentration of the inhibitor increases. Maximum value of inhibition efficiency of 91% was obtained for inhibitor concentration of 300 mg L^{-1}.

The obtained polarization curves indicate that the addition of the bio-copolymer influences the kinetics (decrease in reaction rate) but does not modify the mechanism of the cathodic process [22].

From a thermodynamic point of view, at the acidic pH (1 MHCl) and E_{corr} (ca. −450 mV vs. SCE) of the investigated system, both the oxygen reduction reaction and hydrogen evolution reaction could potentially take place as the iron Pourbaix diagram states [23].

As for the kinetic aspects related to the cathodic Tafel slope calculation, it is apparent that slope values are markedly different from the theoretical ones expected in the case of pure hydrogen evolution reaction. Tafel slope is in fact known to be 118, 39, or 29.5 mV/decade when the discharge reaction, electrochemical desorption reaction, or recombination reaction is rate-determining, respectively [44]. This might mean that more than one reaction is taking place in the considered potential range.

However, it is shown in Table 3 that Tafel slopes do not remarkably change with addition of different concentrations of inhibitor, thus confirming the hypothesis that the inhibitor does not significantly change the cathodic reaction mechanisms. Therefore, the use of the polarization curves as a tool for deriving quantitative parameters, such as inhibition efficiency, can be justified since it is based on relative calculations (using current densities values in the presence and absence of inhibitor) and does not imply an absolute evaluation.

Considering the anodic reaction, C-Mn steel corrodes in acid medium according to an electrochemical process by anodic dissolution of iron. This phenomenon might be accentuated if other parameters intervene, such as microbiological activity or mechanical stresses. The almost constant value of E_{corr} at a first sight might give an indication that the inhibitor is of mixed type acting by adsorption mechanism. However, the cathodic branches trend, due to inhibitor concentration increase, and the overlap of all the anodic branches points to the inhibitor being a cathodic one. This is also in agreement with results obtained at different temperatures. In fact, data showed that the corrosion inhibition mechanism might be more complex than a simple physisorption process on the steel surface.

4.6.2. Electrochemical Impedance Spectroscopy

According to results reported in Figure 10 and Table 4, the impedance diagrams show a semi-circle, indicating that the charge transfer process mainly controls the steel corrosion. By increasing the concentration of St63Gly37, the charge transfer resistance increases and double-layer capacitance (*Cdl*) decreases. A minimum value of *Cdl b* of 31.64 μFcm^{-2} was obtained for inhibitor concentration of 300 mg L^{-1}. The decrease in *Cdl* can be explained by the sorption of inhibitor compound on steel surface leading to the protection from the aggressiveness of HCl. The kinetics of the cathodic hydrogen evolution reaction is decreased and the chlorides are prevented from crossing the protective barrier [26,27]. The resulting inhibition efficiency (E_{Rt}) increases progressively when the concentration of the inhibitor increases. A maximum value of 90% was obtained for inhibitor concentration of 300 mg L^{-1}. These obtained values are in agreement with results obtained by weight loss and polarization method.

5. Conclusions

C-Mn steel in 1 M HCl solution was investigated to study the corrosion behavior in the temperature range 25–50 °C with and without a glycerin-grafted starch as a bio-copolymer working as a corrosion inhibitor.

Results obtained by the weight loss method showed that the inhibition efficiency increases by increasing the inhibitor concentration. Optimum concentrations of inhibitor equal to 300 mg L^{-1} enabled a maximum efficiency of 94.25%.

The corrosion process and the inhibition efficiency were found to be significantly dependent on the temperature and concentration of inhibitor. The values of inhibition efficiency, obtained using the weight loss method in the experimented temperature range, show that higher temperatures might favor the inhibitor sorption onto the steel surface. Results indicated that the corrosion inhibition mechanism might be more complex than a simple physisorption process on the steel surface. This might be explained in terms of chemisorption of polymer on steel surface.

The obtained values of corrosion potential and corrosion current density, E_{corr} and i_{corr}, obtained by potentiodynamic polarization, are in agreement with the weight loss method: The corrosion current densities decrease when the concentration of the inhibitor increases. An analysis of data indicates that the inhibitor might be of the cathodic type.

The decrease in double-layer capacitance values, obtained by EIS method, follows a decrease similar to that obtained for i_{corr} by the polarization method. The decrease in C_{dl} can be explained by the sorption of inhibitor compound on steel surface leading to the protection from the aggressiveness of HCl.

Results obtained in this study might be developed to supplement corrosion protection in real field applications, injecting the optimum quantities of corrosion inhibitor without any harmful effect on the natural environment and human health.

Author Contributions: Investigation, B.B.; Methodology, A.M.; Supervision, A.B.; Writing–original draft, S.L.; Writing–review & editing, L.T. and S.M.

Funding: This research was funded by Direction Générale de la recherché scientifiqueet du développement technologique, Algerian Ministry of Scientific Research.

Acknowledgments: The authors are very grateful to the Algerian Gas Transport TRC for their great help to make a part of this work by providing us with some steel samples. They also thank all researchers at the LAEPO Research Laboratory from Tlemcen University for their experimental contribution during the corrosion test, and the equipment support from the Algerian University and Society. They also appreciate the contribution of the Department of Environmental Science and Policy from Università degli Studi di Milano.

Conflicts of Interest: The authors declare no conflicts of interest. The funders had no role in the design of the study; in the collection, analyses, or interpretation of data; in the writing of the manuscript, or in the decision to publish the results.

References

1. Nazarov, A.P.; Thierry, D. Probing of Atmospheric Corrosion of Metals: Carbon Steel. *Prot. Met.* **2004**, *40*, 377–388. [CrossRef]
2. Jones, T. Steel Industry and the Environment: Technical and Management Issues, The Institute. 1997. Available online: https://books.google.it/books/about/Steel_Industry_and_the_Environment.html?id=9DkL7pfXu30C&redir_esc=y (accessed on 29 July 2019).

3. Choi, H.J.; Warnken, D.; Al Beheiri, F.I.; Al-Bannai, N.S. Field Corrosion Assessment of L80 Carbon Steel Downhole Production Tubing at Khuff Gas Wells. 2006. Available online: https://www.onepetro.org/conference-paper/NACE-06653 (accessed on 29 July 2019).
4. Belahcene, B.; Benmoussat, A.; Mansri, A.; Doghmane, M.Z. Analyze and modeling of damage behavior of a C0.12%Mn1.02%Si0.29% HLE steel solicited in selected physicochemical medium. *J. Mater. Environ. Sci.* **2015**, *6*, 1765–1770.
5. Bianchi, G.; Mazza, F. Corrosione e Protezione dei Metalli, Associazione Italiana di Metallurgia. 2000. Available online: https://www.libreriacortinamilano.it/scheda-libro/francesco-mazza-giuseppe-bianchi/corrosione-e-protezione-dei-metalli-9788885298354-317855.html (accessed on 19 October 2018).
6. Fiori-Bimbi, M.V.; Alvarez, P.E.; Vaca, H.; Gervasi, C.A. Corrosion inhibition of mild steel in HCL solution by pectin. *Corros. Sci.* **2015**, *92*, 192–199. [CrossRef]
7. Li, Y.; Zhao, P.; Liang, Q.; Hou, B. Berberine as a natural source inhibitor for mild steel in 1 M H_2SO_4. *Appl. Surf. Sci.* **2005**, *252*, 1245–1253. [CrossRef]
8. Marzorati, S.; Verotta, L.; Trasatti, S.P. Green Corrosion Inhibitors from Natural Sources and Biomass Wastes. *Molecules* **2018**, *24*, 48. [CrossRef] [PubMed]
9. Vermeirssen, E.L.; Dietschweiler, C.; Werner, I.; Burkhardt, M. Corrosion protection products as a source of bisphenol A and toxicity to the aquatic environment. *Water Res.* **2017**, *123*, 586–593. [CrossRef]
10. Obot, I.; Onyeachu, I.B.; Kumar, A. Sodium alginate: A promising biopolymer for corrosion protection of API X60 high strength carbon steel in saline medium. *Carbohydr. Polym.* **2017**, *178*, 200–208. [CrossRef]
11. Christopher, G.; Kulandainathan, M.A.; Harichandran, G. Biopolymers nanocomposite for material protection: Enhancement of corrosion protection using waterborne polyurethane nanocomposite coatings. *Prog. Org. Coat.* **2016**, *99*, 91–102. [CrossRef]
12. Chetouani, A.; Medjahed, K.; Sid-Lakhdar, K.; Hammouti, B.; Benkaddour, M.; Mansri, A. Poly(4-vinylpyridine-poly(3-oxide-ethylene) tosyle) as an inhibitor for iron in sulphuric acid at 80 °C. *Corros. Sci.* **2004**, *46*, 2421–2430. [CrossRef]
13. Mansri, A.; Bouras, B. Synergistic effect of AM-4VP-9 copolymer and iodide ion on corrosion inhibition of mild steel in 1 M H_2SO_4. *Res. Chem. Intermed.* **2014**, *2*, 252–271. [CrossRef]
14. Mobin, M.; Khan, M.A.; Parveen, M. Inhibition of mild steel corrosion in acidic medium using starch and surfactants additives. *J. Appl. Polym. Sci.* **2011**, *121*, 1558–1565. [CrossRef]
15. Bello, M.; Ochoa, N.; Balsamo, V.; López-Carrasquero, F.; Coll, S.; Monsalve, A.; González, G. Modified cassava starches as corrosion inhibitors of carbon steel: An electrochemical and morphological approach. *Carbohydr. Polym.* **2010**, *82*, 561–568. [CrossRef]
16. Petroleum Institute. API 5L: Specification for Line Pipe, n.d. Available online: https://law.resource.org/pub/us/cfr/ibr/002/api.5l.2004.pdf (accessed on 19 October 2004).
17. Amara Zenati, A.; Benmoussat, A.; Bourdim, A. Localized corrosion study of X60 pipeline steel in soil solution and inhibitive effect by disodic hydrogeno-phosphate. *J. Mater. Environ. Sci.* **2015**, *6*, 1896–1901.
18. Thygesen, L.G.; Løkke, M.M.; Micklander, E.; Engelsen, S.B. Vibrational microspectroscopy of food. Raman vs. FT-IR. *Trends Food Sci. Technol.* **2003**, *14*, 50–57. [CrossRef]
19. Fringant, C.; Desbrières, J.; Rinaudo, M. Physical properties of acetylated starch-based materials: Relation with their molecular characteristics. *Polymer* **1996**, *37*, 2663–2673. [CrossRef]
20. Tizzotti, M.J.; Sweedman, M.C.; Tang, D.; Schaefer, C.; Gilbert, R.G. New ^{1}H NMR Procedure for the Characterization of Native and Modified Food-Grade Starches. *J. Agric. Food Chem.* **2011**, *59*, 6913–6919. [CrossRef]
21. Umoren, S.; Obot, I.B.; Akpabio, L.; Etuk, S.; Umoren, S. Adsorption and corrosive inhibitive properties of Vigna unguiculata in alkaline and acidic media. *Pigment Resin Technol.* **2008**, *37*, 98–105. [CrossRef]
22. Kartsonakis, I.A.; Stanciu, S.G.; Matei, A.A.; Hristu, R.; Karantonis, A.; Charitidis, C.A. A comparative study of corrosion inhibitors on hot-dip galvanized steel. *Corros. Sci.* **2016**, *112*, 289–307. [CrossRef]
23. Zeng, M.; Li, Y. Recent advances in heterogeneous electrocatalysts for the hydrogen evolution reaction. *J. Mater. Chem. A* **2015**, *3*, 14942–14962. [CrossRef]
24. Raja, P.B.; Fadaeinasab, M.; Qureshi, A.K.; Rahim, A.A.; Osman, H.; Litaudon, M.; Awang, K. Evaluation of Green Corrosion Inhibition by Alkaloid Extracts of Ochrosia oppositifolia and Isoreserpiline against Mild Steel in 1 M HCl Medium. *Ind. Eng. Chem. Res.* **2013**, *52*, 10582–10593. [CrossRef]

25. El-Etre, A. Inhibition of acid corrosion of carbon steel using aqueous extract of olive leaves. *J. Colloid Interface Sci.* **2007**, *314*, 578–583. [CrossRef] [PubMed]
26. Loto, R.T. Corrosion inhibition effect of non-toxic α-amino acid compound on high carbon steel in low molar concentration of hydrochloric acid. *J. Mater. Res. Technol.* **2017**, 1–10. [CrossRef]
27. Fu, J.-J.; Li, S.-N.; Cao, L.-H.; Wang, Y.; Yan, L.-H.; Lu, L.-D. l-Tryptophan as green corrosion inhibitor for low carbon steel in hydrochloric acid solution. *J. Mater. Sci.* **2010**, *45*, 979–986. [CrossRef]
28. Keny, S.; Kumbhar, A.; Thinaharan, C.; Venkateswaran, G. Gallic acid as a corrosion inhibitor of carbon steel in chemical decontamination formulation. *Corros. Sci.* **2008**, *50*, 411–419. [CrossRef]
29. Bentiss, F.; Lagrenee, M.; Traisnel, M.; Mernari, B.; Elattari, H. 3,5-bis(n-Hydroxyphenyl)-4-amino-1,2,4-triazoles and 3,5-bis(n-aminophenyl)-4-amino-1,2,4-triazoles: A new class of corrosion inhibitors for mild steel in 1m HCl medium. *J. Appl. Electrochem.* **1999**, *29*, 1073–1078. [CrossRef]
30. Emregül, K.; Hayvalı, M. Studies on the effect of vanillin and protocatechualdehyde on the corrosion of steel in hydrochloric acid. *Mater. Chem. Phys.* **2004**, *83*, 209–216. [CrossRef]
31. Gapsari, F.; Soenoko, R.; Suprapto, A. Bee Wax Propolis Extract as Eco-Friendly Corrosion Inhibitors for 304SS in Sulfuric Acid. *Int. J. Corros.* **2015**, *2015*, 1–10. [CrossRef]
32. Jokar, M.; Farahani, T.S.; Ramezanzadeh, B. Electrochemical and surface characterizations of morus alba pendula leaves extract (MAPLE) as a green corrosion inhibitor for steel in 1 M HCl. *J. Taiwan Inst. Chem. Eng.* **2016**, *63*, 436–452. [CrossRef]
33. Larif, M.; Elmidaoui, A.; Zarrouk, A.; Zarrok, H.; Salghi, R.; Hammouti, B.; Oudda, H.; Bentiss, F. An investigation of carbon steel corrosion inhibition in hydrochloric acid medium by an environmentally friendly green inhibitor. *Res. Chem. Intermed.* **2013**, *39*, 2663–2677. [CrossRef]
34. Chevalier, M.; Roos, C.; Tomi, F.; Sutour, S.; Lebrini, M. Corrosion Inhibition of C38 Steel in 1 M HCl Media by Alkaloidic Extract from Xylopia Frutescens Amazonian Tree. *ECS Trans.* **2015**, *64*, 1–28. [CrossRef]
35. Bouanis, F.; Jama, C.; Traisnel, M.; Bentiss, F. Study of corrosion resistance properties of nitrided carbon steel using radiofrequency N_2/H_2 cold plasma process. *Corros. Sci.* **2010**, *52*, 3180–3190. [CrossRef]
36. Fares, M.M.; Maayta, A.; Al-Qudah, M.M. Pectin as promising green corrosion inhibitor of aluminum in hydrochloric acid solution. *Corros. Sci.* **2012**, *60*, 112–117. [CrossRef]
37. Ating, E.; Umoren, S.; Udousoro, I.; Ebenso, E.; Udoh, A.; Umoren, S. Leaves extract of Ananas sativum as green corrosion inhibitor for aluminium in hydrochloric acid solutions. *Green Chem. Lett. Rev.* **2010**, *3*, 61–68. [CrossRef]
38. Bentiss, F.; Traisnel, M.; Lagrenee, M. The substituted 1,3,4-oxadiazoles: A new class of corrosion inhibitors of mildsteel in acidic media. *Corros. Sci.* **2000**, *42*, 127–146. [CrossRef]
39. Belkaid, S.; Tebbji, K.; Mansri, A.; Chetouani, A.; Hammouti, B. Poly(4-vinylpyridine-hexadecyl bromide) as corrosion inhibitor for mild steel in acid chloride solution. *Res. Chem. Intermed.* **2012**, *38*, 2309–2325. [CrossRef]
40. Czepirski, L.; Balys, M.R.; Komorowska-Czepirska, E. Some generalization of Langmuir adsorption isotherm. *Internet J. Chem.* **2000**, *3*, 14–1099.
41. Dada, A.O.; Olalekan, A.P.; Olatunya, A.M.; Dada, O.J.I.J.C. Langmuir, Freundlich, Temkin and Dubinin—Radushkevich Isotherms Studies of Equilibrium Sorption of Zn^{2+} Unto Phosphoric Acid Modified Rice Husk. *IOSR J. Appl. Chem.* **2012**, *3*, 38–45.
42. Bouklah, M.; Attayibat, A.; Kertit, S.; Ramdani, A.; Hammouti, B. A pyrazine derivative as corrosion inhibitor for steel in sulphuric acid solution. *Appl. Surf. Sci.* **2005**, *242*, 399–406. [CrossRef]
43. Mansri, A.; Bendraoua, A.; Bouras, B. Polyacrylamide-clay microcomposite as corrosion inhibitor for mild steel in 1m hydrochloric acid solution. *J. Mater. Environ. Sci.* **2016**, *7*, 808–819.
44. Mansfeld, F. Recording and Analysis of AC Impedance Data for Corrosion Studies. *Corrosion* **1981**, *37*, 301–307. [CrossRef]

applied sciences

MDPI

Article

The Protection Role of Cysteine for Cu-5Zn-5Al-1Sn Alloy Corrosion in 3.5 wt.% NaCl Solution

Kebede W. Shinato †, Feifei Huang †, Yanpeng Xue, Lei Wen and Ying Jin *

National Center for Materials Service Safety, University of Science and Technology Beijing, Beijing 100083, China
* Correspondence: yjin@ustb.edu.cn; Tel.: +86-17-8010-50028
† These two authors contribute equally to this manuscript.

Received: 4 July 2019; Accepted: 8 August 2019; Published: 17 September 2019

Abstract: In this work, the corrosion mechanism of a Cu-5Zn-5Al-1Sn alloy was examined in a 3.5 wt.% NaCl solution. At the same time, the effect of a cysteine inhibitor was also investigated through a multi-analytical approach. Electrochemical results suggested that inhibition efficiency increased with the increase of cysteine concentration. From potentiodynamic polarization (PD) analysis, a decrease in corrosion current and corrosion potential shift toward a more negative direction was observed. The potential difference between the blank and inhibited surface was found to be 46 mV, which is less than 85 mV, revealing a mixed type inhibition effect of cysteine for the Cu-5Zn-5Al-1Sn alloy. The inhibition mechanism of cysteine (Cys) and the effect of alloying elements were investigated by fitting experimental impedance data according to a projected equivalent circuit for the alloy/electrolyte interface. A Langmuir adsorption isotherm was proposed to explain the inhibition phenomenon of cysteine on the Cu-5Zn-5Al-1Sn alloy surface. Surface morphology observation confirmed that the Cu-5Zn-5Al-1Sn alloy was damaged in 3.5 wt.% NaCl solution and could be inhibited by using the cysteine inhibitor. The impact of alloying elements on the corrosion mechanism was further examined by surface analysis techniques such as X-Ray photoelectron spectroscopy (XPS)/Auger spectra, the results of which indicated that the corrosion inhibition was realized by the adsorption of the inhibitor molecules at the alloy/solution interface.

Keywords: corrosion inhibitor; corrosion mechanism; cysteine; thin film

1. Introduction

In recent years, research on copper alloy corrosion have garnered more attention in the industrial sector since copper and its alloys form a number of industrially important materials [1]. Based on alloying elements, the important properties range from good thermal properties, electrical conductivities, formability, and visual appearance, to good corrosion resistance [2–6]. Cu-5Zn-5Al-1Sn (Cu5Zn5Al1Sn) consists of 89 wt.% copper, 5 wt.% aluminum, 5 wt.% zinc, and 1 wt.% tin, with the latter usually used for the front covering due to its shiny and golden appearance [7]. The features of the microstructure of the Cu5Zn5Al1Sn alloy have been studied by comparing it with pure copper [8]. The microstructure of Cu5Zn5Al1Sn owns numerous properties that are promising from a corrosion resistance perspective. These properties consist of meaningfully smaller grain size (2.1 μm versus 10.4 μm) and a higher segment of coherent twin boundaries [8]. Smaller grain size and a closer, well-ordered grain boundary illustrate the relatively high corrosion resistance of the alloy [8–10]. The Cu5Zn5Al1Sn alloy forms relatively poorly soluble oxide films with its constituent metals in the presence of chlorides when exposed to long-term outdoor and short-term laboratory conditions [11]. Even though corrosion resistant behavior of this golden alloy is much better than that of other comparable alloys, it is found to be easily corroded in acidic media. Similarly to Cu metal, Cu_2O and $Cu_2(OH)_3Cl$ are the principal corrosion products produced from the Cu5Zn5Al1Sn alloy in marine outdoor conditions, with the latter largely present at the outmost surface of both Cu and the Cu5Zn5Al1Sn alloy [7,11]. Although a big difference exists in their corrosion rates, the corrosion process is highly administered by the molecular diffusion between the metal surface and the electrolyte [12].

One approach to prevent corrosion of Cu and its alloy is the use of corrosion inhibitors [13,14]. Among the corrosion inhibitors used in practice, cysteine (Cys) is found to be widely employed for Cu protection in various media [15,16]. Cysteine is an amino acid that contains a –SH group in addition to the amino group; this mercaptan group is strongly attracted to copper. A. A. Nazeer et al. have studied the inhibitive effect of cysteine on a Cu10Ni alloy in sulfide containing atmospheres [16]. Based on this study, cysteine can act as a mixed-type inhibitor and cysteine molecules are adsorbed on the alloy surface. The inhibition mechanism of cysteine with copper-based materials is characterized by the formation of a stable Cu (I)–cysteine complex [17]. As confirmed by many researchers, the better corrosion inhibition effect of cysteine relies on its surface adsorption through sulfur atoms [18,19]. I. Milošev et al. on the other hand have studied different amino acids as inhibitors for copper in acidic environments and confirmed that cysteine has higher efficiency. Based on their molecular dynamic simulation results, it has been suggested that the –SH group is in charge of the good protective effect of cysteine [20]. G. M. Abd El-Hafez et al. have also investigated the protective action of methionine, N-acetyl cysteine, and cysteine for a Cu-10Al-5Ni alloy in 3.5 wt.% NaCl solution and they concluded that cysteine showed higher efficiency ascribed to the existence of the mercaptan moiety [21]. On the other hand, our previous study regarding the effect of cysteine on copper metal in corrosive environments also showed its effectiveness on corrosion inhibition.

It can be seen from the above investigated results cysteine can be a very promising corrosion inhibitor for copper and its alloys, and no information has been found on the action of cysteine for the inhibition of Cu5Zn5Al1Sn alloy corrosion. Furthermore, the interaction between the cysteine molecule and the alloy surface in chloride-containing solution has not yet been investigated. In the present paper, the corrosion and corrosion inhibition of the Cu5Zn5Al1Sn alloy in chloride solutions is investigated. The influence of various cysteine concentrations is also taken into account. The performance of the studied inhibitor is evaluated using various electrochemical and surface analysis techniques. The inhibition efficiency, surface behavior, and adsorption mechanism are clearly analyzed and summarized using the obtained experimental results.

2. Materials and Methods

A commercial Cu5Zn5Al1Sn alloy in ther form of a 1 mm thick sheet (89 wt.% Cu, 5 wt.% Zn, 5 wt.% Al, and 1 wt.% Sn, equivalent to 84 at.% Cu, 11 at.% Al, 4.5 at.% Zn, and 0.5 at.% Sn) was obtained from Aurubis (Hamburg, Germany). The specimens were wrapped in epoxy resin, exposing a 1 cm^2 surface area. Before any test, the alloy was ground by silicon carbide paper from #800 to #2000 and then successively polished with diamond paste to 2.5, 1.5, and 0.5 μm, respectively. The diamond polished specimens were rinsed with analytical grade ethanol and dried in air.

3.5 wt.% sodium chloride (NaCl) was used as the corrosive solution. It was made from an analytical grade reagent of NaCl and deionized water. Cysteine (KC90277-100gm, Suzhou tianke Co. Ltd., Suzhou, China) was used as the inhibitor and stock solution of 1×10^{-2} M was prepared by mixing a suitable amount of cysteine in distilled water. Solutions with cysteine concentrations of 1×10^{-3} to 1×10^{-5} M were set from the stock solution using a dilution method.

The electrochemical cell used was a three-electrode glass cell with a capacity of 400 mL. A silver/silver chloride electrode (Ag/AgCl) was used as the reference electrode. All potential values in the manuscript were referred to using Ag/AgCl. A flat platinum sheet (4 cm^2) was used as the counter electrode. An electrochemical workstation Gamry (Reference 600) with a high input resistance of 10^{14} Ω, a current detection limit of 60 pA, and a current resolution of 20 aA was used to conduct electrochemical measurements. Prior to the tests of electrochemical impedance spectroscopy (EIS), the specimen was deepened in the test solution for 60 min. EIS was carried out under potentiostatic conditions in the frequency range of 100 kHz to 0.01 Hz at open circuit potential (OCP) with an amplitude of 5 mV.

The EIS data was analyzed by Zsimpwin software to determine a simulated circuit and respective parameters. Tafel curves were obtained from potentiodynamic polarization study which was performed by varying the electrode potential automatically from −800 mV to +400 mV with respect to OCP at a rate of 1 mV/s.

An FEI Quanta 250 scanning electron microscope was used to observe the surface morphology of the Cu5Zn5Al1Sn alloy specimen with and without the cysteine inhibitor in the 3.5 wt.% NaCl solution. X-ray photoelectron spectroscopy (XPS) and Auger electron spectroscopy (AES) measurements were performed on an ESCALAB 250Xi spectrometer with Al Kα X-ray (1486.6 eV) irradiation as the photo source. Binding energy (BE) was calibrated against the C1s line of the aliphatic carbon contamination set at 284.8 eV. Xpspeak41 software was used for data fitting, which was carried out graphically within the constraints of Gaussian peak shapes.

3. Results

3.1. Potentiodynamic Polarization

Figure 1 presents Tafel polarization curves for the Cu5Zn5Al1Sn alloy exposed to 3.5 wt.% NaCl solution with and without the Cys inhibitor. The trend of polarization curves suggests that a low amount addition of cysteine (10^{-4} M, 10^{-5} M) in 3.5 wt.% NaCl solution moves the curves to the lower current region in the anodic potential route compared with the blank solution. Nevertheless, the higher concentration (10^{-3} M, 10^{-2} M) of the inhibitor in solution causes a shift of the curves to a more cathodic potential region and decreases the corrosion current meaningfully. Consequently, cysteine influences both the cathodic and anodic reactions. Hence, cysteine is shown to act as a mixed-type inhibitor to the Cu5Zn5Al1Sn alloy in 3.5 wt.% NaCl and to retard majorly cathodic corrosion reactions [22–25]. The corrosion current density (Icorr) and corrosion potential (Ecorr) were calculated by fitting the experimental results with Gamry Echem Analyst software 5.61 and listed in Table 1. The inhibition efficiency (μ%) was also obtained by Equation (1) using parameters listed in Table 1 [26].

$$\mu\% = \frac{\text{Icorr} - \text{Icorr(inh)}}{\text{Icorr}} \times 100\% \quad (1)$$

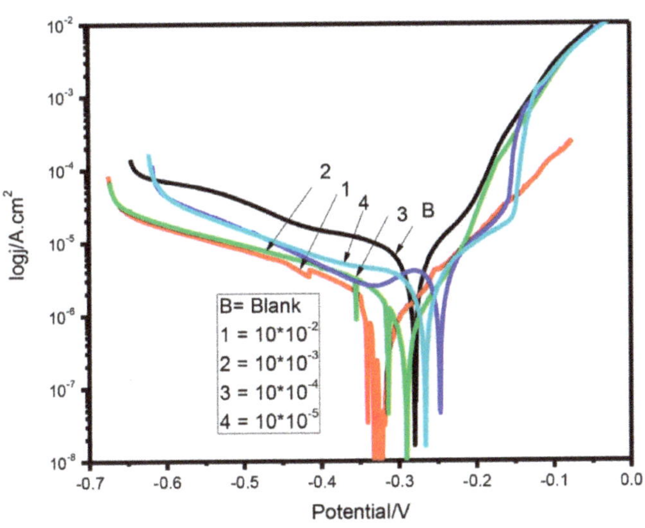

Figure 1. Potentiodynamic polarization curves for the Cu5Zn5Al1Sn alloy after 60 min immersion in 3.5 wt.% NaCl in the absence (Blank) and presence of different concentrations of cysteine (10^{-5} to 10^{-2}).

Table 1. Polarization parameters for the Cu5Zn5Al1Sn alloy measured in 3.5 wt.% NaCl solution in the absence and presence of different concentrations of cysteine.

Cysteine Concentration	E_{corr} (V versus Ag/AgCl)	βa (mV/Decade)	βc (mV/Decade)	I_{corr} (μAcm^{-2})	µ%
Blank (0)	−0.280	69.18	102.3	3.44	-
10^{-5} M	−0.226	77.77	127.2	1.84	46.5
10^{-4} M	−0.247	48.5	63.86	1.78	48.2
10^{-3} M	−0.292	61.23	56.96	0.483	85.9
10^{-2} M	−0.326	23.71	19.49	0.094	97.2

In this equation Icorr (inh) and Icorr are the current densities of the working electrode in the 3.5 wt.% NaCl solution with and without cysteine, respectively.

The Icorr is inversely related to the inhibitor concentration, i.e., an increase in inhibitor concentration results in a lower value of Icorr. This is probably due to the improved protection of the Cu5Zn5Al1Sn alloy by cysteine molecules. Therefore, the mitigation of the Cu5Zn5Al1Sn alloy deterioration in 3.5 wt.% NaCl solution may associated with the development of a protective layer from the adsorbed inhibitor species on the alloy surface [27]. The adsorption layer of cysteine molecules on the alloy surface can hinder the movement of corrosive species, resulting in a reduction in corrosion rate [2,23,24,28,29].

It is evident from Table 1 that both βc and βa are varied on the addition of the inhibitor. From this behavior, it is suggested that both anodic and cathodic reactions are retarded by the protective layer formed by inhibitor molecules and the alloy/electrolyte interface [3,23,29]. In addition, βc values are greater than βa values at all inhibitor concentrations and the variation in βc with the inhibitor concentration is greater than the variations in βa. Both the results give the same evidence that cysteine is more efficient in inhibition of cathodic reactions than that of anodic reactions [3,30]. The efficiency of cysteine increases with its concentration and reaches a maximum value of 97.2% at the concentration 10^{-2} M.

3.2. Electrochemical Impedance Spectroscopy

Nyquist and Bode diagrams of the Cu5Zn5Al1Sn alloy in 3.5 wt.% NaCl solution with and without different amounts of cysteine are illustrated in Figure 2a,b, correspondingly. From Figure 2a, the Nyquist diagram for Cu5Zn5Al1Sn alloy in 3.5 wt.% NaCl solution contains a capacitive loop at intermediate frequency and a straight line in the low frequency region. A straight line at low frequency area may be attributed to the presence of the Walberg constant as a result of the movement of soluble metal species from the alloy surface to the bulk solution [23]. The diameter of the capacitive loop rises in the presence of Cys and the straight line in the low frequency region is removed at higher Cys concentrations, which may be attributed to the inhibitive effect of Cys on the corrosion process [31]. The results of Bode plots (Figure 2b) show that the frequency range with maximum phase becomes larger with increasing Cys concentration. There is a shift in the phase maximum to a lower frequency region and the phase angle is raised to about 80°, in the inhibitor concentration of 10−2 M. These results suggest that the protective ability of the Cu5Zn5Al1Sn alloy is increased with inhibitor concentration, showing that inhibitor particles effectively adsorb on the alloy surface [32,33].

The results of the EIS experiment were fitted with a simulated circuit (Figure 3) and the value of electrical elements that make up the equivalent circuit are presented in Table 2. The equivalent circuit parameters are solution resistance (Rs), film resistance (Rf), film capacitance (Qf), charge transfer resistance (Rct), double layer capacitance (Qdl), and Warburg resistance (W), which result from ionic diffusion of the corrosion product to the electrode surface. The EIS fitting result of the Cu5Zn5Al1Sn alloy in blank 3.5%wt NaCl solution shows a Warburg resistance and a lower value of Rct, as shown in Figure 2a and Table 2, respectively. The Warburg impedance is included to account for the diffusion of corrosion products to the bulk solution and/or dissolved oxygen to the electrode surface. With a lower

Cys concentration, the Warburg impedance remains unchanged, and Rf and Qf are observed to address the film formed between the Cys molecules and alloying elements. The presence of W is attributed to the low inhibition efficiency of the lower concentration of Cys, which may be related to the thinner and unstable film formed which can be easily dissolved to the bulk solution. When more than 10^{-4} M Cys is added to the solution, the Warburg impedances vanish and the Rct values become larger. This shows the presence of high resistance to transfer charge from bulk solution to specimen surface and/or and specimen surface to bulk solution; hence, the diffusion process is controlled. This may be attributed to the development of a thick and protective Cys/metal film on the alloy surface.

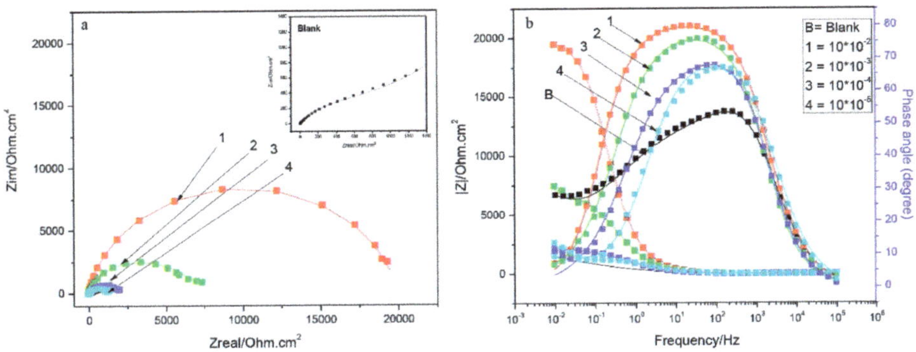

Figure 2. Nyquist (**a**) and Bode (**b**) plots for the Cu5Zn5Al1Sn alloy in 3.5 wt.% NaCl with and without different concentrations of cysteine (10^{-5} to 10^{-2}) (scattered: experimental data; lines: fitting data).

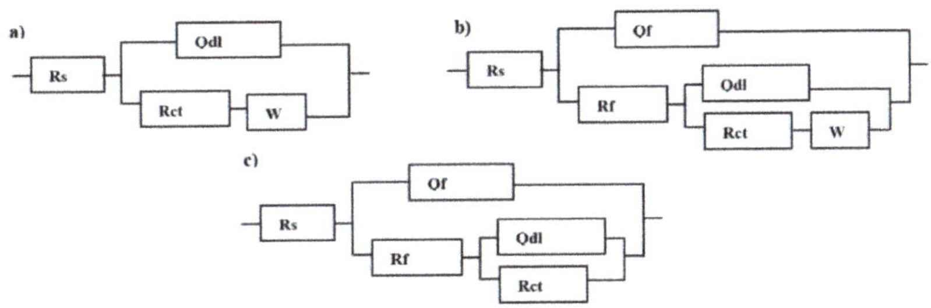

Figure 3. Electrical equivalence circuit for Cu5Zn5Al1Sn alloy in 3.5 wt.% NaCl without R(Q(RW)) (**a**), with 10^{-5} M Cys, R(Q(R(Q(RW)))) (**b**) and 10^{-4}–10^{-2} M Cys, R(Q(R(QR))) (**c**). Legend: solution resistance (Rs), film resistance (Rf), film capacitance (Qf), charge transfer resistance (Rct), double layer capacitance (Qdl), and Warburg resistance (W).

Table 2. Impedance parameters for the Cu5Zn5Al1Sn alloy in 3.5 wt.% NaCl with and without different concentrations of cysteine obtained by using ZSimpWin 3.50.

Cys Conc. (M)	Rs (Ω)	Qf	Rf (Ω)	Qct	Rct (KΩ)	W	Rp (KΩ)	µ%
Blank	2.668			1.55×10^{-4}	692.7	0.0032	0.695	-
10^{-5}	3.224	8.09×10^{-5}	588.1	1.64×10^{-4}	1169	0.0053	1.197	40.76
10^{-4}	2.894	1.05×10^{-4}	212.7	1.61×10^{-4}	1.986		2.201	67.78
10^{-3}	3.345	8.24×10^{-5}	2382	1.10×10^{-4}	6.399		8.784	91.92
10^{-2}	3.067	5.28×10^{-5}	12.02	4.41×10^{-6}	19.940		19.955	96.44

Inhibition efficiency was obtained by using Equation (2) [34–36], the values of which are listed in Table 2. The inhibition efficiency of Cys for Cu5Zn5Al1Sn alloy corrosion in 3.5 wt.% solution is seen to increases with its concentration, and a maximum inhibition efficiency value of 96.44% is recorded for the case of the Cys concentration being 10^{-2} M. The inhibition efficiencies obtained from polarization and EIS investigations are similar and follow the same trend, which implies that the explanation for Cu5Zn5Al1Sn alloy corrosion characteristics on the basis of Icorr and Rp could be well-thought-out as valid and reliable.

$$\mu\% = \frac{\text{Rp(inh)} - \text{Rp}}{\text{Rp(inh)}} \times 100\% \tag{2}$$

where μ, Rp, and Rp (inh) are inhibition efficiency, and polarization resistance for uninhibited and inhibited surfaces, respectively.

The inhibition efficiency results obtained from potentiodynamic (PD) and EIS experiments were used to observe the adsorption behavior of cysteine on the Cu5Z5Al1Sn alloy surface. Therefore, different adsorption isotherms were tested to determine the best suitable model to explain surface phenomena [23,37,38].

$$\frac{C}{\theta} = \frac{1}{\text{Kads}} + C \tag{3}$$

$$\theta = \frac{RT}{b}\text{lnKads} + \frac{RT}{b}\text{lnC} \tag{4}$$

In this equation, C, θ, K_{ads}, and b are inhibitor concentration, the inhibitor's surface coverage, the equilibrium constant related to the interfacial molecular adsorption-desorption phenomena, and a constant relevant to the properties of both adsorbent and adsorbate in the Temkin model, respectively.

Figure 4 shows plots of Langmuir and Temkin adsorption isotherms determined by the respective equations listed in Equations (3) and (4), respectively. As observed in Figure 4a, a slope of 0.99 with $R^2 = 0.99$ derived from EIS and PD results which is close to unity can be observed from a C/θ versus C linear plot for the case of the Langmuir adsorption isotherm. On the other hand, a regression of 0.89 and 0.83 acquired from EIS and PDP, respectively, was obtained from the plot of θ versus ln C based on the Temkin adsorption isotherm (Figure 4b). By comparing the two isotherms, the Langmuir adsorption isotherm, which was seen to have less fitting error, was chosen as the more suitable model to describe the surface inhibition mechanism of cysteine for Cu5Z5Al1Sn alloy corrosion.

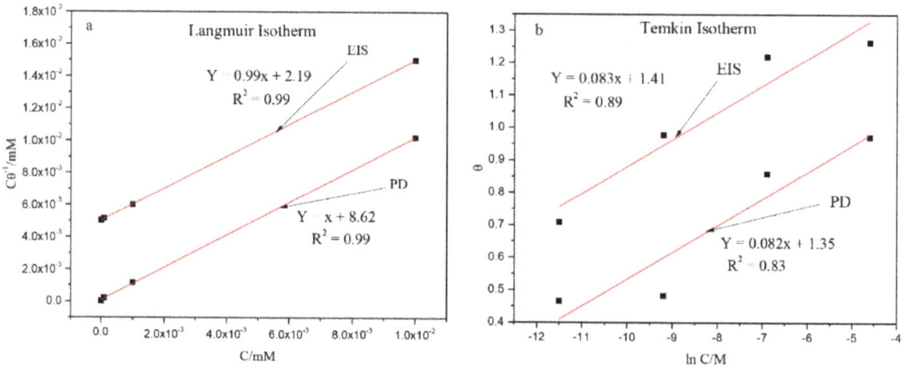

Figure 4. Langmuir (**a**) and Tamkin (**b**) adsorption plots of Cu5Z5Al1Sn alloy in 3.5 wt.% NaCl in the presence of the addition of different concentrations of cysteine obtained by EIS and PD experiments.

3.3. SEM Analysis

Further analysis was performed to examine the surface morphology and elemental characterizations and their effects on corrosion behavior. Figure 5 illustrates the SEM images of the Cu5Zn5Al1Sn alloy exposed to 3.5 wt.% NaCl solution without and with different amounts of Cys. However, the uninhibited copper alloy surface (Figure 5b) is highly corroded and becomes rough as a result of the aggressive attack the from corroding solution. A very different surface morphology is observed in the presence of 10^{-4} M Cys as, shown in Figure 5c. The inhibitor molecules, which partly cover the surface, are seen to be formed. By contrast, Figure 5d does not show any corrosion attack and has almost the same morphology as the unexposed surface, suggesting that the addition of 10^{-2} M Cys leads to a more inhibited corrosion, the surface of which is almost the same as an unexposed polished one (Figure 5a). Thus, it is obvious that corrosive attack is considerably restricted by 10^{-2} M Cys. It is assumed that inhibitor molecules adsorb on the alloy surface and a smoother surface forms compared to the surface treated with the blank 3.5 wt.% NaCl solution.

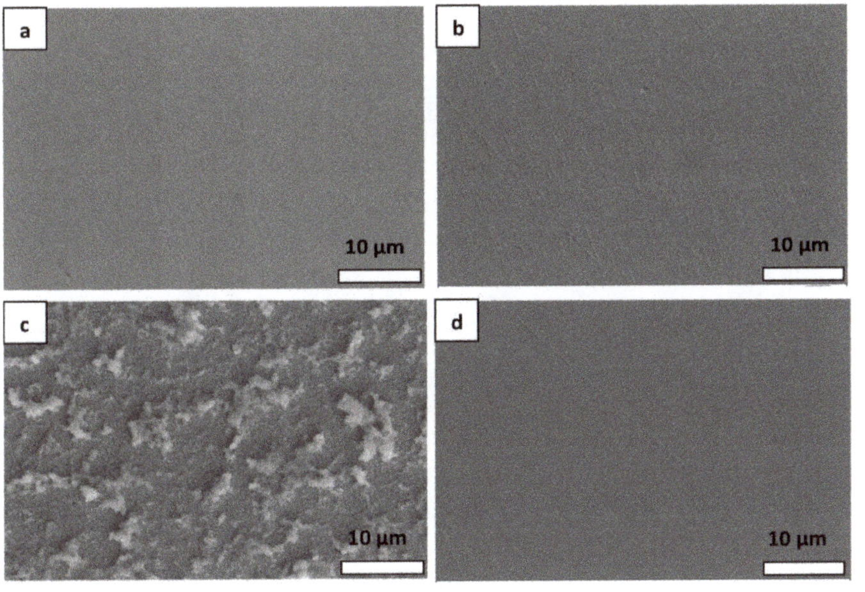

Figure 5. SEM images of Cu-5Zn-5Al-1Sn alloy before (**a**) and after immersion in 3.5 wt.% NaCl, without (**b**) and with 10^{-4} M of cysteine (**c**) and 10^{-2} M cysteine (**d**), respectively.

3.4. XPS and Auger Results

In order to further understand the surface composition of the uninhibited and inhibited copper alloy, high-resolution XPS Cu2p and Auger CuLM2 spectra of the Cu5Zn5Al1Sn alloy surface without and with the addition of various concentrations of cysteine were recorded and are shown in Figure 6. The Cu 2p profiles of Cu(0) and Cu(I) are similar, and as a result it is difficult to use the XPS Cu 2p spectrum alone to differentiate them, whereas the binding energy of the Auger peak of Cu(I) is about 2 eV higher than the Auger peak of Cu(0), meaning the CuLM2 spectrum is necessary to further investigate the valence state of copper. In all cases, there are two deconvoluted peaks of Cu2p 3/2 at different binding energies (Figure 6a). The interpretation of the peaks is different as the reaction taking place on the alloy surface is not same in the blank and Cys-inhibited solutions. In the blank solution, the corrosion products may exist as the form of oxides by reactions with the dissolved oxygen. While in the presence of inhibitors, the inhibitor particles have the opportunity to participate in surface reaction. In blank 3.5 wt.% NaCl solution, the deconvoluted Cu2p peaks at 932.99 and 934.83 eV (Figure 6a) may be assigned to Cu/Cu^{+} and Cu^{2+}, respectively. In addition to the XPS result, the Auger spectrum obtained in the blank solution shows an Auger peak at 570.11 eV, which is a typical characteristic of cuprous ion (Cu^{+}), with a band broadening to the lower binding energy. The band broadening is an indication of the presence of metallic copper, which may be attributed to the uncovered alloy surface. Generally, the XPS and Auger results of the blank solution show the presence of Cu, Cu^{+}, and Cu^{2+} species on the alloy/solution interface, which may be attributed to the uncovered alloy surface.

With regard to the Cu2p 3/2 obtained after immersion in solution with 10^{-4} M Cys, the deconvoluted peaks are located at 932.53 and 934.41 eV, respectively. Similarly, when the surface is inhibited with 10^{-2} M Cys, the Cu2p 3/2 peak is deconvoluted into peaks located at 932.62 and 934.10 eV. The peaks may be assigned to Cu/Cu^{+} and Cu^{2+} ions, respectively, in both cases [39–42]. In the solution containing different concentrations of cysteine ($10^{-4}$and 10^{-2} M), the Auger peaks obtained at 571.05 and 571.02 eV may both be assigned to Cu^{+}. The binding energy of the Cu^{+} peak on the Auger spectra increases in the presence of Cys, which may be associated with the formation of a film containing organic inhibitor molecules typically centered at higher binding energies than the binding energy for the oxide species [43–47]. Both in the absence and presence of Cys, the Cu2p XPS spectra demonstrate the Cu (II) satellite peaks attributed to the presence of copper (II). Furthermore, there is no Auger peak related to Cu^{2+} other than a band broadening in the blank solution, which may be ascribed to the lower content of the Cu^{2+} species. The uninhibited alloy surface and that of the inhibited one with lower Cys concentration (10^{-4} M) have a higher peak intensity, which could be attributed to the poor coverage area of corrosion products and the metal/inhibitor film, respectively [48].

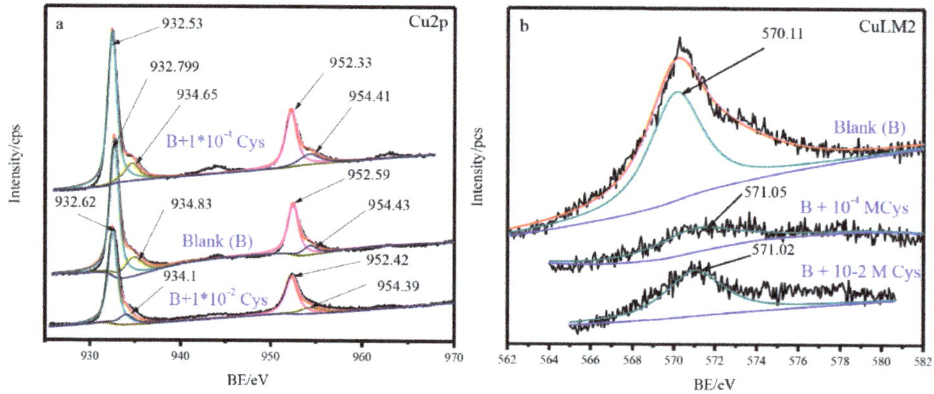

Figure 6. High-resolution XPS Cu2p 3/2 spectra (**a**) and Auger spectra (**b**) of the Cu5Zn5Al1Sn alloy surface without and with addition of inhibitors in 3.5 wt.% NaCl. Legend: BE, binding energy.

High resolution XPS Zn2p spectra (Figure 7a) show two peaks for the alloy surface, both without and with different concentrations of Cys, which may be assigned to Zn2p 3/2 and Zn2p 1/2. The Zn2p 3/2 peak at 1021.75 and 1021.76 eV in the blank and 10^{-4} Cys inhibited surfaces may be assigned to Zn^{2+}, which is associated with the presence of corrosion product [49–51]. However, the peak observed on the surface inhibited with 10^{-2} M Cys is almost negligible due to low Zn^{2+} species on the alloy surface. This phenomenon indicates that the alloy surface is not fully protected by inhibitor molecules in the presence of lower cysteine concentration. Obviously, the dezincification rate of the Cu5Zn5Sn1Al alloy is strongly bounded by the surface film/inhibitor film of the alloy surface. A more dense/integrated surface film leads to a lower dezincification rate. It can also be deduced that the formed inhibited film may rise from other alloying metals of the Cu5Zn5Sn1Al alloy, which will probably be a copper-cysteine film.

The Sn3d peak may be deconvoluted into two different peaks: Sn5/2 and Sn3/2. The peaks centered at 486.65 eV, 486.77 eV, and 486.58 eV correspond to the Sn5/2 peaks, which were obtained in blank solution, with 10^{-4} M Cys and with 10^{-2} M Cys, respectively (Figure 7b). In all cases, the peak values correspond to ionic tin (Sn^{4+}) [39,52–56], which may indicate the formation of the compound of Sn on the alloy surface, meaning, therefore, the contribution of Sn^{2+} to the surface film should not be ignored. In addition to the main Sn3d peaks, satellite peaks at higher binding energies of 499.05 and 499.08 eV appear for the blank and the surface-inhibited surface with lower Cys concentration (10^{-4} M), respectively, which may be attributed to the presence of more tin ion contents on the alloy surface [57].

In the meantime, the atomic ratio shown in Table 3 reveals that the inhibitor free alloy surface contains a greater amount of tin. On the other hand, the alloy surface inhibited with 10^{-2} M Cys contains an almost insignificant amount of SnO_2, which may result in the development of the inhibitor film and prevent the formation of the corresponding metal oxide.

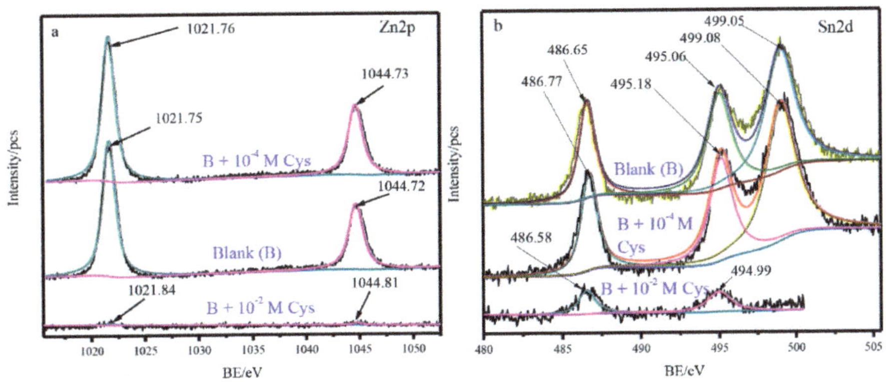

Figure 7. High-resolution XPS of Zn2p (**a**) and Sn3d (**b**) spectra of Cu5Zn5Al1Sn alloy surface without and with addition of inhibitors in 3.5 wt.% NaCl solution.

Table 3. Composition (atomic %) of the surface of Cu5Zn5Al1Sn alloy samples before and after the addition of inhibitors in a 3.5 wt.% NaCl solution.

Element	Atomic %		
	Blank	**Blank + 10^{-4} M Cys**	**Blank + 10^{-2} M Cys**
C1s	37.9	33.54	49.4
N1s	2.77	4.52	7.78
O1s	36.99	37.07	29.38
S2p	0	2.1	3.29
Cu2p	13.22	14.42	9.36
Zn2p	7.98	7.26	0.63
Sn3d	1.14	1.08	0.16

The deconvoluted C1s spectrum illustrated in Figure 8a for the golden alloy in the absence of cysteine may be attributed to three peaks which indicate three chemical forms of the C element present on the alloy surface. The three peaks described above are located at 284.75, 285.73, and 288.59 eV. The component at 284.6 eV is assigned to the non-oxidized carbon containing (C–C) composition. Moreover, the features at BE values of 285.73 eV and 288.59 eV can be assigned to the groups containing a carbon–oxygen bond, i.e., a C–O group like ether or hydroxyl, and adventitious carbon which is usually found on the metal surfaces and results from adsorbed oxidized carbonaceous species from the atmosphere, respectively [44,45,58]. For the case involving the presence of 10^{-4} M cysteine, the C1s spectrum (Figure 7a) also shows three peaks at 288.54, 286.38, and 284.88 eV, and the three deconvoluted peaks for the case of 10^{-2} M Cys are located at 288.68, 286.21, and 284.66 eV, which match with the COO–, C–NH_2, and C– SH groups, respectively [39,40]. The increment of the atomic ratio of carbon (Table 3) of the 10^{-2} M Cys inhibited surface can be ascribed to the existence of the high amount of cysteine on the alloy surface.

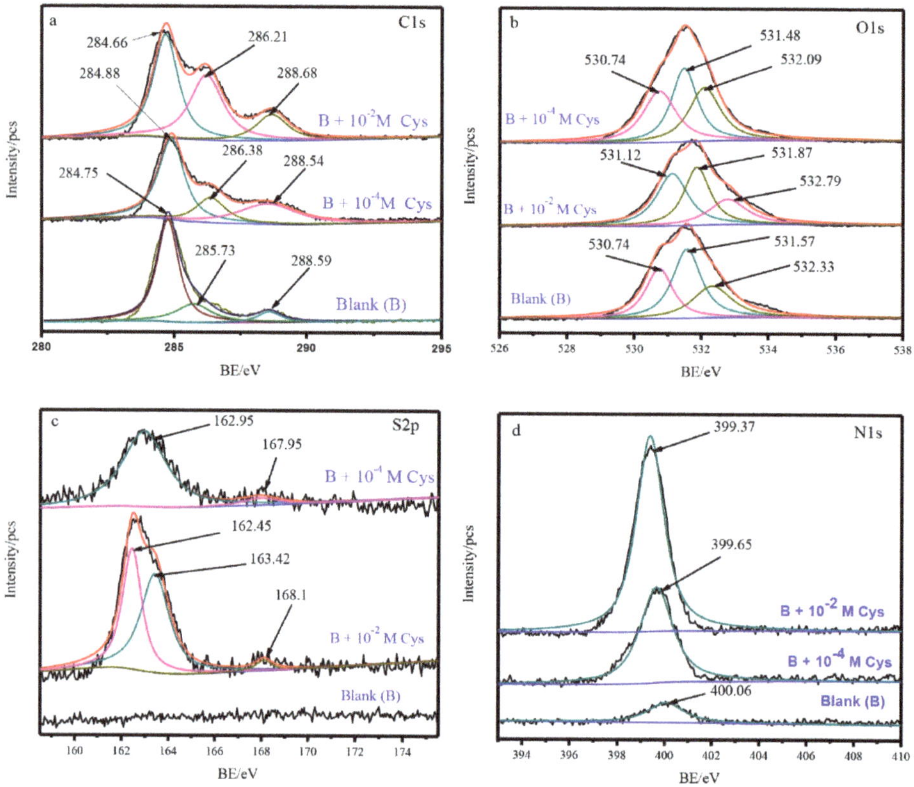

Figure 8. High-resolution XPS C1s (**a**), O1s (**b**), S2p, and N1s spectra of the Cu5Zn5Al1Sn alloy surface before and after addition of inhibitors in 3.5 wt.% NaCl solution (**c**,**d**).

The deconvoluted O1s spectra of the Cu5Zn5Al1Sn alloy in the 3.5 wt.% NaCl solution without and with the Cys inhibitor are shown in Figure 8b. There are three distinct peaks for the blank alloy surface in 3.5 wt.% NaCl solution, with the first peak located at 530.74 eV being attributed to O^{2-}, which could be related to oxygen atoms bound to the constituent metal oxides [48,59,60]. Since the major corrosion product on the outermost part of the alloy is derived from copper, the dominant metal oxide relies on copper [8], whereas the contribution of other constituent metals (Zn, Sn, and Al) cannot be ignored. The presence of Zn and Sn oxides are confirmed by their respective XPS spectrum. On the other hand, it is impossible to obtain peak analysis for Al (neither Al2p nor Al2s can be used) because of the peak overlapping with Cu3p and Cu3s [8]. The second peak observed at 531.57 eV may be ascribed to OH^-, which can be associated with the occurrence of hydrous copper, zinc, tin, and aluminum oxides.

The third peak at 532.33 eV may be assigned to atmospheric oxygen, which is similar to the carbon contamination [42]. With the addition of 10^{-4} M of Cys, the O1s peak is deconvoluted into three peaks located at 530.74, 531.48, and 532.09 eV, which could be assigned to metal oxides, hydrous metal oxides, and C=O or –CON arising from cysteine molecules, respectively. The presence of metal oxide can be ascribed to partial coverage of the metal surface by the inhibitor film [42,61,62]. The O1s spectrum of the sample containing 10^{-2} M Cys may be deconvoluted into three peaks at 531.12, 531.87, and 532.79 eV, which can be attributed to C-O from carboxyl and C=O or -CON, respectively [39]. In contrast to the uninhibited and the 10^{-4} M Cys inhibited alloy surface, the alloy inhibited with the high Cys (10^{-2} M) concentration shows no peak associated with the presence of metal oxide. It is indicated that the alloy surface is fully covered by the inhibitor film after the addition of 10^{-2} M cysteine [63].

Figure 8c shows the deconvoluted S2p spectrum of Cu5Zn5Al1Sn alloy in the absence and presence of cysteine. There is no peak found on the S2p spectrum of the blank Cu5Zn5Al1Sn alloy in 3.5 wt.% NaCl attributed to the nonappearance of any sulfur species. In the presence of 10^{-4} M Cys, two peaks located at 162.95 and 167.95 eV can be recognized, and for the case of 10^{-2} M Cys, the three deconvoluted peaks are located at 162.45, 163.42, and 168.1 eV. The lower binding energies (162.95 and 162.45) may be assigned to the metal–sulfur interaction and the upper ones (167.95 and 168.10) are ascribed to the -SH group of the cysteine molecule [42,61,63,64]. The deconvoluted peaks at 162.45 and 163.42 eV obtained from the 10^{-2} M Cys inhibited surface result from band broadening and because of the presence of more sulfur content they are considered to be peaks derived from the same component [43,65]. The intensity of S2p peaks and percentage ratio of S (Table 3) increases with cysteine concentration, suggesting that the efficiency of cysteine is concentration-dependent and more Cys molecules are adsorbed on the Cu5Zn5Al1Sn alloy surface at the higher inhibitor concentration.

The N1s spectrum for the uninhibited/inhibited surface has only one deconvoluted peak, as shown in Figure 8d, for all cases. The peak at 400.02 eV obtained from the blank solution can be assigned to N raised from the atmosphere during sample preparation. The peaks at 399.65 and 399.37 eV are obtained from 10^{-4} M and 10^{-2} M, respectively, can be assigned to the secondary nitrogen (–NH) attributed to the presence of organic matrix belonging to cysteine [39]. The peak beyond 400 eV in the N1s spectra of the inhibited surfaces is absent, which shows that there is no -$N^{+}H^{-}$ group [66]. Therefore, the above phenomenon indicates that the adsorption site of cysteine on the surface of the studied alloy does not rely on the N group. An atomic ratio of nitrogen (Table 3) in a surface inhibited with 10^{-2} M Cys confirms that the greater amount of inhibitor species are adsorbed on the alloy surface, and hence a better protection effect is displayed.

4. Discussion

The corrosion characteristics of the Cu5Zn5Al1Sn alloy in 3.5 wt.% NaCl solution were examined by different electrochemical and surface analysis techniques. It has been proven from electrochemical tests that the corrosion of the Cu5Zn5Al1Sn alloy in 3.5 wt.% NaCl solution decreases on the addition of the cysteine inhibitor (Figure 1; Figure 2). The effect of cysteine concentration was also investigated and the highest inhibition efficiencies of 97.2% and 96.44% were obtained at a cysteine concentration of 10^{-2} M based on potentiodynamic and EIS investigations, respectively. Morphological analysis based on SEM shows a better protected alloy surface in the presence of the highest concentration (10^{-2} M) of cysteine. Furthermore, composition and the inhibition mechanism were further determined by carrying out elemental analysis using XPS techniques.

From potentiodynamic testing it was determined that the current density decreases upon addition of the inhibitor, which can be accredited to the inhibitive effect of cysteine. In addition, the corrosion potential was observed to shift to the negative direction for the inhibited samples, with a potential difference between the blank and inhibited samples of less than 85 mV, which can be ascribed to the mixed inhibitive effect of cysteine for the Cu5Zn5Al1Sn alloy in 3.5 wt.% NaCl solution and which has a greater effect to retard cathodic reaction [33,67]. A diversity of adsorption isotherms like Langmuir, Temkin, Freundlich, and Frumkin are often used to model experimental results to understand surface reactions [23,37]. An adsorption model to describe the inhibition mechanism of cysteine of the Cu5Zn5Al1Sn alloy in 3.5 wt.% NaCl with different concentrations of cysteine was generated using the data obtained by PDP and EIS experiments. In this paper, the Langmuir and Tamkin adsorption isotherms (Figure 4) have been used to find out the most appropriate model to describe the inhibition mechanism of cysteine for the copper alloy. The Langmuir adsorption isotherm is proposed as a suitable model to describe the inhibition mechanism of cysteine on the Cu5Zn5Al1Sn alloy surface, which can be accredited to the adsorption of inhibitor particles on the active site of the Cu5Zn5Al1Sn alloy substrate [68,69].

In the blank solution, the appearance of Warburg impedance in the simulated circuit of the EIS result (Figure 3) reflects the diffusion process of the ionized alloy molecules from the alloy surface to the bulk solution, or the diffusion process of dissolved oxygen from the wholesale solution to the superficial of the alloy [70,71]. Composition analysis does help to study the corrosion phenomena of the Cu5Zn5Al1Sn alloy in 3.5 wt.% NaCl solution. A Cu2p 1/2 peak at 932.79 eV and Auger CuLMM peak at 570.11 eV with band broadening to low BE direction are observed and may be attributed to the presence of both metallic and ionic copper as corrosion products. There are other alloying elements present on the alloy surface which illustrate their involvement in the corrosion process. As can be seen from Table 3, the ionic ratio of Zn and Sn ions in the blank solution are higher than that for the inhibited surface. This can be ascribed to the ionization of Zn and Sn species in blank solution; they become bounded by the inhibiter film after the addition of cysteine. The damaged surface can also be observed from the SEM image (Figure 5b), which strengthens the results of the electrochemical and XPS experiments.

A relatively protected surface appears after the addition of a small amount of cysteine. The electrochemical experiments (PDP and EIS) show the inhibitive effect of cysteine (Figures 1 and 2), which is confirmed by the decrease in current density and an increase in impedance in the potentiodaynamic polarization and EIS curves, respectively. On the other hand, the inhibition efficiency obtained by electrochemical parameters of both potentiodynamic polarization and EIS is seen to increase with the increase in inhibitor concentration. There are S and N constituents in the inhibited surfaces, indicating the presence of the inhibitor film. However, the atomic ratio of this species is lower compared to a surface which has a high cysteine concentration. As for the alloying elements, the atomic ratio of Cu2p increases and the atomic ratio of Zn and Sn decreases in the presence of 10^{-4} M cysteine, inferring the bounding of the alloy surface by the Cu-inhibitor film.

The atomic ratio of alloying elements other than Cu decreases with the addition of cysteine and becomes negligible at higher cysteine concentration (10^{-2} M), which shows that the inhibitor film is mainly formed by the dominant alloy constituent (Cu) [8] and the inhibitor. In addition, the surface inhibited with 10^{-2} M cysteine concentration shows an almost similar morphology, with an unexposed polished surface surmising a good and almost full coverage of the alloy by cysteine molecules by forming a metal/inhibitor film. It is difficult to obtain XPS data for Al due to peak overlap with Cu3s and Cu3p. Schematic representation of the probable corrosion mechanism of Cu5Z5Al1Sn alloy in 3.5 wt.% NaCl solution with and without the Cys inhibitor is illustrated in Figure 9. As mentioned above, the existence of alloying elements on the blank surface can be attributed to the ionization of these metals in 3.5 wt.% NaCl forming the respective metal oxide and hydroxide (Figure 9a) [11]. On the other hand, the removal of some alloying elements, especially Zn and Sn in the presence of low concentration of the inhibitor, indicates the alloy surface is partially protected by an adsorption film between the dominant metal ion (Cu^+, Cu^{2+}) and inhibitor molecules (Cu(I)-Cys, Cu(I)-Cys-Cu(II)-Cys). With the increase of Cys concentration to 10^{-2} M, the main alloying element existing in the surface film is copper (Table 3), and it is believed that the continuous surface inhibited film formed between the copper ion and Cys molecules has existed on the Cu5Zn5Al1Sn alloy surface (Figure 9c). Therefore, better corrosion efficiency is obtained when 10^{-2} M Cys is added into the 3.5 wt.% NaCl solution.

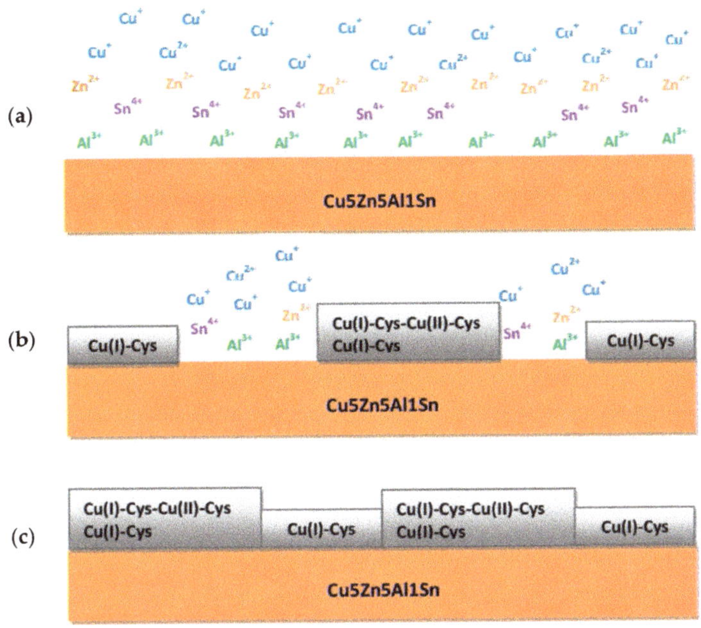

Figure 9. Schematic representation of the corrosion mechanism of the Cu5Zn5Al1Sn alloy in 3.5 wt.% NaCl solution before (**a**) and after addition of 10^{-4} M (**b**) and 10^{-2} M (**c**) cysteine.

5. Conclusions

In this work, the corrosion behavior of a Cu5Zn5Al1Sn alloy was examined in a 3.5 wt.% NaCl solution in the absence and presence of cysteine. Potentiodynamic polarization studies showed that cysteine acts as a mixed-type inhibitor during the corrosion process. The inhibition efficiency increased with an increase in the concentration of cysteine and reached 97.2% in the presence of 10^{-2} M cysteine. Copper was the main corrosion product formed on the Cu5Zn5Al1Sn alloy in 3.5 wt.% NaCl solution. The outermost surface also contained to a small extent Zn- and Sn-rich constituents, as confirmed by XPS investigation. The content of alloy constituents other than Cu decreased in the presence of cysteine and became negligible at a cysteine concentration of 10^{-2} M. The inhibition mechanism relies on the formation of a binding Cu(I)-Cys film and/or Cu(I)-Cys-Cu(II)-Cys film on the alloy surface, which protect the alloy from corrosion attack.

Author Contributions: Conceptualization, K.W.S., F.H., and Y.J.; data curation, K.W.S. and F.H.; formal analysis, K.W.S. and F.H.; investigation, K.W.S. and F.H.; methodology, K.W.S., F.H., and Y.X.; project administration, Y.J.; resources, Y.J.; software, K.W.S. and F.H.; supervision, Y.X., L.W., and Y.J.; writing—original draft, K.W.S.; writing—review & editing, F.H., Y.X., and Y.J.

Funding: This research was funded by the Development and Reform Committee of PRC, (no. YYXM-1412-0001), Fundamental Research Funds for the Central Universities (no. FRF-TP-16-040A1) and 111 Project, (no. B12012).

Conflicts of Interest: The authors declare no conflict of interest. The funders had no role in the design of the study; in the collection, analyses, or interpretation of data; in the writing of the manuscript, or in the decision to publish the results.

References

1. Yu, Y.; Yang, D.; Zhang, D.; Wang, Y.; Gao, L. Anti-corrosion film formed on HAl77-2 copper alloy surface by aliphatic polyamine in 3 wt.% NaCl solution. *Appl. Surf. Sci.* **2017**, *392*, 768–776. [CrossRef]

2. Tan, B.; Zhang, S.; Qiang, Y.; Guo, L.; Feng, L.; Liao, C. A combined experimental and theoretical study of the inhibition effect of three disulfide-based flavouring agents for copper corrosion in 0.5M sulfuric acid. *J. Colloid Interface Sci.* **2018**, *526*, 268–280. [CrossRef] [PubMed]
3. Tasić, Ž.Z.; Petrović, M.M.B.; Radovanović, M.B.; Simonović, A.T.; Antonijević, M.M. Cephradine as corrosion inhibitor for copper in 0.9% NaCl solution. *J. Mol. Struct.* **2018**, *1159*, 46–54. [CrossRef]
4. Sayed, G.H.; Azab, M.E.; Anwer, K.E.; Raouf, M.A.; Negm, N.A. Pyrazole, pyrazolone and enaminonitrile pyrazole derivatives: Synthesis, characterization and potential in corrosion inhibition and antimicrobial applications. *J. Mol. Liq.* **2018**, *252*, 329–338. [CrossRef]
5. Wang, Y.; Yu, Y.; Zhang, J.; Gao, L.; Feng, L.; Zhang, D. Click-assembling triazole membrane on copper surface via one-step or two-steps and their corrosion inhibition performance. *Appl. Surf. Sci.* **2018**, *427*, 1120–1128. [CrossRef]
6. Zhang, X.; Odnevall, W.I.; Leygraf, C. Mechanistic studies of corrosion product flaking on copper and copper-based alloys in marine environments. *Corros. Sci.* **2014**, *85*, 15–25. [CrossRef]
7. Horton, D.J.; Ha, H.; Foster, L.L.; Bindig, H.J.; Scully, J.R. Tarnishing and Cu Ion release in Selected Copper-Base Alloys: Implications towards Antimicrobial Functionality. *Electrochim. Acta* **2015**, *169*, 351–366. [CrossRef]
8. Chang, T.; Odnevall, W.I.; Jin, Y.; Leygraf, C. The golden alloy Cu-5Zn-5Al-1Sn: A multi-analytical surface characterization. *Corros. Sci.* **2018**, *131*, 94–103. [CrossRef]
9. Lapeire, L.; Lombardia, E.M.; Verbeken, K.; De Graeve, I.; Terryn, H.; Kestens, L. Structural dependence of gold deposition by nanoplating in polycrystalline copper. *J. Mater. Sci.* **2014**, *49*, 3909–3916. [CrossRef]
10. Martinez-Lombardia, E.; Maurice, V.; Lapeire, L.; De Graeve, I.; Verbeken, K.; Kestens, L.; Marcus, P.; Terryn, H. In situ scanning tunneling microscopy study of grain-dependent corrosion on microcrystalline copper. *J. Phys. Chem.* **2014**, *118*, 25421–25428. [CrossRef]
11. Chang, T.; Herting, G.; Jin, Y.; Leygraf, C.; Odnevall, W.I. The golden alloy Cu5Zn5Al1Sn: Patina evolution in chloride-containing atmospheres. *Corros. Sci.* **2018**, *133*, 190–203. [CrossRef]
12. Foster, L.L.; Scully, J.R. Corrosion of Cu-5Zn-5Al-1Sn (89% Cu, 5% Zn, 5% Al, 1% Sn) Compared to Copper in Synthetic Perspiration During Cyclic Wetting and Drying: The Fate of Copper. *Corrosion* **2016**, *72*, 1095–1106. [CrossRef]
13. Goyal, M.; Kumar, S.; Bahadur, I.; Verma, C.; Ebenso, E.E. Organic corrosion inhibitors for industrial cleaning of ferrous and non-ferrous metals in acidic solutions: A review. *J. Mol. Liq.* **2018**, *256*, 565–573. [CrossRef]
14. Liu, G.; Huang, Y.; Qu, X.; Xiao, J.; Yang, X.; Xu, Z. Understanding the hydrophobic mechanism of 3-hexyl-4-amino-1, 2,4-triazole-5-thione to malachite by ToF-SIMS, XPS, FTIR, contact angle, zeta potential and micro-flotation. *Colloids Surf. A Physicochem. Eng. Asp.* **2016**, *503*, 34–42. [CrossRef]
15. El-Deab, M.S. Interaction of cysteine and copper ions on the surface of iron: EIS, polarization and XPS study. Mater. *Chem. Phys.* **2011**, *129*, 223–227. [CrossRef]
16. Nazeer, A.A.; Allam, N.K.; Fouda, A.S.; Ashour, E.A. Effect of cysteine on the electrochemical behavior of Cu10Ni alloy in sulfide polluted environments: Experimental and theoretical aspects. *Mater. Chem. Phys.* **2012**, *136*, 1–9. [CrossRef]
17. Radovanovic, M.B.; Petrovic, M.B.; Simonovic, A.T.; Milic, S.M.; Antonijevic, M.M. Cysteine as a green corrosion inhibitor for Cu37Zn brass in neutral and weakly alkaline sulphate solutions. *Environ. Sci. Pollut. Res. Int.* **2013**, *20*, 4370–4381. [CrossRef]
18. Kazansky, L.P.; Selyaninov, I.A.; Kuznetsov, Y.I. Adsorption of 2-mercaptobenzothiazole on copper surface from phosphate solutions. *Appl. Surf. Sci.* **2012**, *258*, 6807–6813. [CrossRef]
19. Ghelichkhah, Z.; Sharifi-Asl, S.; Farhadi, K.; Banisaied, S.; Ahmadi, S.; Macdonald, D.D. L-cysteine/polydopamine nanoparticle-coatings for copper corrosion protection. *Corros. Sci.* **2015**, *91*, 129–139. [CrossRef]
20. Milosev, I.; Pavlinac, J.; Hodoscek, M.; Lesar, A. Amino acids as corrosion inhibitors for copper in acidic medium: Experimental and theoretical study. *J. Serb. Chem. Soc.* **2013**, *78*, 2069–2086. [CrossRef]
21. El-Hafez, G.M.A.; Badawy, W.A. The use of cysteine, N-acetyl cysteine and methionine as environmentally friendly corrosion inhibitors for Cu–10Al–5Ni alloy in neutral chloride solutions. *Electrochim. Acta* **2013**, *108*, 860–866. [CrossRef]
22. Sabet, B.K.; Dehghanian, C. Adsorption behavior of 1H-benzotriazole corrosion inhibitor on aluminum alloy 1050, mild steel and copper in artificial seawater. *J. Environ. Chem. Eng.* **2018**, *6*, 1613–1624. [CrossRef]

23. Tasić, Ž.Z.; Mihajlović, M.B.P.; Radovanović, M.B.; Antonijević, M.M. Electrochemical investigations of copper corrosion inhibition by azithromycin in 0.9% NaCl. *J. Mol. Liq.* **2018**, *265*, 687–692. [CrossRef]
24. Karthik, G.; Sundaravadivelu, M. Investigations of the inhibition of copper corrosion in nitric acid solutions by levetiracetam drug. *Egypt. J. Pet.* **2016**, *25*, 481–493. [CrossRef]
25. Khadom, A.A.; Yaro, A.S. Mass transfer effect on corrosion inhibition process of copper–nickel alloy in hydrochloric acid by Benzotriazole. *J. Saudi Chem. Soc.* **2014**, *18*, 214–219. [CrossRef]
26. Vastag, G.; Shaban, A.; Vraneš, M.; Tot, A.; Belić, S.; Gadžurić, S. Influence of the N-3 alkyl chain length on improving inhibition properties of imidazolium-based ionic liquids on copper corrosion. *J. Mol. Liq.* **2018**, *264*, 526–533. [CrossRef]
27. Al-Sabagh, A.M.; El Basiony, N.M.; Sadeek, S.A.; Migahed, M.A. Scale and corrosion inhibition performance of the newly synthesized anionic surfactant in desalination plants: Experimental, and theoretical investigations. *Desalination* **2018**, *437*, 45–58. [CrossRef]
28. Sabet, B.K.; Dehghanian, C.; Yari, S. Corrosion inhibition of copper, mild steel and galvanically coupled copper-mild steel in artificial sea water in presence of 1H-benzotriazole, sodium molybdate and sodium phosphate. *Corros. Sci.* **2017**, *126*, 272–285. [CrossRef]
29. Srivastava, M.; Tiwari, P.; Srivastava, S.K.; Kumar, A.; Ji, G.; Prakash, R. Low cost aqueous extract of Pisum sativum peels for inhibition of mild steel corrosion. *J. Mol. Liq.* **2018**, *254*, 357–368. [CrossRef]
30. Kong, D.; Dong, C.; Ni, X.; Man, C.; Xiao, K.; Li, X. Insight into the mechanism of alloying elements (Sn, Be) effect on copper corrosion during long-term degradation in harsh marine environment. *Appl. Surf. Sci.* **2018**, *455*, 543–553. [CrossRef]
31. Prakashaiah, B.G.; Vinaya, K.D.; Anup, P.A.; Nityananda, S.A.; Amitha, R.B.E. Corrosion inhibition of 2024-T3 aluminum alloy in 3.5% NaCl by thiosemicarbazone derivatives. *Corros. Sci.* **2018**, *136*, 326–338. [CrossRef]
32. Jiang, L.; Qiang, Y.; Lei, Z.; Wang, J.; Qin, Z.; Xiang, B. Excellent corrosion inhibition performance of novel quinoline derivatives on mild steel in HCl media: Experimental and computational investigations. *J. Mol. Liq.* **2018**, *255*, 53–63. [CrossRef]
33. Boyapati, V.A.R.; Kanukula, C.K. Corrosion Inhibition of Cu-Ni (90/10) Alloy in Seawater and Sulphide-Polluted Seawater Environments by 1,2,3-Benzotriazole. *ISRN Corros.* **2013**, *2013*, 703929. [CrossRef]
34. Deyab, M.A. Corrosion inhibition of heat exchanger tubing material (titanium) in MSF desalination plants in acid cleaning solution using aromatic nitro compounds. *Desalination* **2018**, *439*, 73–79. [CrossRef]
35. Negm, N.A.; Migahed, M.A.; Farag, R.K.; Fadda, A.A.; Awad, M.K.; Shaban, M.M. High performance corrosion inhibition of novel tricationic surfactants on carbon steel in formation water: Electrochemical and computational evaluations. *J. Mol. Liq.* **2018**, *262*, 363–375. [CrossRef]
36. Chaubey, N.; Singh, V.K.; Quraishi, M.A. Corrosion inhibition performance of different bark extracts on aluminium in alkaline solution. *J. Assoc. Arab Univ. Basic Appl. Sci.* **2018**, *22*, 38–44. [CrossRef]
37. Edison, T.N.J.I.; Atchudan, R.; Pugazhendhi, A.; Lee, Y.R.; Sethuraman, M.G. Corrosion inhibition performance of spermidine on mild steel in acid media. *J. Mol. Liq.* **2018**, *264*, 483–489. [CrossRef]
38. Raman, K.; Hansung, K.; Reddicherla, U.; Ompal, S.Y.; Gurmeet, S. Comprehensive adsorption characteristics of a newly synthesized and sustainable anti-corrosion catalyst onmild steel surface exposed to a highly corrosive electrolytic solution. *J. Mol. Liq.* **2018**, *268*, 37–48.
39. Wagner, C.D.; Riggs, W.M.; Davis, L.E.; Moulder, J.F.; Muilenberg, G.E. (Eds.) *Handbook of X-ray Photoelectron Spectroscopy*; Perkin-Elmer Corporation: Waltham, MN, USA, 1979; pp. 41–127.
40. Wang, T.; Wang, J.; Wu, Y. The inhibition effect and mechanism of l-cysteine on the corrosion of bronze covered with a CuCl patina. *Corros. Sci.* **2015**, *97*, 89–99. [CrossRef]
41. Wang, Z.; Gong, Y.; Jing, C.; Huang, H.; Li, H.; Zhang, S. Synthesis of dibenzotriazole derivatives bearing alkylene linkers as corrosion inhibitors for copper in sodium chloride solution: A new thought for the design of organic inhibitors. *Corros. Sci.* **2016**, *113*, 64–77. [CrossRef]
42. Huang, H.; Wang, Z.; Gong, Y.; Gao, F.; Luo, Z.; Zhang, S. Water soluble corrosion inhibitors for copper in 3.5 wt% sodium chloride solution. *Corros. Sci.* **2017**, *123*, 339–350. [CrossRef]
43. Milošev, I.; Kovačević, N.; Kovač, J.; Kokalj, A. The roles of mercapto, benzene and methyl groups in the corrosion inhibition of imidazoles on copper: I. Experimental characterization. *Corros. Sci.* **2015**, *98*, 107–118. [CrossRef]

44. Finšgar, M. EQCM and XPS analysis of 1,2,4-triazole and 3-amino-1,2,4-triazole as copper corrosion inhibitors in chloride solution. *Corros. Sci.* **2013**, *77*, 350–359. [CrossRef]
45. Finšgar, M. 2-Mercaptobenzimidazole as a copper corrosion inhibitor: Part II. Surface analysis using X-ray photoelectron spectroscopy. *Corros. Sci.* **2013**, *72*, 90–98. [CrossRef]
46. Finšgar, M.; Merl, D.K. An electrochemical, long-term immersion, and XPS study of 2-mercaptobenzothiazole as a copper corrosion inhibitor in chloride solution. *Corros. Sci.* **2014**, *83*, 164–175. [CrossRef]
47. Finšgar, M. X-ray excited Auger Cu $L_3L_4,5M_4,5$ spectra measured at low take-off angles as a fingerprint for a Cu-organics connection. *J. Electron Spectrosc. Relat. Phenom.* **2018**, *222*, 10–14. [CrossRef]
48. Qiang, Y.; Fu, S.; Zhang, S.; Chen, S.; Zou, X. Designing and fabricating of single and double alkyl-chain indazole derivatives self-assembled monolayer for corrosion inhibition of copper. *Corros. Sci.* **2018**, *140*, 111–121. [CrossRef]
49. Koitaya, T.; Shiozawa, Y.; Yoshikura, Y.; Mukai, K.; Yoshimoto, S.; Torii, S. Electronic states and growth modes of Zn atoms deposited on Cu(111) studied by XPS, UPS and DFT. *Surf. Sci.* **2017**, *663*, 1–10. [CrossRef]
50. Morozov, I.G.; Belousova, O.V.; Ortega, D.; Mafina, M.K.; Kuznetcov, M.V. Structural, optical, XPS and magnetic properties of Zn particles capped by ZnO nanoparticles. *J. Alloy. Compd.* **2015**, *633*, 237–245. [CrossRef]
51. Sun, J.; Yarmolenko, M.A.; Rogachev, A.A.; Rogachev, A.V.; Jiang, X.; Gorbachev, D.L. Investigation of structural properties of electron-beam deposition of zinc oxide coatings doped with copper. *Surf. Interfaces* **2017**, *6*, 24–32. [CrossRef]
52. Babu, B.; Neelakanta, R.I.; Yoo, K.; Kim, D.; Shim, J. Bandgap tuning and XPS study of SnO_2 quantum dots. *Mater. Lett.* **2018**, *221*, 211–215. [CrossRef]
53. Xu, C.; Zhang, L.; An, Y.; Wang, X.; Xu, G.; Chen, Y. Promotional synergistic effect of Sn doping into a novel bimetallic Sn-W oxides/graphene catalyst for selective oxidation of alcohols using aqueous H_2O_2 without additives. *Appl. Catal. A Gen.* **2018**, *558*, 26–33. [CrossRef]
54. Hutchison, M.J.; Zhou, P.; Ogle, K.; Scully, J.R. Enhanced Electrochemical Cu Release from Commercial Cu-Sn Alloys: Fate of the Alloying Elements in Artificial Perspiration. *Electrochim. Acta* **2017**, *241*, 73–88. [CrossRef]
55. Zatsepin, D.A.; Zatsepin, A.F.; Boukhvalov, D.W.; Kurmaev, E.Z.; Gavrilov, N.V. Sn-loss effect in a Sn-implanted α-SiO_2 host-matrix after thermal annealing: A combined XPS, PL, and DFT study. *Appl. Surf. Sci.* **2016**, *367*, 320–326. [CrossRef]
56. Sangaiya, P.; Jayaprakash, R. Tuning effect of Sn doping on structural, morphological, optical, electrical and photocatalytic properties of iron oxide nanoparticles. *Mater. Sci. Semicond. Process.* **2018**, *85*, 40–51. [CrossRef]
57. Moscu, A.; Theodoridi, C.; Cardenas, L.; Thieuleux, C.; Motta-Meira, D.; Agostini, G. CO dissociation on Pt-Sn nanoparticles triggers Sn oxidation and alloy segregation. *J. Catal.* **2018**, *359*, 76–81. [CrossRef]
58. Finšgar, M.; Merl, D.K. 2-Mercaptobenzoxazole as a copper corrosion inhibitor in chloride solution: Electrochemistry, 3D-profilometry, and XPS surface analysis. *Corros. Sci.* **2014**, *80*, 82–95. [CrossRef]
59. Appa, R.B.V.; Narsihma, R.M. Formation, characterization and corrosion protection efficiency of self-assembled 1-octadecyl-1H-imidazole films on copper for corrosion protection. *Arab. J. Chem.* **2017**, *10*, S3270–S3283. [CrossRef]
60. Wang, Z.; Gong, Y.; Zhang, L.; Jing, C.; Gao, F.; Zhang, S. Self-assembly of new dendrimers basing on strong π-π intermolecular interaction for application to protect copper. *Chem. Eng. J.* **2018**, *342*, 238–250. [CrossRef]
61. Wagner, A.J.; Wolfe, G.M.; Fairbrother, D.H. Reactivity of vapor-deposited metal atoms with nitrogen-containing polymers and organic surfaces studied by in situ XPS. *Appl. Surf. Sci.* **2003**, *219*, 317–328. [CrossRef]
62. Zhang, D.Q.; Gao, L.X.; Zhou, G.D. Synergistic effect of 2-mercapto benzimidazole and KI on copper corrosion inhibition in aerated sulfuric acid solution. *J. Appl. Electrochem.* **2003**, *33*, 361–366. [CrossRef]
63. Krzysztof, S.; Roman, D.; Zofia, P.; Stefan, W. transformation of nitrogen structures in carbonization of model compounds determined by XPS. *Carbon* **1995**, *33*, 1383–1392.
64. Wang, C.; Luo, X.; Jia, Z. Linkage, charge state and layer of L-Cysteine on copper surfaces. *Colloids Surf. B Biointerfaces* **2017**, *160*, 33–39. [CrossRef]
65. Ismail, K.M. Evaluation of cysteine as environmentally friendly corrosion inhibitor for copper in neutral and acidic chloride solutions. *Electrochim. Acta* **2007**, *52*, 7811–7819. [CrossRef]

66. Vinothkumar, K.; Sethuraman, M.G. Corrosion inhibition ability of electropolymerised composite film of 2-amino-5-mercapto-1,3,4-thiadiazole/TiO_2 deposited over the copper electrode in neutral medium. *Mater. Today Commun.* **2018**, *14*, 27–39. [CrossRef]
67. Mahdavian, M.; Tehrani-Bagha, A.R.; Alibakhshi, E.; Ashhari, S.; Palimi, M.J.; Farashi, S.; Javadiane, S.; Ektefae, F. Corrosion of mild steel in hydrochloric acid solution in the presence of two cationic gemini surfactants with and without hydroxyl substituted spacers. *Corros. Sci.* **2018**, *137*, 62–75. [CrossRef]
68. Roland, T.L.; Olukeye, T. Corrosion inhibition properties of the synergistic effect of 4-hydroxy-3-methoxybenzaldehyde and hexadecyltrimethylammoniumbromide on mild steel in dilute acid solutions. *J. King Saud Univ. Eng. Sci.* **2018**, *30*, 384–390.
69. Hany, M.A.; Ahmed, M.A.; Bahaa, E.M. Investigation of adsorption and inhibition effects of some novel anil compounds towards mild steel in H_2SO_4 solution: Electrochemical and theoretical quantum studies. *J. Electroanal Chem.* **2015**, *758*, 135–147.
70. Yang, W.; Li, T.; Zhou, H.; Huang, Z.; Fu, C.; Chen, L.; Li, M.; Kuang, Y. Electrochemical and anti-corrosion properties of octadecanethiol and benzotriazole binary self-assembled monolayers on copper. *Electrochim. Acta.* **2016**, *220*, 245–251. [CrossRef]
71. Chen, W.; Hong, S.; Li, H.B.; Li, B.; Luo, H.Q.; Li, M.; Li, N.B. Protection of copper corrosion in 0.5 M NaCl solution by modification of 5-mercapto-3-phenyl-1,3,4-thiadiazole-2-thione potassium self-assembled monolayer. *Corros. Sci.* **2012**, *61*, 53–62. [CrossRef]

applied sciences

MDPI

Article

Thin Coatings of Cerium Oxide Nanoparticles with Anti-Reflective Properties

Laura J. Romasanta [1], Lauriane D'Alençon [2], Sara Kirchner [3] and Christophe Pradère [3] and Jacques Leng [1,*]

1. University of Bordeaux, Laboratory of the Future (CNRS/SOLVAY), 178 avenue du Docteur Schweitzer, 33600 Pessac, France; ljromasanta@gmail.com
2. Solvay, Functional Inorganic Materials Laboratory, 52 rue de la Haie coq, 93300 Aubervillliers, France; Lauriane.Dalencon@solvay.com
3. Institut d'ingénierie et de mécanique, TREFLE, Esplanade des Arts et Métiers, 33400 Talence, France; sara.kirchner1@gmail.com (S.K.); christophe.pradere@u-bordeaux.fr (C.P.)

* Correspondence: jacques.leng-exterieur@solvay.com; Tel.: +33-556-464-749

Received: 30 July 2019; Accepted: 5 September 2019; Published: 16 September 2019

Abstract: Cerium oxide, in addition to its catalytic properties, is also known for its optical properties such as ultraviolet (UV) radiation filtering and a relatively high refractive index ($n > 2$), which makes it an excellent candidate for multifunctional coatings. Here, we focus on the optical properties of thin deposits ($\lesssim$2 µm) of densely packed CeO_2 nanoparticles, which we assemble using two evaporation-based techniques: convective self-assembly (CSA, a type of very slow blade-coating) to fabricate large-scale coatings of controllable thickness—from tens of nanometres to a few micrometers—and microfluidic pervaporation which permits us to add some micro-structure to the coatings. Spectroscopic ellipsometry yields the refractive index of the resulting nano-porous coatings, which behave as lossy dielectrics in the UV-visible regime and loss-less dielectrics in the visible to infra-red (IR) regime; in this regime, the fairly high refractive index ($\approx$1.8) permits us to evidence thickness-tunable anti-reflection on highly refractive substrates, such as silicon, and concomitant enhanced transmissions which we checked in the mid-IR region.

Keywords: Cerium oxide nanoparticles; anti-reflection; self-assembly; microfluidics; convective self-assembly

1. Introduction

Cerium oxide nanoparticles have been applied in diverse fields including catalysis, luminescence, and nanomedicine, etc. [1–3]. Here, we investigate their optical behaviour when assembled as coatings onto a substrate. As cerium oxide (CeO_2) has a fairly high optical refractive index, we expect that dense coatings made out of CeO_2 nanoparticles will also exhibit a high refractive index.

We follow a low-tech approach to engineer simple coatings out of dispersions of CeO_2 nanoparticles. The coatings are realized using evaporation-based techniques out of water-based dispersions, making the processes fairly appealing in terms of energy consumption and toxicity: Their fabrication requires virtually no external pressure, no toxic or hazardous gases, no elevated temperatures, no etching, no vacuum, and no toxic solvents. The coatings we obtain have simple yet non trivial structures and they perform very well from an optical point of view, in particular for anti-reflection on highly refractive materials, which we demonstrate on silicon as a case study.

First, we re-demonstrate that the evaporation-based blade-coating method is efficient for producing coatings of controllable thickness, in this case, out of CeO_2 dispersions. These results conform to the pioneering work of O. Velev and co-workers [4,5] who coated SiO_2 dispersions on glass and silicon. The very slow regime of blade-coating, were evaporation competes favourably with the casting velocity enhances the formation of a well structured, thin to thick deposit. We also add some complexity via micro-patterned structures using an evaporation-based micro moulding technique. We measure the refractive index of these coatings which behave as lossy dielectrics in the ultraviolet (UV)-visible regime and loss-less dielectrics in the visible to infra-red (IR) regime; in the latter regime, the fairly high refractive index ($\approx$1.8) permits us to demonstrate thickness-tunable anti-reflection properties on substrates with a high refractive index such as silicon substrates, and concomitant enhanced transmission which we tested in the mid-IR region, thereby covering a large spectral range.

2. Thin and Structured Coatings of Densely Packed CeO_2 Nanoparticles

We use two different techniques that permit us to assemble an initially dispersed state of nanoparticles into a solid made of the same, densely packed nanoparticles: convective self-assembly (CSA) and microfluidic pervaporation (µ-pervaporation). A detailed description of these methods will be given in Section 4.2.

In brief, CSA is a modified blade-coating technique where the withdrawing velocity of the substrate on which the coating will be deposited is so small that evaporation competes with the film casting rate, so that the dispersion is pre-concentrated at the level of the deposition meniscus; on top of this, it enhances the structure of the deposit [6–9].

µ-pervaporation takes advantage of the extremely well-defined poly(dimethyl siloxane) (PDMS) geometries crafted by soft-lithography [10,11], as well as the ability of some solvents to pervaporate across the elastomer PDMS matrix [12]. Pervaporation induces a concentration mechanism of the solute, which was initially solubilized/dispersed in the solvent, thus leading to the formation of a solid that grows in a neat geometry [13]. It is an 'augmented' version of the moulding of solids into micro-capillaries, a seminal and inspiring piece of work described in [14].

2.1. Convective Self-Assembly

CSA has been extensively used to deposit a variety of objects (from molecules [15] to large colloids [6]) into homogeneous thin coatings on relatively large scales (tens of cm^2). It is possible to control the thickness and to some extent the morphology of the coating via process parameters such as the casting velocity, the concentration of the stock solution [16], temperature, humidity, etc., but also to couple the reaction to the deposition process, for instance for sol-gel coatings [17], and finally to stop and re-run the process (stop-and-go), in order to add more structure to the deposit [18].

Here, we use the simplest version of CSA, namely a continuous and slow deposition of a dispersion at constant volume fraction ϕ_0 at room temperature and with no specific control of the atmosphere (the room in which we work is nevertheless air-conditioned with a constant temperature 21 °C and relative humidity $\approx$0.5). The air flow over the evaporating zone is not controlled and is left to natural convection; it has been demonstrated that for the deposition colloids and unlike the case of polymers, it is *not* a crucial parameter [7]. The only control parameter we use is the casting velocity v; we deposit the same grade of CeO_2 nanoparticles with batches coming at two different typical sizes: $\approx$4 and $\approx$40 nm (average diameter, see Section 4.1).

The overview of our deposition campaigns is summarized in Figure 1 with the coating thickness, the morphology of the coatings, and their optical appearance. In the velocity range of deposition we studied ($v < 100$ µm.s^{-1}), the decreasing trend of h against v is a clear signature of the CSA regime and with a comparable behaviour for the two sizes of nanoparticles studied here; it contrasts with the Landau-Levich regime where h is expected to scale like $h \sim v^{2/3}$ [15].

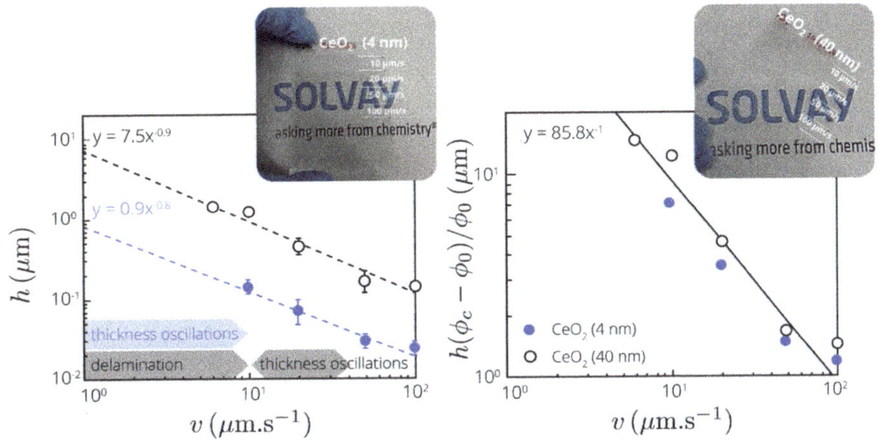

Figure 1. (**Left**): Thickness of coatings of CeO_2 nanoparticles deposited at different coating speeds v; each point averages at least seven independent experiments. The different morphologies of the sample are zoned at the bottom left for the two systems (blue 4 nm, grey 40 nm) and flat films are obtained otherwise, see Figure 2 top; the dashed lines represent the fitting with a power law. (**Right**): Rescaled data considering the initial volume fraction ϕ_0 and a dry volume fraction $\phi_c = 0.61$. The solid line is a fit according to Equation (3). Inserts: Macroscopic views of the samples highlighting the high degree of transparency.

However, the coating is not always homogeneous and flat, see Figure 2 top and the grey and blue zones in Figure 1, bottom left. Instead, for the large particles and at low deposition velocity (v $\lesssim$ 40 µm.s^{-1}) the coating exhibits thickness oscillations. Notice this is definitely not stick-and-slip [19] but instead oscillations around a mean (non zero) thickness with no return to the bare substrate. At even lower velocity (v $\lesssim$ 10 µm.s^{-1}), the coating delaminates, probably because it becomes thick, tough and brittle [20]. For the small-size particles, we observe only thickness oscillations below v $\lesssim$ 10 µm.s^{-1} (blue zone in Figure 1, bottom left).

The same data can be re-scaled using a simple mass-conservation assumption (Equation (3) simply states that all the particles coming from the bulk reservoir are accumulated into the final deposit via evaporation), that takes into account both the initial and final volume fractions, ϕ_0 and ϕ_c respectively. Using $\phi_c = 0.61$ a value that we obtained from ellipsometry, which is in agreement with a close packing of a polydisperse hard-sphere [21], the data reasonably collapse and follow the expected trend $h \sim v^{-1}$ indeed.

In all cases but the ones where the coatings delaminate, the coated substrate appears perfectly transparent with no visible effect of surface undulations and no haze (upper inserts in Figure 1), even though optical microscopy reveals the defects (Figure 2 top). Low resolution scanning electron microscopy (SEM) of the surface of some samples may show evidence of the granulometry of the particles (Figure 2 bottom left) but a cross-section of a selected sample demonstrates a beautiful, flat, and large-scale structure (Figure 2 bottom right).

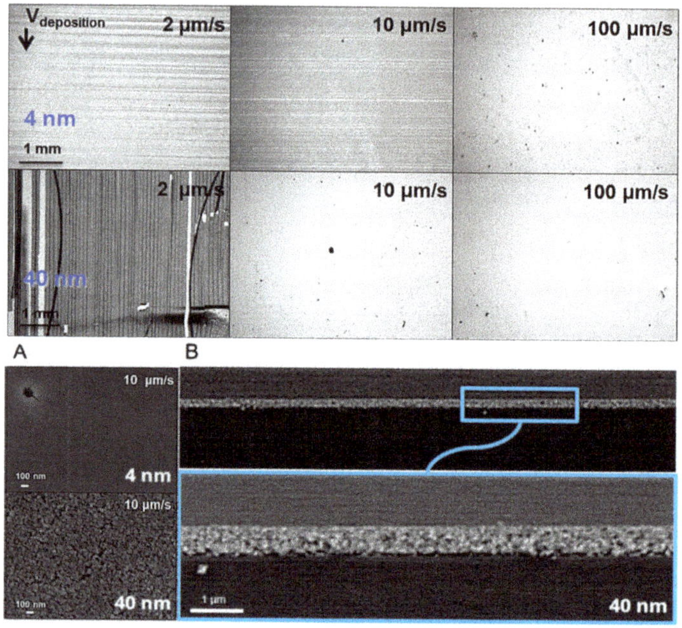

Figure 2. (**Top**): Representative optical microscopy images (bright field) for the coatings deposited at different deposition speeds using CeO_2 sols with nanoparticle size of 4 nm (first row) and 40 nm (second row); the morphologies of the coating are either with thickness oscillations and possible delamination, or flat (see also Figure 1 left). (**Bottom**): Scanning electron microscopy (SEM) (SEM-FEG HR JEOL 6700F) images of (**A**) the surface of coatings deposited at different speeds using CeO_2 sols with nanoparticle size of (top) 4 nm and (bottom) 40 nm and (**B**) of a cross-section of a coating deposited at 20 μm.s^{-1} (40 nm particles).

2.2. Microfluidic Pervaporation

Microfluidic pervaporation is a templated self-assembly method [22] where a PDMS mould—the template—is used to guide the growth of a solid made of nanoparticles. In turn, the template also serves as a pump via the pervaporation mechanism described more in detail in Section 4.2. The process is self-powered: We fill the microfluidic template with a dilute dispersion and pervaporation extracts the solvent, which in turn concentrates the dispersion until a solid nucleates, which then progressively grows. It can be a long process as the growth velocity v_g can be understood on a simple volume conservation assumption, very similar to Equation (3): $v_g \sim v_e\phi_0/(\phi_c - \phi_0)$ where v_e is an evaporation velocity associated with the pervaporation mechanism [12,23]. It offers little flexibility as it is mainly governed by the geometry of the template, i.e., surface exchange for pervaporation, yet the growth velocity is still largely tunable via ϕ_0. We typically see $v_g \approx 0.2$ μm.s^{-1} which permits us to fill the entire length of the template in about one day. It is slow mostly because the stock solution is dilute, $\phi_0 \approx 1\%$, and much slower than CSA ($v \approx 10^1 - 10^2$ μm.s^{-1}, Figure 1) because the evaporation proceeds across the PDMS template, not directly in air.

In general, the solid we obtain nicely replicates the shape of the mould [13]. However, we discovered fairly recently that there are cases where the solid significantly deforms the template during its growth [24]. It is likely due to a poro-elastic mechanism related to the depression generated inside the material, originating from the suction mechanism during the evaporation-induced growth [25].

We deliver here the very same observation: Whereas the mould has a nominal thickness of $\approx$3 μm, the final material barely reaches h $\approx$ 300 nm, see Figure 3. We understand it on the basis of the small size of particles, here ≈4 nm, which generates a massive pressure drop inside the material during the growth of the solid upon evaporation that tends to collapse the soft PDMS template. It is somewhat disappointing as we targeted thick deposits, but the final material turns out to be fairly flat (Figure 3). Altogether, this method mainly leads to arbitrarily structured coatings which are impossible to obtain with CSA only and where the structure could bring some additional function.

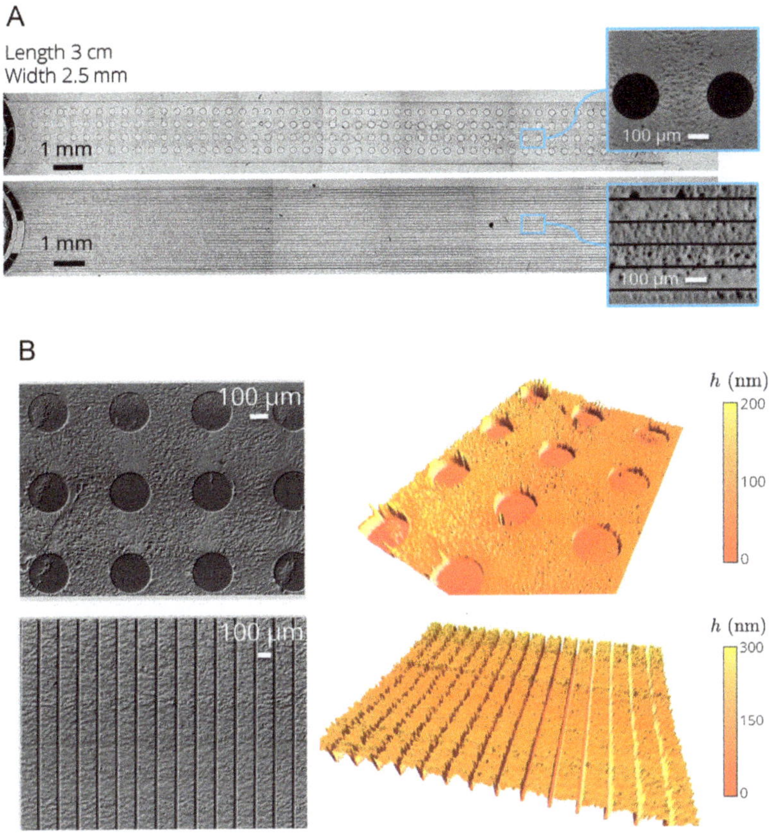

Figure 3. (**A**) Materials engineered with µ-pervaporation made of densely packed CeO_2 nanoparticles, structured with two different designs. In each case, the large figure shows the entire material through a tiled image obtained from optical microscopy in bright field mode and the insert is a SEM view of the top surface of the material. In the upper case, the thickness is about 180 nm; in the bottom case, the thickness is about 260 nm (measured with optical profilometry). (**B**) Optical microscopy (**left**) and 3D reconstruction from optical profilometry (**right**) of the micro-structured materials.

3. Optical Features of CeO_2 Nanoporous Coatings

3.1. Refractive Index

Variable angle spectroscopic ellipsometry (VASE) was used to measure the refractive index of the coatings and we limited ourselves to coatings made of the thinnest grade of nanoparticles where the final material does significantly scatter the light (empirical observation). VASE provides ellipsometric raw data (here ψ and Δ angles) that need to be parametrized in order to extract the complex refractive index $\tilde{n} = n + ik$ [26].

The optimisation approach requires knowledge of the refractive index $\bar{n}_{NP}$ of CeO_2 nanoparticles, their packing density ϕ_c inside the coating ($1 - \phi_c$ being the void fraction or porosity), and the thickness of the coating h. We then compute the ellipsometric angles of the coating with a refractive index calculated via the Bruggeman effective-medium approximation [27] of a flat nano-porous film with no roughness and deposited of a semi-infinite silicon substrate with an oxidized top surface, namely a 2 nm thick silica layer.

To guide the input parameters, we used the coating thickness measured by mechanical profilometry (h = 18 $\pm$ 3 nm), the packing density ϕ_c = 0.64 of a random close packing structure, and the refractive index $\bar{n}_{NP}$ based on literature data [28] and parametrized according to [29]. The optimisation is then performed through the Horiba software and eventually yields ϕ_c = 0.61, h = 15 nm, a model of $\bar{n}_{NP}$ (not shown), and ultimately, a very good agreement with ellipsometric raw data (Figure 4 left). The refractive index of CeO_2 nanoparticles is comparable yet smaller in magnitude than bulk cerium oxide, which could be attributed to a synthetic route that is likely to lead to a material different from bulk CeO_2.

We thus obtain a satisfactory measurement of the refractive index of the nanoporous coating of CeO_2 on a specific range of photon energy (1 − 5 eV, i.e., λ = 1250 − 250 nm, Figure 4): $\bar{n}$ shows a significant absorption of light in the UV-visible regime up to $\lambda \approx 500$ nm where the material becomes a loss-less dielectric with $n \approx 1.75$, $k \approx 0$. We assume it remains true even up to the mid-IR range, for instance λ = 4 µm, which we will also use to characterize the transmittance of the coating in this regime.

Also, we assume that this measurement obtained on a specific coating is intrinsic and thus holds, whatever the thickness of a homogeneous coating made out of the same grade of particles.

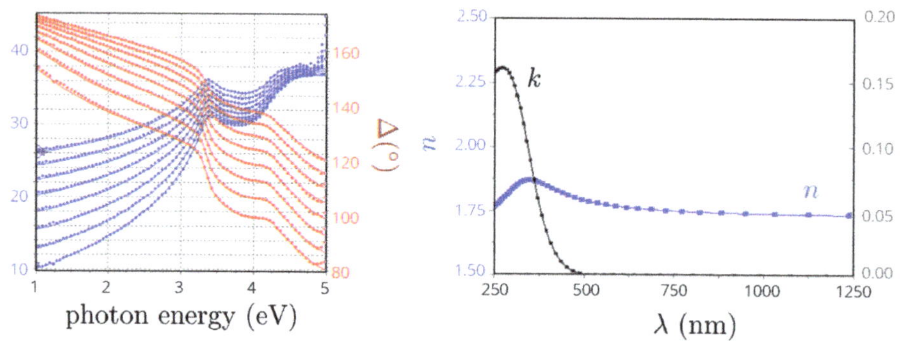

Figure 4. **(Left)**: Examples of ellipsometric angles ψ and Δ collected for eight angles over a given range of energies (symbols) and best parametrization (solid lines). **(Right)**: Complex refractive index ($\bar{n} = n + ik$) of a CeO_2 coating obtained out of these ellipsometric measurements.

3.2. *Reflection Features on Silicon Substrates*

Light reflection is due to a refractive index mismatch at an interface between two different media. The reflectance $R = I_R/I_0$ is the ratio between the reflected intensity I_R over the incident intensity of light I_0 and can be calculated using the Fresnel equation at normal incidence as follows: $R = [(n_1 - n_3)(n_1 + n_3)]^2$ where the n_is stand for the refractive indices of the two media, and which also holds when the refractive indices are complex values. As an example, silicon in air ($n_1 = 1$, $n_3 > 3$) displays a reflectance of around $R > 40\%$ in the visible to near-infrared range, see the dashed blue line in Figure 5.

An anti-reflective coating (ARC) with an adequate refractive index and thickness can lower this light reflection at an interface between two different media. The simplest ARC with a refractive index n_2 leading to $R = 0$ at normal incidence and at a single wavelength of incident radiation λ is obtained at a thickness $h = \lambda/(4n_2)$ when $n_2 = (n_1 n_3)^{1/2}$. More precisely, such a result is obtained through the calculation of the multi-layer reflection coefficient $R = |r|^2$ based on r, the amplitude Fresnel coefficient of a three-layer stack where layer 1 stands for air (refractive index $n_1 = 1$, semi-infinite), layer 2 stands for the coating, and layer 3 is the semi-infinite substrate [30,31]:

$$r = \frac{r_{12} + r_{23}\exp(2i\beta)}{1 + r_{12}r_{23}\exp(2i\beta)}, \tag{1}$$

with $\beta = 2\pi n_2 h/\lambda$ the phase shift in the coating of thickness h, and $r_{12} = (n_1 - n_2)/(n_1 + n_2)$ and $r_{23} = (n_2 - n_3)/(n_2 + n_3)$ the amplitude Fresnel coefficients at interfaces between media 1 and 2, and 2 and 3 respectively. Notice that the n_is may admit complex values. Here, medium 1 is air ($n_1 \equiv 1$), medium 2 is the nanoporous CeO_2 coating (see Figure 4 right for n_2), and layer 3 is the silicon substrate (n_3 found in Reference [32]).

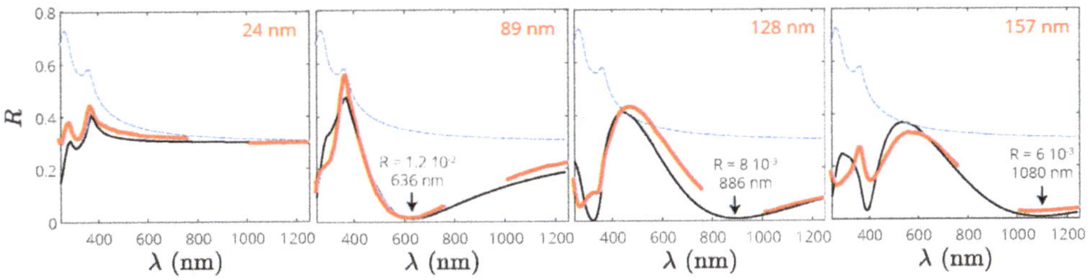

Figure 5. Intensity reflection coefficient R against wavelength for four different coating thickness (values given in red); in red the measurement, in black the calculation according to Equation (1) with no free parameter, in dashed blue the reflection coefficient of a bare silicon substrate. The value of R denoted by the arrow is the best extinction calculated from Equation (1) in the visible to near-infra-red (IR) range.

We performed the characterization of intensity reflectance R under normal incidence of light in the range $\lambda = 250 - 1300$ nm with bench-top, fairly basic equipment; the main limitation of our set-up comes from the illumination and collection of optical fibres which do not work exactly at normal incidence but accommodate some angular opening, which is somewhat detrimental to the fine measurement of R. Figure 5 shows measurements of R *vs* λ when the coating is deposited on the polished side of a silicon wafer. Here, we show the results for coatings with four different thickness (red curves) along with the reference reflectance of the bare silicon substrate (blue dashed curve) and the calculated reflectance, black curve with Equation (1).

In all cases, the reflectance is lowered by the presence of the coating (Figure 5), and for the three thickest coatings, it even exhibits a vanishing reflectance at several wavelengths, which expectingly increase with the thickness of the coating. Owing to the absence of a fitting parameter, the agreement between the experiments and the calculation is satisfactory, especially in the visible to near-IR range. The slight discrepancy that shows for the thickest coating could be due to the fact that we are not working at a perfect normal incidence.

The ARC is particularly efficient for $h = 157$ nm where $R < 1\%$ in a significant range of wavelengths in the near-IR range $\lambda = 1000 - 1300$ nm, Figure 5. In Section 5, we give a systematic mapping of R highlighting the regions where such a coating performs the best in terms of thickness and wavelength.

3.3. Enhanced Transmission on Silicon Substrates

Along with anti-reflection behaviour comes a possible enhanced transmission provided by the ARC. Now, we work with a bare substrate which has two polished sides and which is thick enough (350 μm) to exhibit an incoherent behaviour [30], so that we can neglect the interferences.

We nevertheless take into account the multiple reflections for calculating the total optical path from which the total transmittance follows:

$$T = \frac{(1 - R_{ACS})(1 - R_{AS})}{1 - R_{SCA}R_{AS}} \exp(-\alpha L), \tag{2}$$

with R is the reflectance at the different interfaces where subscripts A, C, and S stand for air, coating, and substrate respectively, $\alpha = 4\pi k_3/\lambda$ is the extinction coefficient in the substrate with k_3 as the imaginary part of the refractive index of the substrate of thickness L. R_{ACS} and R_{SAC} are calculated following the procedure of Equation (1); in the case of a bare substrate, replace R_{ACS} and R_{SCA} with R_{AS}.

We measured the transmission T of a two-sided polished substrate coated with a micro-structured pattern using a spectro-imaging set-up (described in Section 4.4 and in [33,34]). Owing to the imaging capability of the device, it is possible to select the measurement place in a heterogeneous sample such as ours, see Figure 6A with a spatial resolution of about 25 µm in this specific case, but which depends on the magnification. We thus performed a reference measurement on the bare substrate, the result of which is shown in Figure 6B (blue symbols) and which agrees well with the calculated Equation (2) and tabulated values [32] (blue line). We notice a residual peak at $\lambda = 4.2$ µm which we attribute to CO_2 absorption, and which we have difficulty in systemically removing. When the measurement was performed and averaged on a specific part of the pattern—namely the bright lines of Figure 6A—we observed a significant increase of transmittance, see the red symbols of Figure 6B, and which is parametrized with Equation (2) where the thickness h of the coating is left as a free parameter in the range 300–600 nm. It is clear that the experimental transmission is properly framed by the calculated T although the best agreement is found for $h = 400$ nm which does not perfectly match the actual thickness $h = 260$ nm. The slight discrepancy could be due either to the fact that the sample is actually slightly slanted and not perfectly perpendicular to the incident beam, or possibly also that the pixel size of the imaging set-up ($\approx$25 µm) is not very small compared to the width of the pattern $\approx$100 µm), which could induce some 'blurring' in the measured transmittance. Yet, it definitely corroborates that the anti-reflection coating is accompanied with an enhanced transmission; such a coating could thus be used to enhance the silicon transmission when this material is used as an optical component for mid-IR imaging purposes for instance, and is particularly efficient at some specific wavelengths but is obviously wavelength dependent, see Section 5.

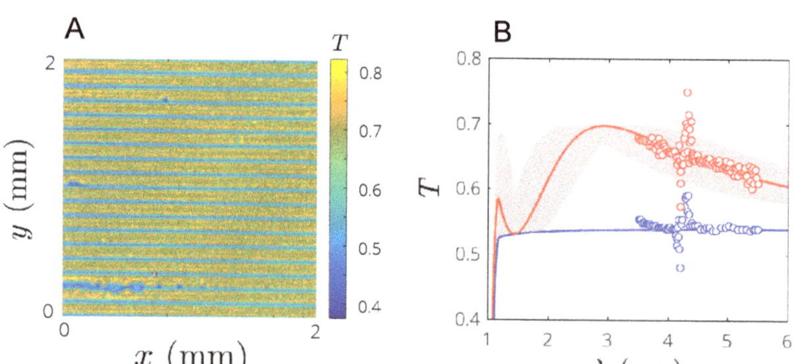

Figure 6. (**A**) Imaging the intensity transmission T at $\lambda = 4$ µm for a micro-structured pattern. (**B**) Spectral transmission T in the mid-IR range. Symbols are the local average of the wavelength dependent transmittance for the bare silicon substrate (blue) and for the pattern (red, average along the bright lines). The solid blue line is the theoretical transmission of an incoherent silicon substrate; the solid red line corresponds to a calculated T with a coating thickness $h = 400$ nm, whereas the pale red zone shows the parametrization of T with h ranging from 300 to 600 nm.

4. Materials And Methods

4.1. Dispersions of CeO_2 Nanoparticles

Fairly monodisperse cerium oxide (CeO_2) nanoparticle dispersions were synthesized and kindly provided by Solvay with two grades: a small diameter (4 ± 1 nm) at volume fraction $\phi_0 = 1.2 \times 10^{-2}$, pH = 1.6, and a large diameter (40 ± 10 nm) at volume fraction $\phi_0 = 5.6 \times 10^{-2}$, pH = 4.2, see Figure 7. These acidic dispersions are formulated in water with acetic acid in order to ensure the long-term colloidal stability of the particles (month/years at 4 °C) with no additional additives such as surfactants. Transmitted-electron microscopy (TEM EOL JEM 2200FS FEG HR 200 kV) observations reveal the polyhedral shape of the particles, along with a significant size range which was otherwise also estimated using dynamic light scattering (values given above). The dispersions are used as such for the assembly of the nanoparticles. Importantly, we also obtained the very same results concerning coatings, using a commercial dispersion available at Sigma-Aldrich (catalog number 289744-500g).

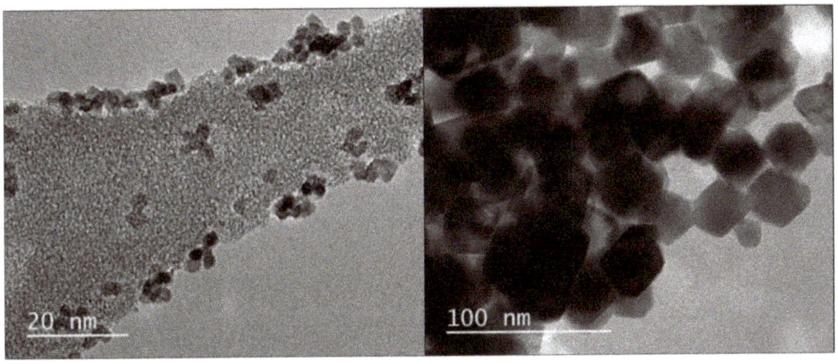

Figure 7. Transmitted-electron microscopy. (TEM) of CeO_2 nanoparticles deposited on the TEM grid, with diameters in the range of 4 ± 1 nm (**left**) and 40 ± 10 nm (**right**).

4.2. Assembly Methods

4.2.1. Films Obtained by Convective Self-Assembly (CSA)

Convective self-assembly (CSA) is a modified blade-coating technique where the withdrawing velocity of the substrate on which the coating will be laid (Figure 8) is so small that evaporation competes advantageously with the film drawing rate: both the meniscus and the wet film (deposited continuously due to the displacement of substrate with good wetting properties) act as suction pumps via evaporation that drag the nanoparticles in this confined zone to eventually build a dense film, which fully dries soon after [7–9]. The final thickness h of the film is controllable via the velocity v, the evaporation rate, the bulk volume fraction ϕ_0 of the dispersion, shape of the meniscus, etc. From mass conservation and as a rough guide [6,35]:

$$h \sim \frac{\phi_0}{\phi_c - \phi_0} \frac{Q_e}{v}, \tag{3}$$

where ϕ_0 is the volume fraction of the dispersion in the reservoir, ϕ_c the one of the final deposit, and Q_e the volumetric evaporation rate per unit of length in the lateral direction, perpendicular to v ($[Q_e] \equiv m^2.s^{-1}$). Interestingly, the coating thickness increases with a decreasing v, which sustains the enhanced role of evaporation in order to pre-concentrate the dispersion at the level of the meniscus.

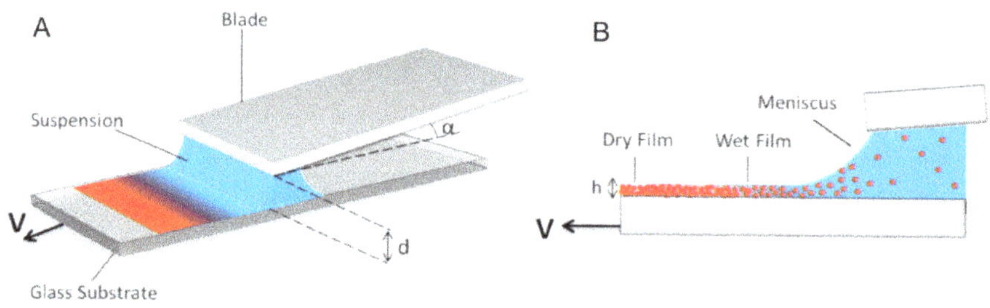

Figure 8. Sketch of the convective self-assembly (CSA) set-up: (**A**) A small volume of a dispersion of nanoparticles (the 'reservoir', in blue) is placed in-between a static blade and a substrate moving at controlled velocity v. (**B**) Side view of the geometry which leads to a deposit, in red, of controllable thickness h drying soon at the level of the meniscus [7,8]. The figures were reprinted with permission from *Langmuir* 2016, 32, 51, 13657–13668. Copyright 2016 American Chemical Society.

This regime differs from the so-called Landau-Levich (LL) regime where the withdrawing velocity v is fast enough that the deposited film remains liquid during deposition and dries afterwards. Here, the liquid film thickness increases with the velocity v due to enhanced viscous dissipation in the liquid film, and consequently, the thickness of the final deposit increases with v.

The transition between CSA and LL regimes as a function of velocity, concentration, evaporation rate, etc., lead to a V-shaped curve for h against v [15]: $h \sim v^{-1}$ at low v and $h \sim v^{2/3}$ at high v. It has been largely explored both experimentally and theoretically but remains a debated issue subject to continuous improvements. In particular, there exist considerable differences for the drying mechanisms between molecular or polymeric solutions and colloidal dispersions; heterogeneous coatings are often observed but remain largely unexplained, e.g., stick-slip sparse coatings or continuous yet undulating coatings (Figure 2, top). Nevertheless, homogeneous coatings are also at stake (Figure 2, bottom).

In practice, a small volume of a dispersion of nanoparticles ($\sim 100\,\mu$L) is placed in between the static blade and the substrate which moves at a velocity v in the range of $1 - 100\,\mu\text{m.s}^{-1}$. Here, both the blade and the substrate are 3×2 square-inches, 1 mm thick glass slides (Marienfield Ref. 11400420) which were super-cleaned via piranha treatment (20 min) followed by air plasma treatment (1 min), both performed in a clean room. Glass slides are exposed to air only at the time of setting up the CSA experiment.

4.2.2. Deposits Engineered with Microfluidic Pervaporation

Microfluidic Pervaporation combines the moulding in micro-capillary methods (MIMIC [14]) with the ability of the poly(dimethyl siloxane) (PDMS), to let solvents pervaporate (permeation followed by evaporation) across it, including water even though the PDMS is hydrophobic [12,36,37]. In the MIMIC method, the PDMS mould contains micro-channels with precise and sometimes complex morphologies engineered with soft-lithography (typical dimensions: thickness 0.5–50 µm, width 10–500 µm, length 1–10 mm) which are filled with a dispersion which then dries, mostly from the openings. It leads to a solid material made out of the dispersion which nicely replicates the shape of the micro-channels. Including thin membranes to the PDMS mould [12] (Figure 9) enhances tremendously the drying via permeation across the membrane, and permits us to build solidified materials out of (possibly ultra-) dilute dispersions [13,38–40]. Additionally, the versatility of soft-lithography leads to a rich variety of mould's topographies into which the growth of a material is made possible.

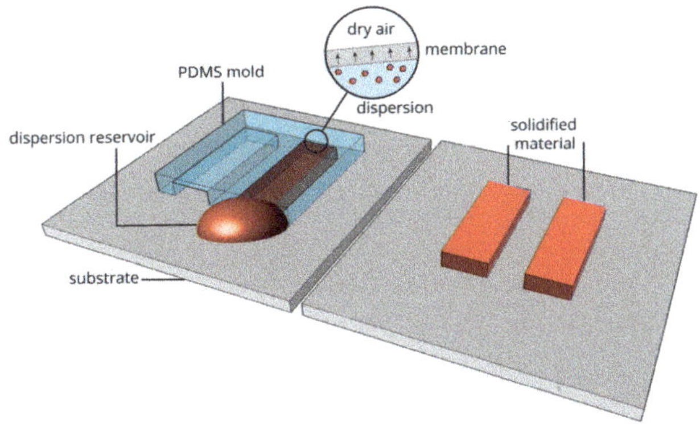

Figure 9. Sketch of the μ-pervaporation technique leading to micro-structured deposits: A poly(dimethyl siloxane) (PDMS) mould (**left**) is designed with channels initially filled with a dilute dispersion; solvent spontaneously permeates across the elastomer, concentrates the nanoparticles, and leads to a solid material out of the dispersion upon removal of the mould (**right**).

Indeed, we fabricated here two types of structured moulds: a series of long and thin parallel channels (width 100 μm, height 3 μm, length 30 mm), and a large structure consisting of a single channel (width 2.5 mm, height 3 μm, length 30 mm) with poles preventing the collapse of the PDMS mould. This structures are suggestive to the potential of the μ-pervaporation for engineering structures that go beyond simple coatings, e.g., gratings for instance.

4.3. Substrates

We use mostly used glass substrates for high-throughput coating campaigns and switched to silicon substrates for measuring the optical properties of the coatings on a high-index substrate (BT Electronics, one-sided polished 1 mm thick and two-sided polished 0.345 mm thick).

4.4. Optical Methods

Optical performance was measured as follows: Spectroscopic ellipsometry permitted us to extract the refractive index of the coatings; reflectance in the visible-to-near IR range leads to the reflection coefficient evidencing anti-reflection in some cases; mid-IR provided us with the transmittance of the samples.

The variable angles spectroscopic ellipsometer we used was a HORIBA Jobin Yvon UVISEL working here in the 250–1300 nm range and we analysed eight angles in the range 55–75°. We worked only on the thinnest grade of nanoparticles (4 nm) as the material made out of the largest ones scatters too much light.

The reflection measurements were performed at 'normal' incidence with basic bench-top equipment from Avantes. The sample was illuminated via the same co-axial fibre, from which the light emerging from the sample is also collected (Avantes reference FCR-7UVIR400-2-BX/ME). Neither the illumination nor the collection are actually at perfect normal incidence: The illumination fibre is not collimated (numerical aperture 0.22). The lamp shines at the spectral range of 200–2500 nm, but due to the two detectors we use, there is an unfortunate detection gap around 800–1000 nm.

Transmission in the mid-IR range was measured on a specific system, namely fast infrared imaging spectroscopy (FIIST [33,34]), which permits us to image the optical transmission in the range of 2–6 μm.

5. Appendix: Optical Performances

5.1. Calculated Reflectance

Here, the reflectance R of a nanoporous coating of CeO_2 is calculated on silicon substrates as described in the main text as a function of its thickness h with the measured refractive index (Figure 4) and over a large range of wavelengths. Results are shown in Figure 10 where we set a threshold to R at 5% in order to best highlight the performance zones in terms of anti-reflection; the mapping is split in two zones: small h in the visible to near-IR range and large h in the mid-IR range. In all cases, the near-zero R shows as blue cones with quite narrow bands and which are clearly sometimes degenerated, namely several regions with $R \approx 0$ show as a function of λ for a single h.

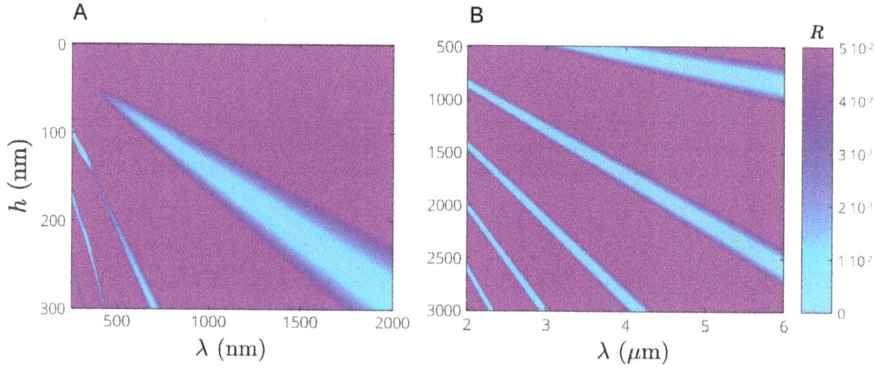

Figure 10. Reflectance R calculated for a nanoporous CeO_2 coatings of variable thickness h deposited on a silicon substrate. (**A**): Small thickness h in the visible and near-IR range; (**B**): Large thickness in the mid-IR range.

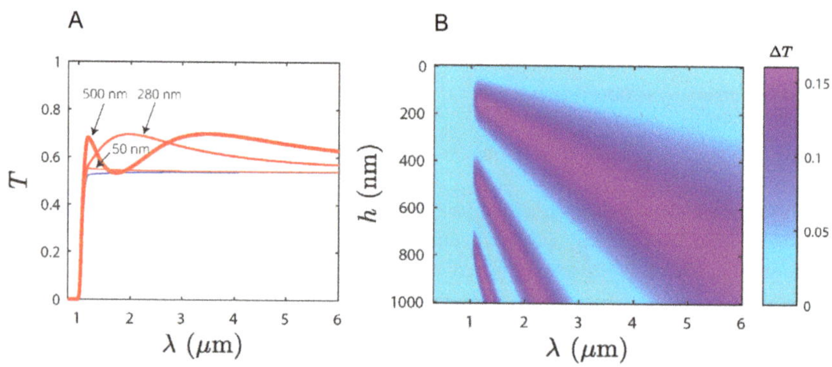

Figure 11. (**A**): Transmittance T calculated for a nanoporous CeO_2 coatings of variable thickness h (red curves) deposited on a silicon substrate (blue curve). (**B**): Transmittance excess $\Delta T = T_{Si+coat} - T_{Si}$ as a function of wavelength and coating thickness h.

5.2. Calculated Transmittance

The same type of calculation was performed for transmittance T on double-sided polished silicon substrates in the range of mid-IR, see Figure 11. In part A, a few examples of T against λ are given for three different coating thicknesses (30, 280, and 500 nm, red curves) and are compared to the transmittance of the bare substrate in blue. Obviously, the silicon substrate is opaque ($T = 0$) below some λ which actually depends on its thickness, here $\lambda \approx 1$ μm for a 325 μm thick substrate.

Above this opacity cut-off, the substrate is basically homogeneously semi-opaque (e.g., $T \approx 0.57$ at 4 μm) while adding a thin coating can dramatically enhance the transmittance. Part B of Figure 11 shows a mapping of the increase of transmittance that the coating provides, namely $\Delta T = T_{Si+coat} - T_{Si}$, against h and λ. In the mid-IR range, up to 15% of transmittance enhancement can be observed; obviously, this enhancement depends significantly on λ but can nevertheless be of some interest for monochromatic processes such as imaging.

6. Conclusions

We demonstrated that low-cost, low-tech techniques such as blade coating out of aqueous based dispersions offer an interesting alternative for producing thin, homogeneous, and well controlled nanoporous coatings. While this has been known for years, we demonstrated here the benefit of using dispersions of CeO_2, an oxide possessing a high refractive index and which results in nanoporous coatings with a high refractive index. These coatings are then interesting candidates for use as an intermediate layer between air and substrates, such as silicon with a even higher refractive index for providing, for instance, anti-reflection features. Eventually, we believe the UV-blocking of the coating could be of potential interest for protecting organic sub-layers and we think the catalytic capabilities of CeO_2 could add even more functionalities to the final material.

Author Contributions: J.L. and L.J.R. conceived the experiments, analysed and modelled the data, and wrote the paper; L.J.R. performed experiments with the help of S.K. and C.P. for mid-IR imaging; L.D.A. provided nanoparticle samples.

Funding: The authors thank funding from Solvay and Agence Nationale de la Recherche (ANR) within the EVAPEC grant 13-BS09-0010-01. The financial support from the CPER CAMPUSB project funded by the French state and the Région Nouvelle-Aquitaine is gratefully acknowledged.

Acknowledgments: The authors thank J.-B. Salmon, Ch. Loussert, M. [2] Lehtihet for fruitful help and support.

Conflicts of Interest: The authors declare no conflict of interest.

Abbreviations

The following abbreviations are used in this manuscript:

IR	Infra-Red
UV	Ultra-Violet
SEM	Scanning Electron Microscopy
TEM	Transmission Electron Microscopy

References

1. Das, S.; Dowding, J.M.; Klump, K.E.; McGinnis, J.F.; Self, W.; Seal, S. Cerium oxide nanoparticles: Applications and prospects in nanomedicine. *Nanomedicine* **2013**, *8*, 1483–1508. [CrossRef]
2. Younis, A.; Chu, D.; Li, S. Cerium Oxide Nanostructures and their Applications. In *Functionalized Nanomaterials*; InTech: London, UK, 2016.
3. Dhall, A.; Self, W. Cerium Oxide Nanoparticles: A Brief Review of Their Synthesis Methods and Biomedical Applications. *Antioxidants* **2018**, *7*, 97. [CrossRef] [PubMed]
4. Prevo, B.G.; Hwang, Y.; Velev, O.D. Convective Assembly of Antireflective Silica Coatings with Controlled Thickness and Refractive Index. *Chem. Mater.* **2005**, *17*, 3642–3651. [CrossRef]
5. Prevo, B.G.; Hon, E.W.; Velev, O.D. Assembly and characterization of colloid-based antireflective coatings on multicrystalline silicon solar cells. *J. Mater. Chem.* **2007**, *17*, 791–799. [CrossRef]
6. Dimitrov, A.S.; Nagayama, K. Continuous Convective Assembling of Fine Particles into Two-Dimensional Arrays on Solid Surfaces. *Langmuir* **1996**, *12*, 1303–1311. [CrossRef]
7. Doumenc, F.; Salmon, J.B.; Guerrier, B. Modeling Flow Coating of Colloidal Dispersions in the Evaporative Regime: Prediction of Deposit Thickness. *Langmuir* **2016**, *32*, 13657–13668. [CrossRef] [PubMed]

8. Loussert, C.; Doumenc, F.; Salmon, J.B.; Nikolayev, V.S.; Guerrier, B. Role of Vapor Mass Transfer in Flow Coating of Colloidal Dispersions in the Evaporative Regime. *Langmuir* **2017**, *33*, 14078–14086. [CrossRef] [PubMed]
9. Joshi, K.; Gilchrist, J.F. Estimation of drying length during particle assembly by convective deposition. *J. Colloid Interface Sci.* **2017**, *496*, 222–227. [CrossRef] [PubMed]
10. Qin, D.; Xia, Y.; Whitesides, G.M. Soft lithography for micro- and nanoscale patterning. *Nat. Protoc.* **2010**, *5*, 491–502. [CrossRef] [PubMed]
11. Gale, B.; Jafek, A.; Lambert, C.; Goenner, B.; Moghimifam, H.; Nze, U.; Kamarapu, S. A Review of Current Methods in Microfluidic Device Fabrication and Future Commercialization Prospects. *Inventions* **2018**, *3*, 60. [CrossRef]
12. Leng, J.; Lonetti, B.; Tabeling, P.; Joanicot, M.; Ajdari, A. Microevaporators for Kinetic Exploration of Phase Diagrams. *Phys. Rev. Lett.* **2006**, *96*, 084503. [CrossRef] [PubMed]
13. Angly, J.; Iazzolino, A.; Salmon, J.B.; Leng, J.; Chandran, S.P.; Ponsinet, V.; Désert, A.; Le Beulze, A.; Mornet, S.; Tréguer-Delapierre, M.; et al. Microfluidic-Induced Growth and Shape-Up of Three-Dimensional Extended Arrays of Densely Packed Nanoparticles. *ACS Nano* **2013**, *7*, 6465–6477. [CrossRef] [PubMed]
14. Kim, E.; Xia, Y.; Whitesides, G.M. Micromolding in Capillaries: Applications in Materials Science. *J. Am. Chem. Soc.* **1996**, *118*, 5722–5731. [CrossRef]
15. Le Berre, M.; Chen, Y.; Baigl, D. From Convective Assembly to Landau Levich Deposition of Multilayered Phospholipid Films of Controlled Thickness. *Langmuir* **2009**, *25*, 2554–2557. [CrossRef] [PubMed]
16. Kumnorkaew, P.; Gilchrist, J.F. Effect of Nanoparticle Concentration on the Convective Deposition of Binary Suspensions. *Langmuir* **2009**, *25*, 6070–6075. [CrossRef]
17. Grosso, D. How to exploit the full potential of the dip-coating process to better control film formation. *J. Mater. Chem.* **2011**, *21*, 17033. [CrossRef]
18. Farcau, C.; Sangeetha, N.M.; Moreira, H.; Viallet, B.; Grisolia, J.; Ciuculescu-Pradines, D.; Ressier, L. High-Sensitivity Strain Gauge Based on a Single Wire of Gold Nanoparticles Fabricated by Stop-and-Go Convective Self-Assembly. *ACS Nano* **2011**, *5*, 7137–7143. [CrossRef]
19. Bodiguel, H.; Doumenc, F.; Guerrier, B. Stick-Slip Patterning at Low Capillary Numbers for an Evaporating Colloidal Suspension. *Langmuir* **2010**, *26*, 10758–10763. [CrossRef]
20. Smith, M.I.; Sharp, J.S. Effects of Substrate Constraint on Crack Pattern Formation in Thin Films of Colloidal Polystyrene Particles. *Langmuir* **2011**, *27*, 8009–8017. [CrossRef]
21. Farr, R.S.; Groot, R.D. Close packing density of polydisperse hard spheres. *J. Chem. Phys.* **2009**, *131*, 244104. [CrossRef]
22. Whitesides, G.M. Self-Assembly at All Scales. *Science* **2002**, *295*, 2418–2421. [CrossRef]
23. Salmon, J.B.; Leng, J. Application of microevaporators to dynamic exploration of the phase diagram. *J. Appl. Phys.* **2010**, *107*, 084905. [CrossRef]
24. Laval, C.; Poulin, P.; Salmon, J.B. Investigation of the dynamics of growth of polymer materials obtained by combined pervaporation and micro-moulding. *Soft Matter* **2016**, *12*, 1810–1819. [CrossRef]
25. Bouchaudy, A.; Salmon, J.B. Drying-induced stresses before solidification in colloidal dispersions: In situ measurements. *Soft Matter* **2019**, *15*, 2768–2781. [CrossRef]
26. Tompkins, H.G.; Irene, E.A. (Eds.) *Handbook of Ellipsometry*; Springer: Berlin/Heidelberg, Germany, 2005.
27. Oh, T.S.; Tokpanov, Y.S.; Hao, Y.; Jung, W.; Haile, S.M. Determination of optical and microstructural parameters of ceria films. *J. Appl. Phys.* **2012**, *112*, 103535. [CrossRef]
28. Vangelista, S.; Piagge, R.; Ek, S.; Sarnet, T.; Ghidini, G.; Martella, C.; Lamperti, A. Structural, chemical and optical properties of cerium dioxide film prepared by atomic layer deposition on TiN and Si substrates. *Thin Solid Films* **2017**, *636*, 78–84. [CrossRef]
29. Yvon, H.J. *New Amorphous Diserpsion Formula*; Technical Report; Horiba Jobin Yvon: France, 2012. Available online: http://www.horiba.com/fileadmin/uploads/Scientific/Downloads/OpticalSchool_CN/TN/ellipsometer/New_Amorphous_Dispersion_Formula.pdf (accessed on 15 June 2019).
30. MacLeod, H.A. Antireflection Coatings. In *Thin-Film Optical Filters, Fourth Edition*; CRC Press: Boca Raton, FL, USA, 2010; pp. 105–184.
31. Kats, M.A.; Blanchard, R.; Genevet, P.; Capasso, F. Nanometre optical coatings based on strong interference effects in highly absorbing media. *Nat. Mater.* **2013**, *12*, 20–24. [CrossRef]

32. RefractiveIndex.INFO. Optical Constants of Si (Silicon) after Aspnes and Studna, 1983 (n, k for $\lambda = 0.21$–0.83 μm). 2019. Available online: https://refractiveindex.info/?shelf=main&book=Si&page=Aspnes (accessed on 4 May 2019).
33. Romano, M.; Ndiaye, C.; Duphil, A.; Sommier, A.; Morikawa, J.; Mascetti, J.; Batsale, J.; Servant, L.; Pradere, C. Fast infrared imaging spectroscopy technique (FIIST). *Infrared Phys. Technol.* **2015**, *68*, 152–158. [CrossRef]
34. Kirchner, S.; Narinsamy, S.; Sommier, A.; Romano, M.; Ryu, M.; Morikawa, J.; Leng, J.; Batsale, J.C.; Pradère, C. Calibration Procedure for Attenuation Coefficient Measurements in Highly Opaque Media Using Infrared Focal Plane Array (IRFPA) Spectroscopy. *Appl. Spectrosc.* **2018**, *72*, 177–187. [CrossRef]
35. Dimitrov, A.S.; Nagayama, K. Steady-state unidirectional convective assembling of fine particles into two-dimensional arrays. *Chem. Phys. Lett.* **1995**, *243*, 462–468. [CrossRef]
36. Verneuil, E.; Buguin, A.; Silberzan, P. Permeation-induced flows: Consequences for silicone-based microfluidics. *Europhys. Lett. (EPL)* **2004**, *68*, 412–418. [CrossRef]
37. Randall, G.C.; Doyle, P.S. Permeation-driven flow in poly(dimethylsiloxane) microfluidic devices. *Proc. Natl. Acad. Sci. USA* **2005**, *102*, 10813–10818. [CrossRef] [PubMed]
38. Baron, A.; Iazzolino, A.; Ehrhardt, K.; Salmon, J.B.; Aradian, A.; Kravets, V.; Grigorenko, A.N.; Leng, J.; Le Beulze, A.; Tréguer-Delapierre, M.; et al. Bulk optical metamaterials assembled by microfluidic evaporation. *Opt. Mater. Express* **2013**, *3*, 1792. [CrossRef]
39. Gómez-Graña, S.; Fernández-López, C.; Polavarapu, L.; Salmon, J.B.; Leng, J.; Pastoriza-Santos, I.; Pérez-Juste, J. Gold Nanooctahedra with Tunable Size and Microfluidic-Induced 3D Assembly for Highly Uniform SERS-Active Supercrystals. *Chem. Mater.* **2015**, *27*, 8310–8317. [CrossRef]
40. Gomez-Graña, S.; Le Beulze, A.; Treguer-Delapierre, M.; Mornet, S.; Duguet, E.; Grana, E.; Cloutet, E.; Hadziioannou, G.; Leng, J.; Salmon, J.B.; et al. Hierarchical self-assembly of a bulk metamaterial enables isotropic magnetic permeability at optical frequencies. *Mater. Horiz.* **2016**, *3*, 596–601. [CrossRef]

applied sciences

MDPI

Article

Enhancement of Corrosion Resistance of Aluminum 7075 Surface through Oil Impregnation for Subsea Application

Youngkyun Seo [1], Jung-Yeul Jung [2,*], Jihoon Chung [3] and Sangmin Lee [3]

1 Offshore Industries R & BD Center, Korea Research Institute of Ships & Ocean Engineering (KRISO), Geoje 53201, Korea
2 Maritime Safety and Environmental Research Division, Korea Research Institute of Ships & Ocean Engineering (KRISO), Daejeon 34103, Korea
3 School of Mechanical Engineering, Chung-Ang University, Seoul 06974, Korea
* Correspondence: jungjy73@kriso.re.kr

Received: 6 August 2019; Accepted: 29 August 2019; Published: 9 September 2019

Abstract: This study investigated the corrosion resistance of oil impregnated anodic aluminum oxide (AAO) surfaces of aluminum 7075 for subsea application. Although aluminum 7075 has high strength, it is scarcely used in the subsea industry because of its corrosion issue. Some treatment of aluminum 7075 is required for subsea application. In this study not only a plate shape but also a cylindrical shape were investigated because a cylindrical shape is frequently used in the subsea industry for electronic device housing. Contact angles of bare aluminum and oil impregnated AAO surfaces of aluminum 7075 were measured after a salt spray test and a pressure test. The results showed that the contact angle of bare aluminum was considerably decreased after the salt spray test, whereas the oil impregnated AAO surface presented a relatively high contact angle after the salt spray test and the pressure test. These results showed that the corrosion resistance of aluminum 7075 could be enhanced by oil impregnation on the AAO surface, and thus can be utilized in the subsea industry.

Keywords: aluminum 7075; anodizing; oil-impregnation; corrosion resistance; salt spray test

1. Introduction

Aluminum alloy is one of the most widely used materials in the subsea industry because of its low cost, excellent thermal conductivity, good strength, acceptability for short-term development, and low density. The main reason for use in the subsea industry is the excellent thermal conductivity. The generated heat in subsea electronic equipment should be removed to prevent thermal deformation. The produced heat can be eliminated by high thermal conductivity materials by exchanging the heat with the surroundings.

There are eight series of aluminum alloys depending on its composition elements. The broadly used aluminum alloys in the subsea industry are 6000 series. The 6000 series has moderate strength, and is resistant to corrosion. The 7000 series has the highest strength among all the series, but it requires some treatment for the corrosion issue. Although the 7000 series has higher strength than the 6000 series, the 6000 series is more frequently used in the subsea industry.

Several methods have been suggested to prevent aluminum corrosion such as electrodeposition of cerium or silane films, cathodic protection using an Mg-rich coating, organic coating, conversion coating, and polymer coating [1–6]. Anodizing (Anodic Aluminum Oxide, AAO) is considered one of the most practical methods for corrosion protection. Anodizing forms a nanoporous oxide layer on the top of the aluminum surface. The well-established nanoporous structure increases the corrosion resistance. The dimensions of the nanoporous structure like pore diameter, interpore distance, and thickness of the nanoporous oxide layer can be controlled by changing the type and concentration of the acid electrolyte solution, anodic voltage, anodizing time, and temperature [7–13]. The anodizing time affects the pore depth, and the anodic voltage and the temperature of the electrolyte solution influence the pore diameter and the interpore distance [14]. S.-K. Hwang et al. [13] experimentally showed that the anodizing time determines the pore depth. The pore diameter is an important characteristic of the nanoporous oxide layer [15,16]. The pore diameter can be enlarged through an etching process [12,17].

A corrosive medium can permeate the pores in the produced nanoporous structure [18–20]. Several solutions have been suggested to prevent absorbing the corrosive medium. One method is to fill the pores with solid-state oxide materials such as hydrothermal, dichromate, and nickel-salt [18,21,22]. Another solution is the coating because the hydrophobic coatings on the hydrophilic metallic surface make the surface superhydrophobic [23–27]. A further solution is oil impregnation.

Oil impregnation into the nanoporous oxide layer is suggested for corrosion protection and self-healing to withstand external damages or local defects using high purity aluminum [28]. Several studies have shown that the retention of oil on the surface resisted ingress of water or organic liquid [29,30]. The marine industry has employed oil impregnation into the porous structure to prevent corrosion of the material in a marine environment using high purity aluminum, 5000 series aluminum and low alloy steel [31–34]. Some studies have indicated that the oil impregnated nanoporous oxide layer is likely to lose the oil under a dynamic environment [35].

Although several studies have investigated oil impregnation into the nanoporous oxide layer using pure aluminum or 6000 series aluminum for corrosion protection, no studies have analyzed the oil impregnated nanoporous oxide layer using 7000 series aluminum, to our best knowledge. There are no studies on conducting a pressure test for the oil-impregnated nanoporous oxide to investigate the influence of high-pressure conditions. Therefore, this study investigated the corrosion resistance of aluminum 7075 surfaces with an oil impregnated nanoporous oxide layer to enhance its corrosion resistance for subsea application. The structure of this study is as follows. The experiment for the oil impregnated nanoporous oxide layer is described including information on the materials, preparation of the AAO surface, oil impregnation on the AAO surface, a salt spray test, a pressure test, and characterization in Section 2. The results and discussion are indicated in Section 3. Finally, the conclusion is presented.

2. Experimental

2.1. Material

Commercial aluminum alloy 7075 was employed as the material for the AAO surface because it is widely used in the 7000 series aluminum. Components of aluminum 7075 are aluminum (87.1–91.4 wt%), zinc (5.1–6.1 wt%), magnesium (2.1–2.9 wt%), copper (1.2–2 wt%), chromium (0.18–0.28 wt%), iron (Max 0.5 wt%), silicon (Max 0.4 wt%), manganese (Max 0.3 wt%), titanium (max 0.2 wt%), and other total (max 0.15 wt%). Its density is approximately 2.81 g/cc. The ultimate tensile and the tensile yield strengths are 572 MPa and 503 MPa, respectively.

2.2. Preparation of the AAO Surface

Not only a plate shape but also a cylindrical shape were prepared for the experiment in this study because the cylindrical shape is frequently used in the subsea industry as housing for an electronic device. The AAO devices for the plate and cylindrical shapes are shown in Figures 1 and 2, respectively. The following describes the details of the AAO preparation for both shapes.

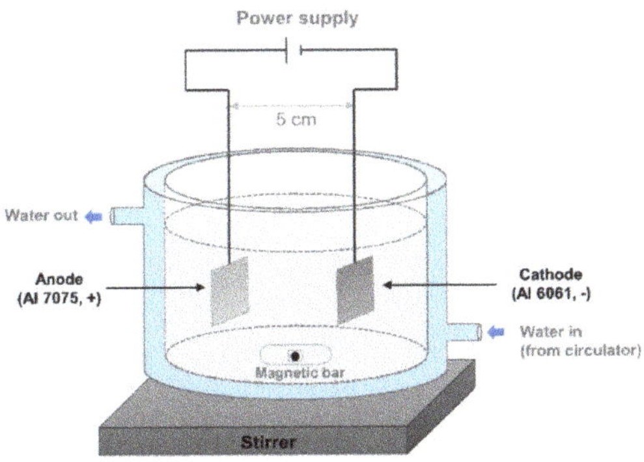

Figure 1. Anodic aluminum oxide device for the plate shape.

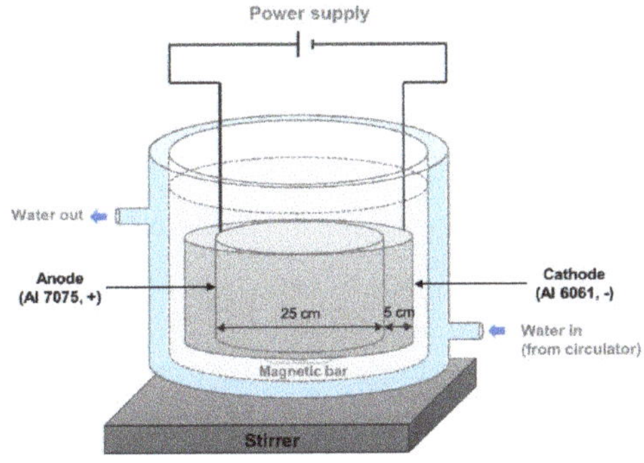

Figure 2. Anodic aluminum oxide device for the cylindrical shape.

2.2.1. Plate shape

All the samples were first rinsed with ethanol and isopropyl alcohol, and then they were cleaned using deionized water in a sonicator for 5 min. The anode was aluminum 7075, and the cathode was aluminum 6061. The distance between the anode and the cathode was fixed at approximately 5 cm as shown in Figure 1. A DC power supply (DRP-92305DU power supply) provided a constant voltage of 40 V, and the temperature was maintained by a circulator (Lab. Companion. RW-0525G) at 25 °C. The solution was 0.3 M oxalic acid solution, and it was stirred by a magnetic bar at a constant speed. The anodization was conducted for 3 h to generate 15 μm pore depth. After the anodizing, widening was conducted for 1 h under the same condition with anodizing to enlarge the pore diameter. The next step was oil impregnation. Oil (rust preventive oil, NP-7) was deposited on the nanoporous structure. The oil used in this study was a lubricant, which mostly consists of paraffin.

2.2.2. Cylindrical shape

The cylindrical shape's anode was installed as shown in Figure 2 and the cylindrical shape's cathode was also built for homogeneous anodizing. First, the samples were rinsed with ethanol and isopropyl alcohol. The anode was commercial aluminum 7075, and the cathode was aluminum 6061. The diameter of the cylindrical anode was 25 cm, and that of the cylindrical cathode was 35 cm. The distance between the anode and the cathode was adjusted to be about 5 cm. A DC power supply (PNCYS. EP-10010) provided a constant voltage of 40 V, and the temperature was maintained by a circulator (Lab. Companion. RW-0525G) at 25 °C. The anodizing time was 3 h, and the solution was 0.3 M oxalic acid solution—the same condition as the anodizing condition of the plate shape. Finally, the oil was impregnated on the surface of the AAO.

2.3. Salt Spray Test

The salt spray test was performed to investigate the corrosion effect on the surface directly. The (bare) aluminum and the oil impregnated AAO surfaces of aluminum 7075 were analyzed for comparison. This salt spray test complies with 'Standard: KSA–KS D 9502'. This standard contains a salt spray test for the anodic oxidation of metallic materials and various coatings. It conforms with ISO 9227:2006 (Corrosion tests in artificial atmospheres—salt spray tests). A 5% concentration of brine was utilized for the salt spray test, and the temperature of the test was maintained at 35 °C. The test was performed for 720 h for the oil impregnated AAO surface of aluminum 7075.

2.4. Pressure Test

This study carried out a pressure test to investigate the surface of the oil impregnated AAO of aluminum 7075 under high-pressure conditions. As the subsea equipment is installed in the seabed, it is exposed to the high-pressure environment—when going down 10 m below sea level, the pressure increases by approximately 1 bar. Since the surface of the oil impregnated AAO can be damaged under the high-pressure condition, a pressure test was required. Figure 3a,b shows the facility for the high-pressure test, which complies with 'MIL-STD-810G' (Department of defense test method standard, part two-laboratory test methods, 512.5 Immersion) and 'ABS NOTICE' (Rules for building and classing underwater vehicles, systems and hyperbaric facilities, Section 3 general requirement and safeguards, 3 proof testing, 3.1 hydrostatic test). Two cylindrical samples, which had 80 mm (diameter) × 150 mm (length), were prepared (Figure 3c). The pressure of the facility was gradually increased to 600 bar for 6000 s as shown in Figure 4. The samples were examined at 600 bar for 1800 s.

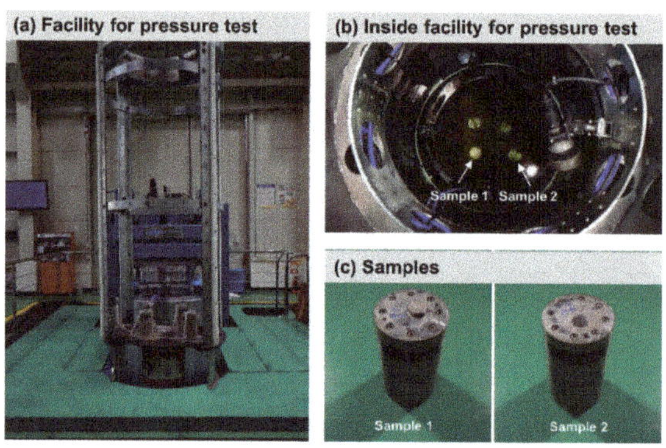

Figure 3. Facility for pressure test and samples.

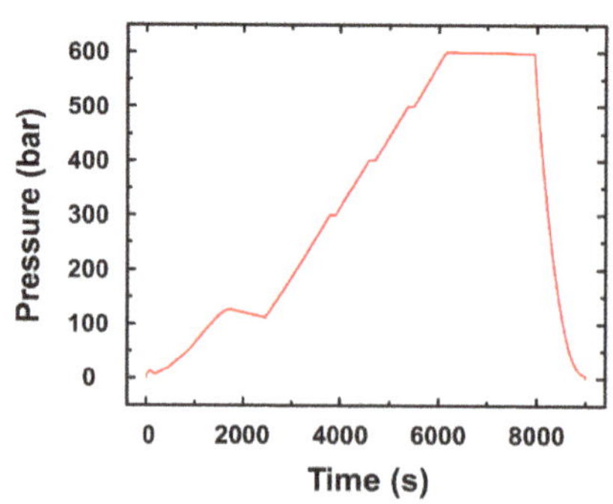

Figure 4. Pressure with time in pressure test facility.

2.5. Measurement

The contact angle of a water droplet on the surface was measured to analyze the corrosion resistance by the wettability. The water droplets were placed on the surface by a microsyringe and then the measurement of the contact angle was conducted for each sample at least five times. The mean value was used as the result value. Low contact angle (high wettability) means that a water droplet is spread over a relatively wide area of the surface. It can be easily attacked by corrosion due to its large interaction area. In contrast, a water droplet having a high contact angle interacts with a small region on the surface (low wettability). Field Emission Scanning Electron Microscope (FE-SEM, JSM-7100F) was utilized for SEM images.

3. Results and Discussion

3.1. Plate Shape

3.1.1. FE-SEM Image of Plate Shape

Figure 5 shows the FE-SEM images of AAO and AAO with 1-h widening. Before widening, the pore diameter was about 35 nm, and the pore depth approximately 15 μm. The uniform surface is shown in the microscale image (Figure 5a), and the nanoporous structure is indicated in the nanoscale image (Figure 5b). After 1 h widening, a crack was generated in the microscale image (Figure 5c) and the nanoporous structure was collapsed in the nanoscale image (Figure 5d). The widening process was not considered in this study due to the collapse of the nanoporous structure.

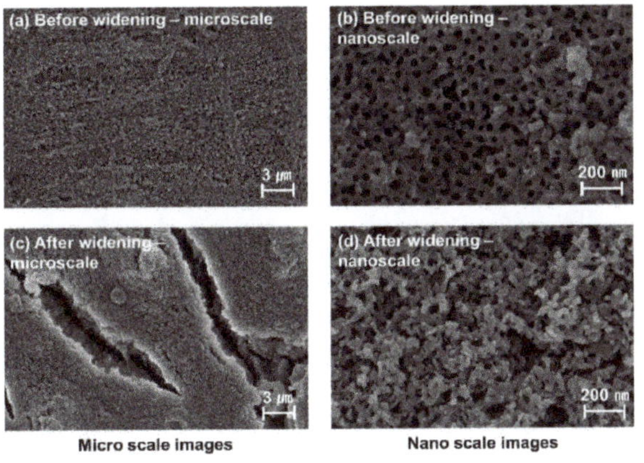

Figure 5. FE-SEM images of anodic aluminum oxide with widening of aluminum 7075.

3.1.2. Salt Spray Test

Figure 6 shows the results of the salt spray test for bare aluminum and AAO-O (oil impregnation on anodic aluminum oxide) surfaces of aluminum 7075. Corrosion was indicated after 5 h in the case of bare aluminum, whereas the corrosion was not generated in the case of the AAO-O after 720 h. The salt spray test was stopped after 100 h in the bare aluminum case as the corrosion was intensively examined after 100 h. In the case of AAO-O, the test was continuously conducted up to 720 h. The results of the salt spray test showed that the oil impregnation on the AAO surface of aluminum 7075 could enhance the corrosion resistance.

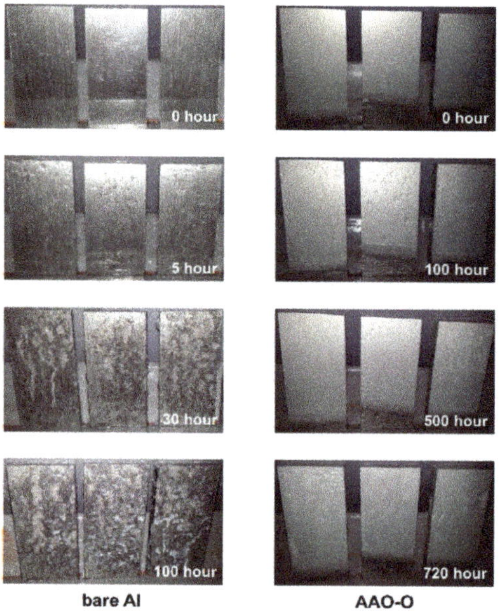

Figure 6. Results of salt spray test for bare aluminum and AAO-O (oil impregnation on anodic aluminum oxide) surfaces.

3.1.3. Contact Angle of Plate Shape

Figure 7 indicates the contact angles of AAO, AAO-O (oil impregnation), and AAO-O-SST (salt spray test). The contact angles of bare aluminum and bare aluminum-SST (salt spray test) were also investigated for comparison. The contact angle of a water droplet on the bare aluminum was 74.4° ± 5.8°, and it was significantly decreased to 24.8° ± 5.6° after the salt spray test. The bare aluminum is susceptible to corrosion. After anodizing, the contact angle was dropped to 21.0° ± 1.7°. It showed that the nanoporous structure enhanced the wettability of water by increasing surface roughness. After oil impregnation on the AAO surface, the contact angle was increased. The contact angle of a water droplet on AAO-O-SST was 81.1° ± 1.9°. Although the salt spray test was conducted on the oil-impregnated AAO surface, the contact angle of the water droplet on AAO-O-SST was significantly increased.

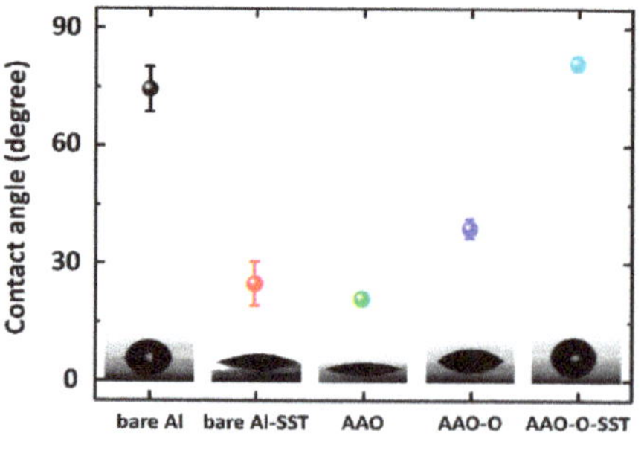

Figure 7. Contact angles of bare aluminum, bare aluminum-SST (salt spray test), AAO (anodic aluminum oxide), AAO-O (oil impregnation), and AAO-O-SST (salt spray test) surfaces of aluminum 7075.

Figure 8 presents the schematic cross-section diagram of bare aluminum, bare aluminum-SST (salt spray test), AAO, AAO-O (oil impregnation), AAO-O-SST (salt spray test) surfaces. Bare aluminum indicated a relatively flat surface, whereas the roughness was considerably increased after the salt spray test as shown in Figure 6. The results of the contact angle test showed that the contact angle of bare Al was decreased after the salt spray test. A surface of high roughness generally presents a low contact angle. Oxidized aluminum by the salt spray test increased the roughness of the surface, and the increased roughness reduced the contact angle. After anodizing, a nanoporous structure was generated. The produced nanoporous structure by anodizing decreased the contact angle. After oil impregnation on the AAO surface, the contact angle was increased. We assumed that the oil entirely covered the surface of the AAO. The contact angle on the surface of the AAO-O-SST was significantly increased after the salt spray test. It could be imagined that some oil remained on the nanoporous structure after the salt spray test as shown in Figure 8e. If there was no oil remaining on the surface of the AAO-O-SST, the contact angle would be similar to the contact angle of the AAO. When the oil layer was completely covered on the top of the surface of the AAO-O-SST, the contact angle would be similar to the contact angle of the AAO-O. The contact angle of the AAO-O-SST had a similar contact angle to the bare aluminum because the exposed nanoporous structure held the droplet.

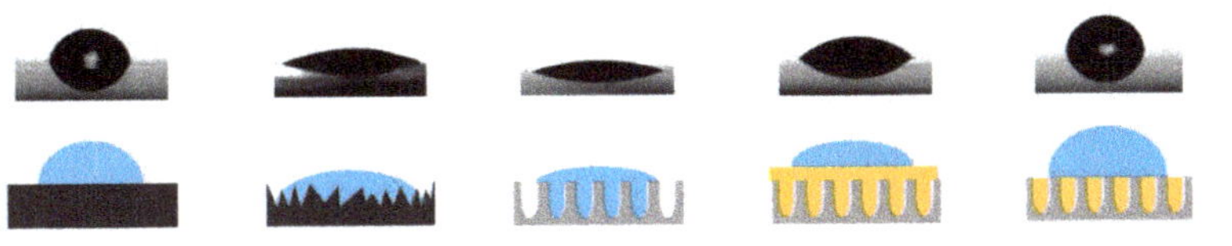

Figure 8. Schematic cross-section diagram of bare aluminum, bare aluminum-SST (salt spray test), AAO (anodic aluminum oxide), AAO-O (oil impregnation), AAO-O-SST (salt spray test) surfaces.

This study analyzed the interfacial tension between the proposed surfaces and water. The interfacial tension shows the adhesive force between the liquid phase of one substance and the liquid, solid or gas state of another element. When there is high interfacial tension, the water spreads on the surface. The high interfacial tension indicates a hydrophilic surface, whereas little interfacial tension shows a hydrophobic surface. Therefore, the high interfacial tension poses a low contact angle, and the low interfacial tension has a high contact angle. The interfacial tension was calculated by proposed equations from previous studies.

We employed the formulas in the Smith et al. study [30] to calculate the interfacial tension between the water and the proposed surfaces. The formula for the interfacial tension between the aluminum surfaces and the water is indicated in Equation (1). The formula for the interfacial tension between the AAO surfaces and the water is shown in Equation (2). Equation (3) shows the formula for the interfacial tension between the AAO-O-SST surfaces and the water. Since we assumed that the oil on the surface of the AAO-O-SST was thoroughly infused, the interfacial tension between the aluminum and the oil was not taken into account.

$$\gamma_{sw} = \sigma_s + \sigma_w - 2\sqrt{\sigma_s \cdot \sigma_w} \tag{1}$$

$$\gamma_{Aw} = r\gamma_{sw} \tag{2}$$

$$\gamma_{ASw} = f\gamma_{sw} + (1-f)\gamma_{ow} \tag{3}$$

where γ_{sw} is interfacial tension between the solid (aluminum) and the liquid (water), γ_{ow} is that between the oil and the water, γ_{Aw} is that between the AAO surfaces and the water, and γ_{ASw} is that between the AAO-O-SST surfaces and the water. σ_s is the surface free energy of the solid, and σ_w is the surface tension of the liquid. r is the roughness factor which is the ratio of the total surface area to the projected area of the solid in contact with liquid, and f is the fraction of the projected area of the surface that is occupied by the solid.

The roughness factor (r) was calculated by Equation (4).

$$r = 1 + \frac{2\pi \cdot \frac{a}{2} \cdot h}{(a+b)^2} \tag{4}$$

where a, b, and h are the pore diameter, edge-to-edge spacing, and the height, respectively. a and b were 30 and 70 nm, and h was 15 µm. f is predicted to be 0.65 using the FE-SEM image of the AAO surface.

There are various models to predict the interfacial tension between different substances and phases: Owens–Wendt–Rabel–Kaelble (OWRK), Wu, Oss and Good (acid-base), Fowkes, and Extended Fowkes [36,37]. This study employed the Fowkes model. Although the Extended Fowkes model furthermore consideres polar interactions and a hydrogen bonding fraction as shown in Equation (5), those two fractions are insignificant in these cases. The surface tension of water, aluminum, and oil are 72.8, 45.0, and 25.0 mN/m, respectively. The dispersion contribution of water is 21.8 mN/m.

$$\gamma_{\alpha\beta} = \sigma_\alpha + \sigma_\beta - 2\left(\sqrt{\sigma_\alpha^D \cdot \sigma_\beta^D} + \sqrt{\sigma_\alpha^P \cdot \sigma_\beta^P} + \sqrt{\sigma_\alpha^H \cdot \sigma_\beta^H}\right) \tag{5}$$

where α is aluminum or oil, and β is water. σ^D is the dispersion force contribution, and σ^P is the polar interaction. σ^H indicates the hydrogen bonding fraction.

The diagrams of the proposed surfaces and the interfacial tensions are tabulated in Table 1. The interfacial tension of bare aluminum and AAO-O-SST showed relatively low values, whereas that for AAO noted a considerably high value. As indicated before, the surfaces of aluminum and AAO-O-SST had a relatively high contact angle. Since the high contact angle means low interfacial tension, the calculated results of interfacial tension agreed with the results of the contact angle. The interfacial tension of the AAO-O-SST was slightly less in value than that of bare aluminum, and the contact angle of the AAO-O-SST was slightly higher than that of bare aluminum. The interfacial tensions at AAO had a significantly high value due to the high roughness factor. The deep pore depth caused the high roughness factors.

Table 1. Interfacial tensions between water and bare aluminum, AAO (anodic aluminum oxide), and AAO-O-SST (oil impregnation—salt spray test).

	Bare Aluminum	AAO	AAO-O-SST
Diagram	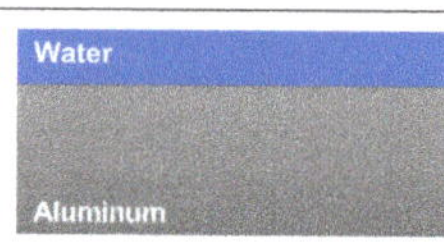	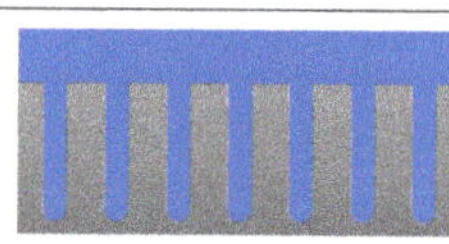	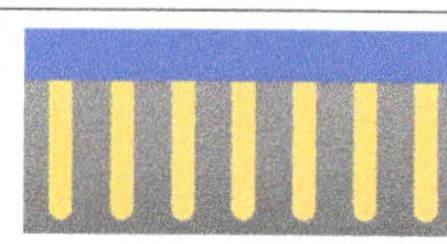
Interfacial tension (mN/m)	55.2	7853.0	53.7

3.2. Cylindrical Shape

Contact Angle of Cylindrical Shape

Figure 9 indicates the contact angles of AAO, AAO-O, and AAO-O-PT (pressure test) for the cylindrical shape of aluminum 7075. After anodizing, the contact angle was decreased. After oil impregnation, the contact angle was increased to 50.0° ± 4.4° like that in the plate shape. The contact angle of the AAO-O-PT was approximately 89.1° ± 13.6°. Although the pressure test was conducted on AAO-O, the contact angle was increased by 39.1°. This result showed that some oil remained in the nanoporous structure, even though the oil on the top of the nanoporous structure was removed. The combination of the nanoporous structure and the oil in the pores caught the water droplet. Therefore, the high contact angle was investigated in the case of oil impregnation.

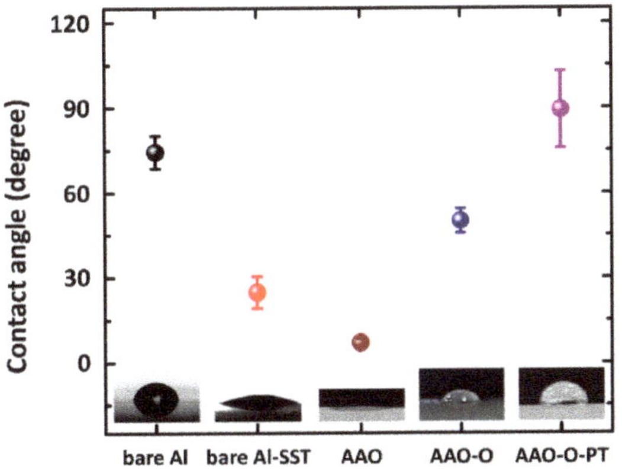

Figure 9. Contact angles of AAO (anodic aluminum oxide), AAO-O (oil impregnation), and AAO-O-PT (pressure test) surfaces for cylindrical shape of aluminum 7075.

3.3. *Comparison of Plate with Cylindrical Shapes*

The cylindrical shape indicated a similar tendency to the plate shape. Although the contact angle of AAO was low in both shapes, they were increased by the oil impregnation. The salt spray test and the pressure test eliminated the covered oil on the top of the nanoporous structure, but could not remove the oil in the pores. The combination of the exposed nanoporous structures and the oil filled in the pores increased the contact angle significantly. These results showed that the oil impregnation on the AAO surface of aluminum 7075 in both shapes increased the corrosion resistance.

4. Conclusions

This study analyzed the corrosion resistance of the oil impregnated AAO surface of aluminum 7075 to suggest a way to improve the corrosion protection for subsea application. The cylindrical shape was investigated as well as the plate shape. The contact angle was measured, and the interfacial tension between the water and the surfaces were calculated using the proposed equations. The salt spray test was conducted in the case of the plate shape, whereas the pressure test was carried out in the case of the cylindrical shape. Although salt spray and pressure tests were conducted, the oil in the pores of the AAO surface remained. The contact angle of the oil impregnated AAO surface was increased after the salt spray and pressure tests. That is, the nanoporous structure and the remaining oil in the pores help to maintain a high contact angle. The calculated interfacial tension coincided with the measurement of the contact angle. Corrosion was not indicated in the oil impregnated AAO surface for the 720 h salt spray test. The impregnated oil in the pores can endure a high-pressure condition (600 bar). Enhanced aluminum 7075 by AAO and oil impregnation can be used for subsea equipment.

Author Contributions: Conceptualization, J.-Y.J.; Methodology, J.C.; Validation, S.L.; Formal Analysis, Y.S.; Writing—Original Draft Preparation, Y.S.; Writing—Review & Editing, J.-Y.J.

Funding: This research was supported by a grant from the Endowment Project of "Development of the Surface Modification Technologies based on Metal-Oxide for Offshore Equipment" (PES9320) and "Development of Technology to Support the Rapid Search and Rescue of Marine Accidents" (PES3130) funded by Korea Research Institute of Ships and Ocean Engineering.

Conflicts of Interest: The authors declare no conflict of interest.

References

1. Wu, L.K.; Liu, L.; Li, J.; Hu, J.M.; Zhang, J.Q.; Cao, C.N. Electrodeposition of cerium (III)-modified bis-[triethoxysilypropyl]tetra-sulphide films on AA2024-T3 (aluminum alloy) for corrosion protection. *Surf. Coat. Technol.* **2010**, *204*, 3920–3926. [CrossRef]
2. Hu, J.M.; Liu, L.; Zhang, J.Q.; Cao, C.N. Electrodeposition of silane films on aluminum alloys for corrosion protection. *Prog. Org. Coat.* **2007**, *58*, 265–271. [CrossRef]
3. Simões, A.M.; Battocchi, D.; Tallman, D.E.; Bierwagen, G.P. SVET and SECM imaging of cathodic protection of aluminium by a Mg-rich coating. *Corros. Sci.* **2007**, *49*, 3838–3849. [CrossRef]
4. Grundmeier, G.; Schmidt, W.; Stratmann, M. Corrosion protection by organic coatings: Electrochemical mechanism and novel methods of investigation. *Electrochim. Acta* **2000**, *45*, 2515–2533. [CrossRef]
5. Cohen, S.M. Review: Replacements for Chromium Pretreatments on Aluminum. *Corrosion* **1995**, *51*, 71–78. [CrossRef]
6. Diggle, J.W.; Downie, T.C.; Goulding, C.W. Anodic oxide films on aluminum. *Chem. Rev.* **1969**, *3*, 365–405. [CrossRef]
7. Masuda, H.; Fukuda, K. Ordered metal nanohole arrays made by a two-step replication of honeycomb structures of anodic alumina. *Science* **1995**, *268*, 1466–1468. [CrossRef]
8. Masuda, H.; Satoh, M. Fabrication of gold nanodot array using anodic porous alumina as an evaporation mask. *Jpn. J. Appl. Phys. Part 2 Lett.* **1996**, *35*. [CrossRef]
9. Masuda, H.; Yada, K.; Osaka, A. Self-Ordering of Cell Configuration of Anodic Porous Alumina with Large-Size Pores in Phosphoric Acid Solution Self-Ordering of Cell Configuration of Anodic Porous Alumina with Large-Size Pores in Phosphoric Acid Solution. *Jpn. J. Appl. Phys.* **1998**, *1340*, 9–12. [CrossRef]
10. Li, A.P.; Müller, F.; Birner, A.; Nielsch, K.; Gösele, U. Hexagonal pore arrays with a 50–420 nm interpore distance formed by self-organization in anodic alumina. *J. Appl. Phys.* **1998**, *84*, 6023–6026. [CrossRef]
11. Schwirn, K.; Lee, W.; Hillebrand, R.; Steinhart, M.; Nielsch, K.; Gösele, U. Self-Ordered Anodic Aluminum Oxide Formed by H2SO4Hard Anodization. *ACS Nano* **2008**, *2*, 302–310. [CrossRef]
12. Zhang, J.; Kielbasa, J.E.; Carroll, D.L. Controllable fabrication of porous alumina templates for nanostructures synthesis. *Mater. Chem. Phys.* **2010**, *122*, 295–300. [CrossRef]
13. Hwang, S.; Jeong, S.; Hwang, H.; Lee, O.; Lee, K. Fabrication of Highly Ordered Pore Array in Anodic Aluminum Oxide. *Korean J. Chem. Eng.* **2002**, *19*, 467–473. [CrossRef]
14. Lee, K.; Lyu, S.; Lee, S.; Kim, Y.S.; Hwang, W. Characteristics and self-cleaning effect of the transparent super-hydrophobic film having nanofibers array structures. *Appl. Surf. Sci.* **2010**, *256*, 6729–6735. [CrossRef]
15. Nelson, S.F.; Lin, Y.Y.; Gundlach, D.J.; Jackson, T.N. Temperature-independent transport in high-mobility pentacene transistors. *Appl. Phys. Lett.* **1998**, *72*, 1854–1856. [CrossRef]
16. Whitney, T.M.; Searson, P.C.; Jiang, J.S.; Chien, C.L. Fabrication and magnetic properties of arrays of metallic nanowires. *Science* **1993**, *261*, 1316–1319. [CrossRef]
17. Choi, D.H.; Lee, P.S.; Hwang, W.; Lee, K.H.; Park, H.C. Measurement of the pore sizes for anodic aluminum oxide (AAO). *Curr. Appl. Phys.* **2006**, *6*, 125–129. [CrossRef]
18. Lee, J.; Jung, U.; Kim, W.; Chung, W. Effects of residual water in the pores of aluminum anodic oxide layers prior to sealing on corrosion resistance. *Appl. Surf. Sci.* **2013**, *283*, 941–946. [CrossRef]
19. Huang, Y.; Shih, H.; Huang, H.; Daugherty, J.; Wu, S.; Ramanathan, S.; Chang, C.; Mansfeld, F. Evaluation of the corrosion resistance of anodized aluminum 6061 using electrochemical impedance spectroscopy (EIS). *Corros. Sci.* **2008**, *50*, 3569–3575. [CrossRef]
20. Moutarlier, V.; Gigandet, M.P.; Normand, B.; Pagetti, J. EIS characterisation of anodic films formed on 2024 aluminium alloy, in sulphuric acid containing molybdate or permanganate species. *Corros. Sci.* **2005**, *47*, 937–951. [CrossRef]
21. Zuo, Y.; Zhao, P.H.; Mao, J.M. The influences of sealing methods on corrosion behavior of anodized aluminum alloys in NaCl solutions. *Surf. Coati. Technol.* **2003**, *166*, 237–242. [CrossRef]
22. González, J.A.; López, V.; Otero, E.; Bautista, A. Postsealing changes in porous aluminum oxide films obtained in sulfuric acid solutions. *J. Electrochem. Soc.* **2000**, *147*, 984–990. [CrossRef]
23. Lu, Z.; Wang, P.; Zhang, D. Super-hydrophobic film fabricated on aluminium surface as a barrier to atmospheric corrosion in a marine environment. *Corros. Sci.* **2015**, *91*, 287–296. [CrossRef]

24. Jeong, C.; Lee, J.; Sheppard, K.; Choi, C.H. Air-Impregnated Nanoporous Anodic Aluminum Oxide Layers for Enhancing the Corrosion Resistance of Aluminum. *Langmuir* **2015**, *31*, 11040–11050. [CrossRef]
25. Wang, P.; Zhang, D.; Qiu, R.; Wan, Y.; Wu, J. Green approach to fabrication of a super-hydrophobic film on copper and the consequent corrosion resistance. *Corros. Sci.* **2014**, *80*, 366–373. [CrossRef]
26. Ishizaki, T.; Masuda, Y.; Sakamoto, M. Corrosion resistance and durability of superhydrophobic surface formed on magnesium alloy coated with nanostructured cerium oxide film and fluoroalkylsilane molecules in corrosive NaCl aqueous solution. *Langmuir* **2011**, *27*, 4780–4788. [CrossRef]
27. Liu, T.; Chen, S.; Cheng, S.; Tian, J.; Chang, X.; Yin, Y. Corrosion behavior of super-hydrophobic surface on copper in seawater. *Electrochim. Acta* **2007**, *52*, 8003–8007. [CrossRef]
28. Lee, J.; Shin, S.; Jiang, Y.; Jeong, C.; Stone, H.A. Oil-Impregnated Nanoporous Oxide Layer for Corrosion Protection with Self-Healing. *Adv. Funct. Mater.* **2017**, *27*, 1606040. [CrossRef]
29. Rykaczewski, K.; Anand, S.; Subramanyam, S.B.; Varanasi, K.K. Cryo-FIB/SEM Investigation of Mechanism of Frost Formation on Lubricant- Impregnated Surfaces. *Microsc. Microanal.* **2013**, *19*, 926–927. [CrossRef]
30. Smith, J.D.; Dhiman, R.; Anand, S.; Reza-Garduno, E.; Cohen, R.E.; McKinley, G.H.; Varanasi, K.K. Droplet mobility on lubricant-impregnated surfaces. *Soft Matter* **2013**, *9*, 1772–1780. [CrossRef]
31. Wang, P.; Zhang, D.; Lu, Z.; Sun, S. Fabrication of Slippery Lubricant-Infused Porous Surface for Inhibition of Microbially Influenced Corrosion. *ACS Appl. Mater. Interfaces* **2016**, *8*, 1120–1127. [CrossRef]
32. Wang, P.; Lu, Z.; Zhang, D. Slippery liquid-infused porous surfaces fabricated on aluminum as a barrier to corrosion induced by sulfate reducing bacteria. *Corros. Sci.* **2015**, *93*, 159–166. [CrossRef]
33. Song, T.; Liu, Q.; Liu, J.; Yang, W.; Chen, R.; Jing, X.; Takahashi, K.; Wang, J. Fabrication of super slippery sheet-layered and porous anodic aluminium oxide surfaces and its anticorrosion property. *Appl. Surf. Sci.* **2015**, *355*, 495–501. [CrossRef]
34. Yang, S.; Qiu, R.; Song, H.; Wang, P.; Shi, Z.; Wang, Y. Slippery liquid-infused porous surface based on perfluorinated lubricant/iron tetradecanoate: Preparation and corrosion protection application. *Appl. Surf. Sci.* **2015**, *328*, 491–500. [CrossRef]
35. Jacobi, I.; Wexler, J.S.; Stone, H.A. Overflow cascades in liquid-infused substrates. *Phys. Fluids* **2015**, *27*. [CrossRef]
36. Shaw, D.J. *Chapter 4 Liquid-gas and liquid-liquid interfaces In Introduction to Colloid and Surface Chemistry*, 2nd ed.; CRC Press: Burlington, VT, USA, 1980; pp. 64–114.
37. Surface Free Energy (SFE), Surface Energy. Available online: https://www.kruss-scientific.com/services/education-theory/glossary/surface-free-energy/ (accessed on 27 August 2019).

applied sciences

MDPI

Article

Poly(phenylene methylene)-Based Coatings for Corrosion Protection: Replacement of Additives by Use of Copolymers

Marco F. D'Elia [1,*], Mirko Magni [2], Stefano P. M. Trasatti [2,*], Thomas B. Schweizer [1], Markus Niederberger [1] and Walter Caseri [1]

1 Department of Materials, ETH Zürich, 8093 Zürich, Switzerland
2 Department of Environmental Science and Policy, University of Milan, 20133 Milan, Italy
* Correspondence: marco.delia@mat.ethz.ch (M.F.D.); stefano.trasatti@unimi.it (S.P.M.T.)

Received: 17 July 2019; Accepted: 27 August 2019; Published: 29 August 2019

Abstract: Poly(phenylene methylene) (PPM) is a thermally stable, hydrophobic, fluorescent hydrocarbon polymer. Recently, blended PPM has been proposed as a valuable anti-corrosion coating material, and, in particular, rheological additives such as external plasticizers resulted crucial to prevent crack formation. Accordingly, to avoid common problems related to the use of external plasticizers, the development of PPM-related copolymer-based coatings containing *n*-octyloxy side chains and their anti-corrosion behavior were explored in this study. The aluminum alloy AA2024, widely employed for corrosion studies, was selected as a substrate, covered with a thin layer of a polybenzylsiloxane in order to improve adhesion between the underlying hydrophilic substrate and the top hydrophobic coating. Gratifyingly, coatings with those copolymers were free of bubbles and cracks. The *n*-octyloxy side-chains may be regarded to adopt the role of a bound plasticizer, as the glass transition temperature of the copolymers decreases with increasing content of alkoxy side-chains. Electrochemical corrosion tests on PPM-substituted coatings exhibited good corrosion protection of the metal surface towards a naturally aerated near-neutrally 3.5% wt.% NaCl neutral solution, providing comparable results to blended PPM formulations, previously reported. Hence, the application of rheological additives can be avoided by use of proper design copolymers.

Keywords: poly(phenylene methylene) coatings; PPM-related copolymer; rheological additive-free polymer formulation; AA2024; corrosion protection; electrochemistry

1. Introduction

Poly(phenylene methylene) (PPM) is a hydrocarbon polymer with the general formula $(C_6H_4[CH_2])_n$. It is structurally located between polyethylene and polyphenylene, consisting of an alternating sequence of phenylene and methylene units (Figure 1a). Remarkably, it exhibits a rather unique combination of material properties. Besides high hydrophobicity [1], it is highly thermally stable (onset of decomposition temperature at 450–470 °C) [2–5] and fluorescent [6]. This optical property, unusual for a non-conjugated polymer such as PPM, was attributed to homoconjugation [6]. The rare phenomenon of homoconjugation can arise under particular geometric conditions when conjugated π-orbital systems interact with each other, even though they are electronically separated by an insulating methylene group [7,8], as illustrated in Figure 1b.

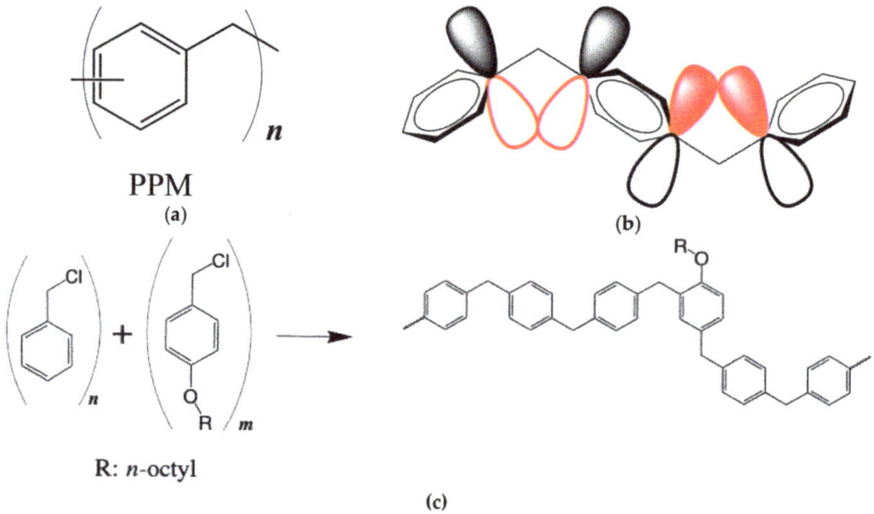

Figure 1. (**a**) Repeating units of poly(phenylene methylene) (PPM), (**b**) schematic representation of homoconjugation in PPM: p-orbitals of phenylene rings overlap even if they are electronically separated by methylene group, (**c**) scheme of the synthesis of random copolymers based on PPM obtained by mixing different fractions (*n* and *m*, $m/(m+n)$ = 5.3% mol/mol or 11.2% mol/mol) of benzyl chloride and its derivative (4-octyloxybenzyl chloride), and structure of a sequence of the resulting random copolymers.

Moreover, importantly, PPM has also been shown to be effective in corrosion protection, however, only when blended with rheological additives, such as polysiloxanes and benzylbutyl phthalate, as an external plasticizer in order to prevent cracking of the surface due to the stiffness of the polymer [1]. Notably, the principal aim of plasticizers is to ameliorate the elasticity and processability of polymers by lowering the second order transition temperature (glass transition temperature), thus decreasing the tendency of coatings to formation of cracks [1,9]. More specifically, external plasticizers are low molar mass compounds which are dispersed in the polymeric matrix spreading the polymer chains apart [10]. Thus, plasticizers depress the polymer–polymer secondary interactions into the polymeric matrix enhancing the mobility of the polymer chains, resulting in a softer material which can be easily deformed. However, due to the weak interactions between polymer chains and external plasticizers, the plasticizer can leave the material matrix by evaporation, migration or extraction, and upon exposure to UV light the plasticizer can degrade and subsequently also initiate degradation of the polymer [11,12]. Further, external plasticizers can increase the erodibility of the coating, thus reducing the material's lifetime [10,12].

However, migration of external plasticizers and the film-forming properties of polymers can be managed by application of polymer-bound molecules which can quasi be regarded as internal plasticizers [13]. In particular, the presence of bulky side chains along the polymer backbone lowers the internal forces between the polymer main chains [13,14].

In this way, we sought to design a copolymer based on phenylene methylene units for corrosion protection, by inserting different fractions of *n*-octyloxy side chains into the PPM backbone. The poly(phenylene methylene) derivative was synthesized by copolymerization of a mixture of benzyl chloride and variable fractions of 4-octylbenzyl chloride in presence of tin tetrachloride as catalyst, analogously to the synthesis of PPM itself [2].

Hence, in this presented work, the preparation of corrosion-resistant PPM-related coatings on pretreated aluminum alloy AA2024 containing different molar fractions of *n*-octyloxy side chains (Figure 1c) were prepared to explore the efficacy of long alkoxy side-chains as a reliable alternative to rheological additives. Accordingly, the ability of PPM derivatives in anti-corrosion protection was examined by coating pretreated aluminum alloy samples (AA2024) and studying their behavior in a naturally aerated near-neutral 0.6 M sodium chloride solution anodic polarization and potentiostatic polarization techniques. Additionally, thermal stability and glass transition temperature of the copolymers were investigated by thermogravimetric analysis (TGA) and differential scanning calorimetry (DSC).

2. Materials and Methods

2.1. Reagents and Solvents

Benzyl chloride stabilized by propylene oxide (99%), tin(IV) chloride, phosphoric acid (85%), thionyl chloride (99%), sodium sulfate (99%) and chloroform were purchased from Sigma Aldrich (Buchs, Switzerland), benzyltriethoxysilane from Fluorochem (Hadfield, UK), 4-hydroxybenzaldehyde (98%) and 1-bromooctane (98%) from abcr (Karlsruhe, Germany), sodium borohydride (98%) and tetrahydrofuran (CROMANORM) from VWR Chemicals BDH (Leuven, Belgium), acetonitrile (99.9%), dichloromethane (99.8%), sodium hydroxide (98.66%) and potassium hydroxide (86%) from Fischer Chemicals (Loughborough, UK), and methanol (98%) from Merck (Darmstadt, Germany).

2.2. Apparatus

^{1}H NMR spectra were recorded on a Bruker AV 300 MHz (Billerica, MA, USA) instrument using $CDCl_3$ as solvent. Chemical shifts (δ) for 1H spectra are expressed in ppm relative to internal Me_4Si as standard. Signals were abbreviated as s, singlet; d, doublet; t, triplet; m, multiplet.

Elemental analysis was performed by the Microanalytic Laboratory of Organic Chemistry (LOC) at ETH Zürich.

Gel permeation chromatography (GPC) analysis were performed by using Viscotek GPC system (Malvern, Worcs, UK) equipped with a pump and degasser (GPCmax VE2001, 1.0 mL min^{-1} flow rate), a detector module (Viscotek 302 TDA) and three columns (2× PLGel Mix-C and 1× ViscoGEL GMHHRN 18055, dimensions 7.5 × 300 mm for each column) using tetrahydrofuran as an eluent.

Rheology measurements were carried out using an Anton-Paar MCR-302 rheometer with parallel plates (Graz, Austria), at a temperature of 120 °C. Complex viscosity was measured collecting 10 points at angular frequencies starting from 0.23–22 rad s^{-1} applying 5% of strain.

For thermogravimetric analysis (TGA), a Mettler Toledo TGA/DSC 3+ STAR^e system instrument (Mettler-Toledo, Schwerzenbach, Switzerland) was used heating the samples from 25 to 1000 °C with a heating rate of 10 °C min^{-1} under nitrogen atmosphere.

Differential scanning calorimetry (DSC) was carried out with a Mettler Toledo DSC822^e instrument (Columbus, OH, USA) using a cooling and heating rate of 10 °C min^1 under nitrogen atmosphere.

Pictures of polymer-coated AA2024 samples were obtained with an optical microscope (Wild Photomakroskop M400, Switzerland) equipped with a digital camera, combined with a portable UV lamp emitting at 395 nm (4 W, Lighting EVER).

Scanning Electron Microscopy (SEM)was performed with a LEO Gemini 1530.

2.3. Synthesis of 4-Octyloxybenzyl Chloride

4-Octyloxybenzyl chloride was synthesized according to the literature [15,16] as follows: 4-hydroxybenzaldehyde (0.16 mol, 20 g) was dissolved together with potassium hydroxide (0.2 mol, 11.12 g) in acetonitrile (270 mL) in a three-necked flask. The reaction was stirred at room temperature and heated to reflux at 90 °C. 1-Bromooctane (0.15 mol, 27 mL) was then added during 1 h and the reaction mixture was vigorously stirred overnight. Thereafter, the mixture was cooled to room temperature and quenched with 250 mL of water, transferred to a separatory funnel and extracted with hexane (200 mL). The organic layer was washed twice with 30 mL of NaOH solution (10% in mass) and finally with water (three times 30 mL). The organic phase was dried over anhydrous sodium sulfate and concentrated in a rotary vapor at 300 mbar (40 °C) giving 4-octyloxy benzaldehyde as a pale-yellow oil (25.1 g, 0.11 mol, yield: 70%). ^{1}H NMR ($CDCl_3$): δ = 0.88 (t, 3H, CH_3), 1.19–1.55 (m, 10 H, 5 CH_2), 1.74–1.84 (m, 2H, CH_2), 4.01 (t, 2H, CH_2O), 6.65–6.98 (d, 2H, Ar), 7.79–7.82 (d, 2H, Ar), 9.85 (s, 1H, COH) ppm.

Afterwards, 4-octyloxy benzaldehyde (25.1 g, 0.107 mol) was dissolved in tetrahydrofuran (THF) (20 mL) and added dropwise to a stirred suspension of $NaBH_4$ (8.1 g, 0.2 mol) in dry THF (140 mL) at 0 °C. The reaction mixture was then warmed up to room temperature and stirred overnight. The reaction mixture was quenched with 50 mL of water and the organic layer was separated, washed with water (three times) and dried over anhydrous sodium sulfate [17]. The solvent was then removed by evaporation at 11 mbar and 40 °C, giving a white solid that was dissolved in 5 mL of THF and poured into 400 mL of water under stirring. The white precipitate was filtered (cellulose filter) and the cake was dried under vacuum (0.7 mbar) to give 4-octyloxybenzyl alcohol as a white powder (0.08 mol, 20.14 g, yield: 80%). ^{1}H NMR ($CDCl_3$): δ = 0.86–0.91 (t, 3H, CH_3), 1.20–1.49 (m, 10 H, 5 CH_2), 1.73–1.82 (m, 2H, CH_2), 3.93–3.95 (t, 2H, CH_2OAr), 4.61 (s, 2H, CH_2OH) 6.87–6.90 (d, 2H, Ar), 7.26–7.29 (d, 2H, Ar) ppm. $C_{15}H_{24}O_2$ (236.35 g mol^{-1}): calcd. C 76.23, H 10.23, found C 76.18, H 10.12.

Subsequently, 4-octcyloxybenzyl alcohol (0.03 mol, 7.08 g) was dissolved in dichloromethane (200 mL) and thionyl chloride (0.036 mol, 4.26 g) was added to the solution dropwise at 0 °C. The mixture was then stirred at the same temperature during 2 h and the reaction was subsequently quenched with water (100 mL). The organic layer was extracted, washed with water (40 mL) and saturated aqueous sodium hydrogen carbonate (140 mL) and dried over anhydrous sodium sulfate. The dichloromethane was evaporated at reduced pressure (100 mbar at 40 °C) obtaining 4-octyloxybenzyl chloride as a yellow pale oil (0.016 mol, 4.16 g, yield: 54%). ^{1}H NMR ($CDCl_3$): δ = 0.89–0.93 (t, 3H, CH_3), 1.31–1.49 (m, 10 H, 5 CH_2), 1.75–1.84 (m, 2H, CH_2), 3.94–3.98 (t, 2H, CH_2OAr), 4.57 (s, 2H, CH_2Cl) 6.87–6.90 (d, 2H, Ar), 7.29–7.32 (d, 2H, Ar) ppm. $C_{15}H_{24}OCl$ (254.80 g): calcd. C 70.71, H 9.10, found C 70.92, H 9.22.

2.4. *Synthesis of the Octyloxy-Containing PPM Derivatives*

Copolymerization of benzyl chloride in presence of 4-octyloxybenzyl chloride was performed with two ratios of the comonomers, 5.3% mol/mol and 11.2% mol/mol of 4-octyloxybenzyl chloride, respectively. First, propylene oxide stabilizer (0.25% w/v) was removed exposing benzyl chloride to vacuum (0.7 mbar) overnight. The removal of propylene oxide stabilizer was verified by ^{1}H NMR spectroscopy by disappearance of signals of propylene oxide at 2.96, 2.73, 2.41, and 1.31 ppm. In the case of 5.3% mol/mol, 1.58 g (1.6 mL, 6.2 mmol) of 4-octyloxybenzyl chloride was added under nitrogen atmosphere to 14.9 g (13.7 mL, 118 mmol) of destabilized benzyl chloride in a 100 mL three-neck flask equipped with a mechanical stirrer. In the case of 11.2% mol/mol, 1.49 g (1.50 mL 5.84 mmol) of 4-octyloxybenzyl chloride were added under nitrogen atmosphere to 6.1 mL (52 mmol) of destabilized benzyl chloride in a 100 mL three-neck flask equipped with a mechanical stirrer. Thereafter, the respective mixtures were heated up to 60 °C and 0.05 mL (0.46 mmol) of $SnCl_4$ was added. The copolymerizations were carried out under nitrogen flow of 0.4–0.5 mL min^{-1} to allow the produced HCl to leave the reaction environment. After 3 h, due to the increase of viscosity, the temperature was risen to 120 °C for 3 h and subsequently to 180 °C for 17 h. During the reaction, the color shifted from deep red after the addition of $SnCl_4$ to clear amber at the end of the reaction. Afterwards, the mixture was cooled down to room temperature and the product was solubilized in 10 mL of THF. This solution was then poured into 400 mL of methanol under vigorous stirring, and after 4 h, the obtained powder was filtered through a cellulose filter and dried under vacuum (10^{-2} mbar) for 12 h, yielding 3.87 g and 13.26 g of product, respectively, containing 13.4% and 6.1% of the octyloxy repeat units, respectively, as obtained from ^{1}H NMR spectra (see below). Accordingly, yields of 63% and 75%, respectively, were calculated in the case of 13.4% and 6.1% octyloxy repeat unit.

2.5. Preparation of Coated AA2024

Sheets of 12 cm in length, 3 cm in width and 4 mm in thickness of high strength aluminum alloy AA2024 (4.3%–4.5% copper, 1.3%–1.5% magnesium, 0.5%–0.6% manganese and less than 0.5% of other elements) were provided by Aviometal s.p.a (Varese, Italy) and used as substrate.

Samples of 4 cm in length were cut and subsequently polished with abrasive papers of 300, 500, 800, 1200, and 4000 grit. Immediately after polishing, the samples were cleaned by immersion in ethanol in an ultrasonic bath (Banderlin, Berlin, Germany) for 5 min. Then AA2024 samples were removed from the ethanol bath and the residual alcohol at the surface was evaporated by means of a flush of nitrogen.

A layer of benzyltriethoxysilane was applied by spin coating (3500 rpm, 30 s) on freshly cleaned AA2024 samples and subsequently heated up to 100 °C for 1 min, whereupon condensation of benzyltriethoxysilane to respective polysiloxanes proceeded [1]. These samples were finally coated with substituted PPM as described in the section Results and Discussion, using about 100 mg of polymer.

2.6. Electrochemical Characterization of Coated AA2024

The anticorrosion ability of the PPM derivatives as thin protective films was studied by means of electrochemistry techniques, carrying out tests on AA2024 samples coated with the two copolymers (6.1% mol/mol and 13.4% mol/mol). The protocol for the coating deposition is described in the Results and Discussion section.

Electrochemical corrosion tests were conducted in a naturally aerated near-neutral simulated marine environment prepared by dissolving 0.6 mol L^{-1} sodium chloride (≥99.0%, Sigma-Aldrich) in MilliQ® water. The pH value was adjusted to 6.7 ± 0.1 by adding few drops of 0.2 mol L^{-1} sodium hydroxide solution to the stock solutions. All the experiments, if not otherwise stated, were carried out at ambient temperature (24 ± 3 °C, with a variation lower than 2 °C during each single run). In all cases, the operative temperature was below the glass transition temperature of PPM derivatives (see below).

The apparatus used for the measurements consisted of a glass cell with a hole (1 cm in diameter) in the middle of the flat bottom part which assures the contact between the coated metallic plate (working electrode, exposed area 0.78 cm^2) and the working solution (0.6 M NaCl). The sealing was guaranteed by a bi-adhesive layer (a2 Soluzioni Adesive, Italy) pressed between the sample and the bottom of the cell. The electrochemical setup also included a platinum coil as counter electrode and an aqueous saturated calomel electrode as reference one (E_{SCE} = 0.242 V vs. SHE). The latter was inserted into a glass double bridge (filled with the same working solution) ending with a Luggin capillary aimed to minimize the ohmic drop between working and reference electrode. No instrumental compensation of the residual ohmic drop was performed.

The electrochemical characterization included both potentiodynamic and potentiostatic methods. The former consisted of a single anodic polarization scan, sweeping the potential from OCP to 2.5 V vs. SCE, at a scan rate of 10 mV min^{-1} (each run lasting ca. 5.5 h). A limit current density of 4 mA cm^{-2} was imposed, thereafter the scan was automatically aborted independently by the achievement of the final potential. The second characterization implies the application of a constant potential to the metallic sample and the recording of the current flow between working and counter electrode. In our experiments, an oxidizing potential of 0 V vs. SCE was applied for 24 h. Potentiodynamic and potentiostatic curves were recorded after an initial delay time of 600 s for assuring the equilibration of the system at OCP.

Some potentiodynamic curves were recorded also at a fixed temperature of 35 °C, just above the glass transition temperature of the modified PPM (see Results and Discussion). For these experiments, a suitable cell surrounded by a jacket filled by a flux of water controlled by a thermostat (Haake CH Fisons coupled to a Haake F3 Fision) was adopted.

3. Results and Discussion

3.1. Synthesis and Structural Characterization

PPM-based copolymers containing *n*-octyloxy side chains were obtained by mixing the comonomers benzyl chloride and 4-octyloxybenzyl chloride at two ratios, 5.3% mol/mol and 11.2% mol/mol, respectively, and polymerization was subsequently carried out under the conditions reported in the Materials and Methods section. For both ratios the number-average molar masses of the resulting copolymers (M_n) amounted to about 2500 g mol^{-1} and the weight-average molar masses (M_w) to about 5400 g mol^{-1}, resulting in polydispersity indices (PDI= M_w/M_n) of about 2.2–2.3 (Table 1). Notably, the fluorescence observed for PPM itself also emerged in the copolymers (Figure 2).

Table 1. Molar masses of copolymers prepared from benzyl chloride and 4-octylbenzyl chloride (6.1% mol/mol and 13.4% mol/mol).

Sample	Mn (g mol^{-1})	Mw (g mol^{-1})	PDI (Mw/Mn)
6.1% octyloxy units	2438	5402	2.21
13.4% octyloxy units	2382	5496	2.3

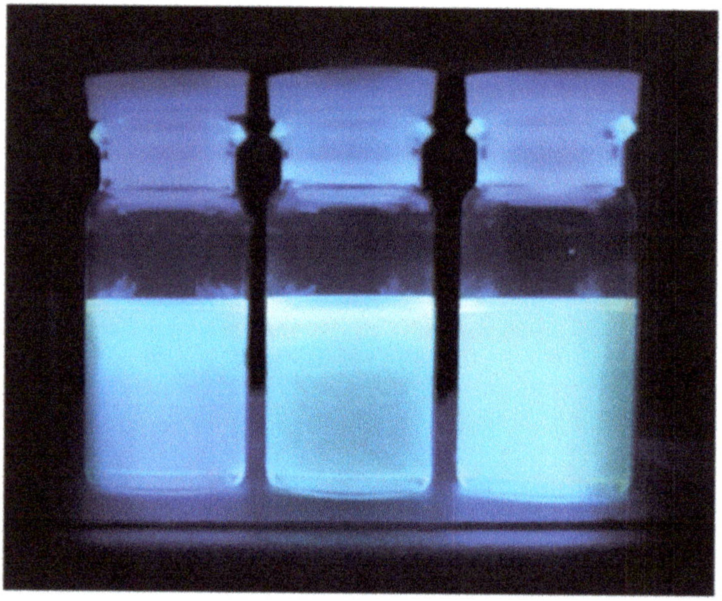

Figure 2. Photographs taken under UV-light illumination of PPM and its derivatives solutions in chloroform (≈0.5% m/m, 2 cm diameter of the vials): PPM (left), copolymer containing 13.4% (center) and 6.1% (right).

The presence of octyloxy groups in the copolymers was assessed with ^1H NMR spectroscopy. With regard to the reported values of PPM [18,19], ^1H NMR spectra of the copolymers showed the aromatic resonances (H_{Ar}) in the typical region of 6.8–7.2 ppm. The position of the bridging methylene signals (H_m) at 3.9 ppm differed from that of the corresponding signals of the CH_2Cl group of the comonomers at 4.5 ppm (Figure 3). Notably, the presence of the signals corresponding to the alkoxy chains (H_{Alk}) in the region of 0.5–1.5 ppm confirms the inclusion of octyloxy side chains into the polymers. Furthermore, the ratio between integrated peak areas of alkyl signals (H_{Alk}) and those of the methylene bridges (H_m) were used to calculate the effective molar composition of the copolymers. The obtained values showed somewhat higher molar ratios of constitutional repeat units of the octyloxy groups (13.4% and 6.1%) than the comonomer ratio employed for the reaction (11.2% and 5.3% respectively). Obviously, polymers with a higher fraction of unsubstituted phenylene methylene units were removed during sample workup with somewhat higher preference.

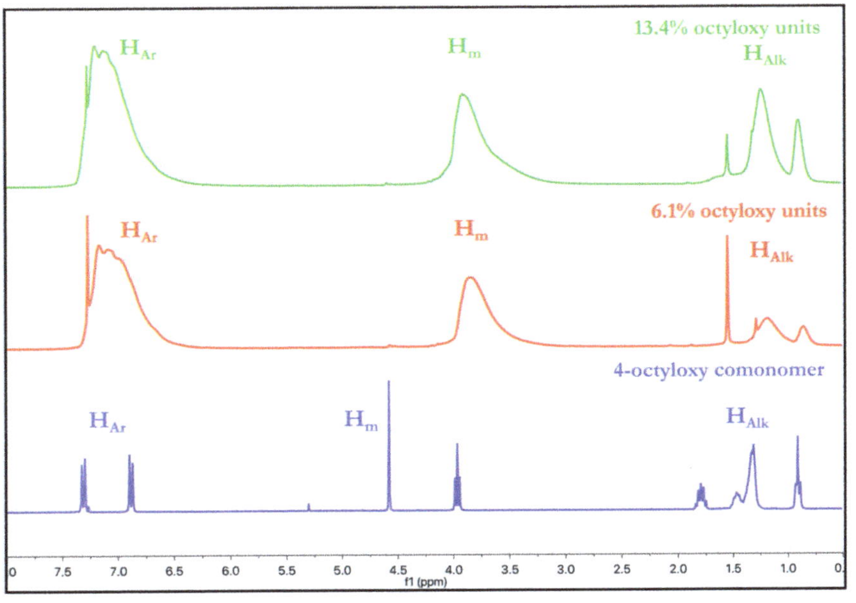

Figure 3. ^{1}H NMR spectra. From the bottom to top: 4-octyloxybenzyl chloride (blue line), copolymer with 6.1% mol/mol (red line), and 13.4% mol/mol 4-octyloxybenzyl-containing constitutional repeat units (green line).

3.2. Thermal Analysis

Thermogravimetric analysis of the polymers carried out under ambient atmosphere revealed an onset of thermal decomposition at about 380 and 400 °C respectively for PPM containing 13.4% mol/mol and 6.1% mol/mol of repeat units with alkoxy side chains, respectively, while the maximum decomposition rates were observed at 482 °C (for 13.4%) and 508 °C for (for 6.1%), compared to 510–515 °C for PPM itself [5,18].

Differential scanning calorimetry (DSC) performed with the copolymers revealed in each case only a single glass transition temperature (Table 2). The glass transition temperature (T_g) of the copolymers decreases as the fraction of the octyloxy side chains along the polymeric backbone increases. The pure PPM of M_n = 2400 g mol^{-1} features $T_g \approx$ 55 °C [2] while the PPM derivative containing 6.1% and 13.4% mol/mol of octyloxy side-chains exhibited T_g values of 48 and 31 °C, respectively (Table 2). This is likely due to the relative difference in mobility of the polymer chains, whose motion is progressively enhanced with the increase of the concentration of octyloxy substituents, leading to the corresponding decrease of T_g [20]. Moreover, the lack of any thermal transition beyond the glass transition temperature indicates that the copolymers are amorphous like PPM itself [6].

Table 2. Glass transition temperatures (T_g) and the difference between the temperature T applied in rheology experiments (120 °C and T_g), of PPM and copoylmers with (6.1% and 13.4% octyloxy groups, respectively.

Sample	T_g (°C)	($T-T_g$) (°C)
PPM	55	65
6.1% octyloxy units	48	71
13.4% octyloxy units	31	89

3.3. Rheology

Dynamic rheological data of molten PPM and the related copolymers were obtained from frequency sweeps over the range of 0.3–22 rad s^{-1} at 5% strain at the temperature of 120 °C (the temperature of coating application). The applied 5% strain was confirmed to be within the linear viscoelastic region by strain sweep measurements. The complex viscosity (η^*) of PPM and the copolymers are shown in Figure 4. The copolymer containing 6.1% mol/mol of *n*-octyloxy repeat units revealed a slight decrease of viscosity compared to PPM itself, while increasing the fraction of units with *n*-octyloxy side-chains to 13.4% mol/mol led a pronounced drop of viscosity (decrease by order of magnitude). It has to be noted that the measurements were performed at the film processing temperature (see below) which is well above the glass transition temperature of the polymers (Table 1). It appears, therefore, in agreement with glass transition temperatures, that *n*-octyloxy side-chains act as spacers between the polymer backbones decreasing the cumulative intermolecular forces along the polymer chains, thus inducing a decrease in viscosity.

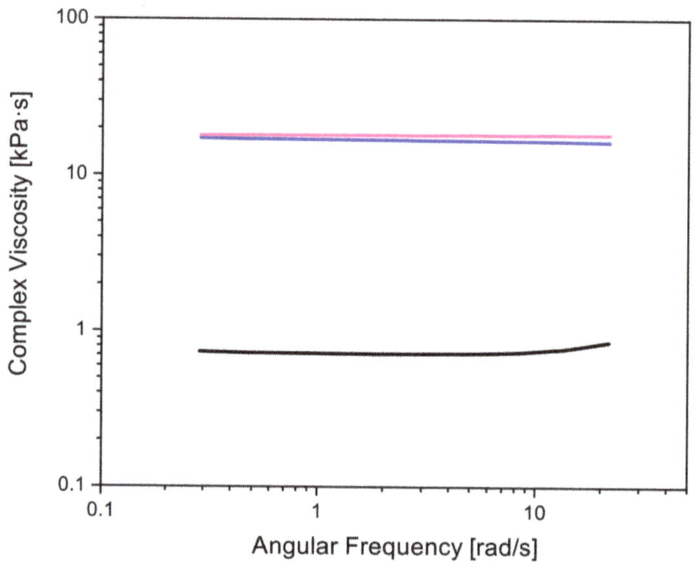

Figure 4. Effect of *n*-octyloxy side chains on the complex viscosity. From top to bottom: PPM (pink line), 6.1% mol/mol octyloxy units (blue line) and 13.4% mol/mol octyloxy units (black line).

3.4. Coatings

Coatings of the copolymers were manufactured by pressing polymer powders onto a silane-pretreated [1] AA2024 specimen, using polyetheretherketone (PEEK) foil to separate the PPM-based polymers from the pressing instrument. Pressing was performed for both copolymers (6.1% mol/mol and 13.4% mol/mol octyloxy units) at a temperature of 120 °C for 30 s. The thickness of the resulting films was about 30 μm (Figures S1 and S2), and the coatings appeared very uniform and homogeneous although no rheological additive was added (Figure 5). Notably, at the processing temperature, the low viscosity of the molten polymer with 13.4% mol/mol octyloxy units enhances self-diffusion within the polymeric matrix, and therefore, during film formation, the cohesion between the polymer chains, and thus, coalescence are promoted [20]. Moreover, polymers which contain side chains that do not strikingly hinder self-diffusion may have a greater cohesive strength than non-branched polymers, based on a firmer anchoring of such macromolecules in the polymeric matrix [20].

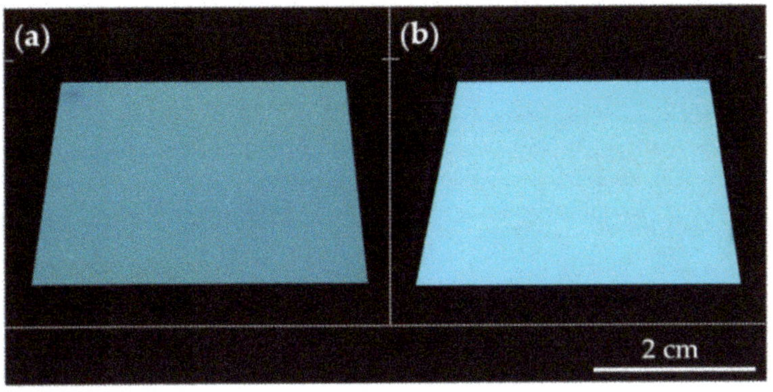

Figure 5. PPM derivatives coatings on polybenzylsiloxane-modified AA2024. (**a**) Surface coated with CO-PPM containing 13.4% of 4-octyloxy side-chains. (**b**) Surface coated with CO-PPM containing 6.1% of 4-octyloxy side-chains.

Remarkably, it is common knowledge that corrosion inhibition may also be favored by the hydrophobic behavior of polymer coatings and the homogeneity of their surface. Advancing and receding contact angles of water on copolymer coatings confirmed the low wettability of the protective films (Table 3). Moreover, the remarkably low contact angle hysteresis (4° and 1°, for 6.1% mol/mol and 13.4% mol/mol octyloxy units) indicates a very smooth and uniform surface (notably, the contact angle hysteresis was even lower than that on AA2024 itself) [1].

Table 3. Contact angles.

Sample	Advancing	Receding
6.1% octyloxy units	104°	100°
13.4% octyloxy units	102°	101°
AA 2024 (1)	55°	41°

1 literature data [1].

3.5. Protective Behavior of PPM-Based Coatings against Corrosion of AA2024

The coatings of the copolymers are an electrical insulating layer deposited on the silane treated surface of aluminum AA2024. Thus, the protective action of the coating against corrosion of the metallic substrate is merely due to a physical barrier effect, aimed to prevent the direct contact between the aggressive environment (i.e., naturally aerated near-neutral 3.5 wt.% NaCl aqueous solution) and the underlying aluminum surface.

Polarization scans are a useful electrochemical method to compare the barrier properties of the here prepared PPM-based coatings containing 6.1% mol/mol and 13.4% mol/mol repeat units with long alkoxy side chains that, extending from the main polymer backbone, act as an internal plasticizer.

Contrary to single-cycle polarization technique (commonly called pitting scan), in which an anodic scan is immediately followed by a backward one, which has been recognized as a valuable tool for the detailed corrosion morphology analysis of aluminum [21–23], anodic polarization provides a faster but more qualitative method to study the corrosion phenomena. In this work, the recorded current density was exploited as a diagnostic parameter to identify the occurrence of localized corrosion.

In the case of the copolymer with 6.1% octyloxy units, excessive porosity (Figure 5b) and/or presence of cracks in the polymer layer resulted into a worst physical barrier effect, in turns detected by a current increase during the polarization test (Figure 6a) due to localized (i.e., pitting) corrosion, thus resulting in aluminum dissolution. This well-known corrosion phenomenon to which Al is susceptible in the presence of chlorides occurs already at OCP for bare aluminum (Figure S3), resulting in a quick increase of the current density to values of many Ma cm^{-2} already at –0.6 V vs. SCE [21–23]. The resulting shift toward more positive potentials for the abrupt increase of the current density for the coated metal (Figure 6a) with respect to the bare one (Figure S3) is due to the barrier-like protective property of the PPM coatings. Therefore, current densities 10^6 times lower (i.e., few Na cm^{-2}) are maintained at potentials even more positive than the OCP and the critical pitting potential of bare AA2024, resulting into a significantly increased resistance towards the aggressive environment.

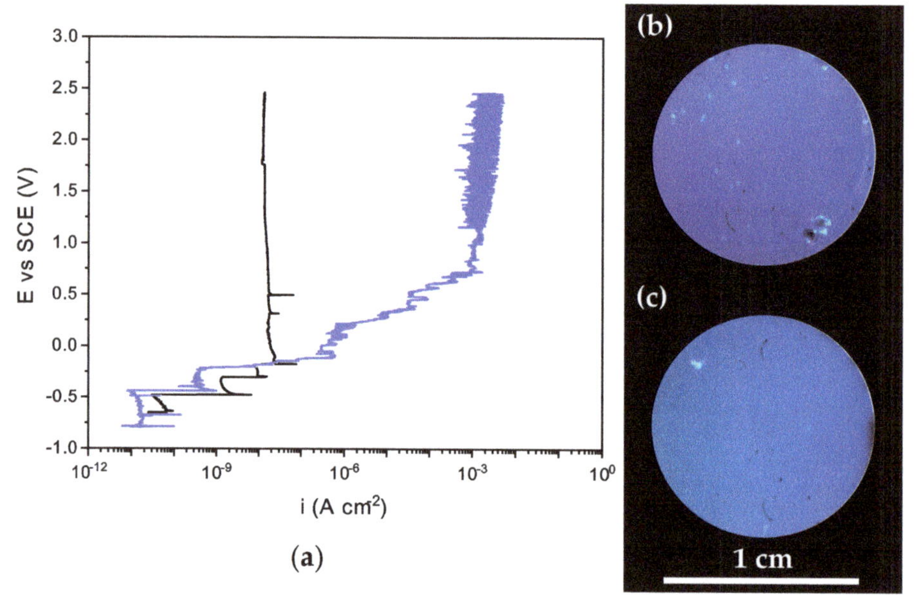

Figure 6. (**a**) Anodic polarization curves for AA2024 coated with copolymer with 6.1% (blue lines) and 13.4% ocytyloxy units (black lines) in naturally aerated near-neutral 0.6 M NaCl solution. Right: Optical microscope pictures (under 395 nm light irradiation) of samples coated with copolmyer containing 6.1% (**b**) and 13.4% octyloxy units (**c**) after polarizations tests.

Best anodic polarization curves recorded for silane treated AA2024 coated with a layer of 6.1% and 13.4% octyloxy units are presented in Figure 6a, where the scans started from OCP (between –0.78 and –0.61 V vs. SCE) as detected after 600 seconds of equilibration between the samples and the solution. For sake of reproducibility, different samples of each copolymer were tested. All data are reported in the Electronic Supporting Information (Figures S4 and S5).

While at the more negative potentials both types of copolymers assure a comparable good protection of the underlying AA2024 (current densities around 10 nA cm^{-2}), the coating prepared with a lower amount of the comonomer exhibited an abrupt increase of the current density starting at ca.

−0.50 V vs. SCE (Figure 6a and Figure S4). This sudden and almost monotonic increase of current (from nA cm^{-2} to mA cm^{-2} within 1 V) is a proof of the limited protection ability of the copolymer with 6.1% octyloxy units, that invariably showed defects after the polarization tests, responsible for the direct metal/solution contact (Figure 6b, and Figures S6 and S7). On the other hand, coatings obtained with the copolymer containing octyloxy units (13.4%) exhibited almost stable current density of 13–30 nA cm^{-2} up to at least 2.5 V vs. SCE (Figure 6a and Figure S5). These results are compatible with very good isolation of the Al surface from the solution by this coating, as a result of an almost complete absence of cracks and holes (Figure 5a, Figure 6c, and Figures S8, and S9) and a low porosity level. To give a clearer quantitative comparison, AA2024 in de-aerated near-neutral 3.5 wt.% NaCl (a less aggressive analog with respect to the aerated solution employed in this study) exhibits passivation current densities around μA cm^{-2} [23], i.e., two orders of magnitude higher than those recorded with the copolymer with 13.4% octyloxy units even applying extremely more oxidizing potentials.

By comparing the performance of the two types of coatings (Figure 6a), it is possible to tentatively attribute the better barrier effect of the copolymer with 13.4% octyloxy units to the higher content of long alkyl side chains that act as a more effective plasticizer by increasing compactness, cohesion, and adhesion ability of the cured films.

As confirmed by DSC (see the previous section), the addition of a comonomer introducing long side chains resulted in a modified PPM with a glass transition temperature (T_g) of 31 °C, lower than pristine PPM of the same molar mass (around 55 °C [1]). Considering that T_g of the copolymers is comparable to the temperatures that an AA2024 manufacture can encounter in real application, additional anodic polarization curves were recorded on aluminum alloy samples coated with the best performing copolymer (13.4% octyloxy groups), by setting the temperature of the 3.5 wt.% NaCl solution at 35 °C, just above the glass transition temperature of the copolymer (Figure S10). The coating, when operating at $T > T_g$, showed, on average, worse corrosion-protective behavior with respect to that at temperatures lower than T_g. The worsening can be attributed to the higher mobility of the polymer matrix that loses, at least partially, its barrier effect by favoring, for example, the permeation of the solution toward the underneath aluminum alloy surface.

Stability of the better performing corrosion-protective coating copolymer (13.4% octyloxy units) was further investigated by applying a constant polarization at 0 V vs. SCE to the aluminum-coated sample for 24 h in naturally aerated near-neutral 0.6 M NaCl solution (Figure 7). Current density, starting from some nA cm^{-2} constantly grew reaching, in three hours, a value around 6 μA cm^{-2} that remained almost stationary for a relatively long period of time (up to 12 h). After some hours characterized by high instability, current density started again increasing progressively up to ca. 100 μA cm^{-2}. The reported trend in the current is coherent with a progressive loss of the barrier effect by the polymeric coating, attributable to both increasing permeation of solution through pores and the formation of defects that grow up in time affecting the integrity of the barrier itself. However, it is apparent that the occurrence of defects is confined to few zones mainly localized close to the borders where some contributions of crevice corrosion can take place (Figure S11). The characteristic "up and down" current spikes are potentially attributable to the initiation of small metastable pits.

The corrosion-protective ability of the best formulated "self-plasticized" polymer coating (i.e., 13.4% octyloxy groups) is in line with the performance already reported for coatings of PPM with addition of siloxanes as an external plasticizer [1], especially in terms of current densities recorded during anodic polarization tests (few nA cm^{-2}). This comparison seems to qualitatively confirm that the design of the self-plasticized PPM approach is a good alternative to the more classical one which is based on the addition of external plasticizers.

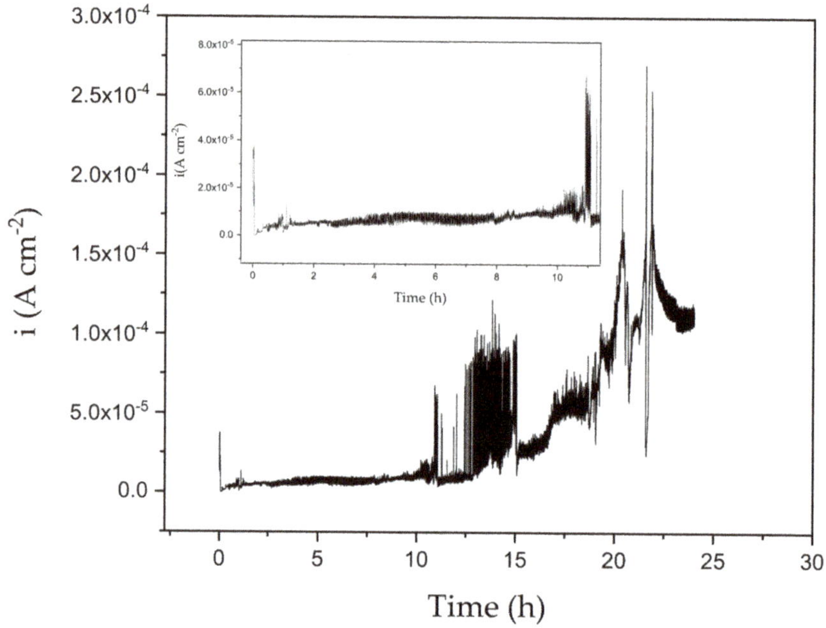

Figure 7. Current density as a function of time for AA2024 coated with a copolymer with 13.4% octyloxy units, polarized at 0 V vs. SCE in naturally aerated near-neutral 0.6 M NaCl solution, duration: 24 h. Inset: Magnification of the first 12 h.

4. Conclusions

Basically, poly(phenylene methylene) (PPM) is a valuable polymer for studies in the area of corrosion prevention coatings due to its thermal stability, hydrophobicity, and fluorescence. The latter facilitates optical detection of inhomogeneities, cracks and other defects caused by pit attacks upon observation under UV-light. However, PPM has to be used with additives, such as plasticizers or polysiloxanes, in order to enable processing into crack-free coating surfaces. Therefore, we aimed at the synthesis of copolymers on the basis of PPM which can be processed to coatings without addition of additives. For this purpose, benzyl chloride and 4-octyloxybenzyl chloride were copolymerized, using tin (IV) chloride as catalyst. Thus, copolymers with 6.1% mol/mol and 13.4% mol/mol octyloxy units were obtained. Notably, fluorescence of PPM was preserved in presence of side-chains on the polymer backbone.

In addition, it was shown that the copolymers are materials with high thermal stability. Further, the alkoxy chains lead to a decrease of the glass transition temperature of the copolymers, confirming the plasticizing effect of the substituent. This finding is line with the viscosity drop at higher fraction of alkyloxy units (13.4% mol/mol).

Gratifyingly, powders of the copolymers can be processed into coatings, providing homogeneous -crack-free surfaces on pretreated AA2024 substrates. The presence of alkoxy side chains along the polymer chains enhances the elasticity and the cohesion of the system, and hence the coating can adapt better to deformation which occurs upon cooling of the coated system prepared by hot pressing.

The isolation of copolymers based on PPM and their corrosion protection ability opens the way for studies on the application of such materials in the area of corrosion protection and monitoring.

Supplementary Materials: The following are available online at http://www.mdpi.com/2076-3417/9/17/3551/s1. Figure S1: SEM image of cross section of AA2024 coated CO-PPM containing 13.4% mol/mol of 4-octyloxy side chains. Figure S2: SEM image of cross section of AA2024 coated CO-PPM containing 6.1% mol/mol of 4-octyloxy side chains. Figure S3: Behavior of uncoated AA2024, in naturally aerated near-neutral 0.6 M NaCl solution.

Figure S4: Reproducibility tests (anodic polarization) for sample AA2024 coated with PPM_6.1%-OcOx. Figure S5: Reproducibility tests (anodic polarization) for sample AA2024 coated with PPM_13.4%-OcOx. Figure S6: SEM image of AA2024 coated CO-PPM containing 6.1% mol/mol of 4-octyloxy side chains. Figure S7: SEM image of cross section of a defect on AA2024 coated CO-PPM containing 6.1% mol/mol of 4-octyloxy side chains after polarization. Figure S8: SEM image of AA2024 coated CO-PPM containing 13.4% mol/mol of 4-octyloxy side chains which shows a uniform and homogenous surface even after polarization (top view). Figure S9: SEM image of cross section of AA2024 coated CO-PPM containing 13.4% mol/mol of 4-octyloxy side chains after polarization. Figure S10: Anodic polarization curves for AA2024 coated with PPM_13.4%-OcOx at 35 °C. Figure S11: optical microscope pictures of PPM_13.4%-OcOx under 24h polarization at 0 V vs. SCE.

Author Contributions: M.F.D. performed the synthesis of the monomers and copolymers. The manufacture of the coatings was performed under supervision of W.C. and M.N.; T.B.S. contributed to the rheology measurements which were performed and designed together with M.F.D.; the corrosion tests were performed by M.M. under supervision of S.P.M.T., and the manuscript was written by M.F.D., W.C. and M.M.

Funding: This research received no external founding.

Acknowledgments: We thank Elena Tervoort for taking SEM images.

Conflicts of Interest: The authors declare no conflict of interest.

References

1. D'Elia, M.F.; Braendle, A.; Schweizer, T.; Ortenzi, M.; Trasatti, S.; Niederberger, M.; Caseri, W. Poly(Phenylene Methylene): A Multifunctional Material for Thermally Stable, Hydrophobic, Fluorescent, Corrosion-Protective Coatings. *Coatings* **2018**, *8*, 274. [CrossRef]
2. Braendle, A.; Schwendimann, P.; Niederberger, M.; Caseri, W. Synthesis and fractionation of poly(phenylene methylene). *J. Polym. Sci. Part A Polym. Chem.* **2018**, *56*, 309–318. [CrossRef]
3. Ellis, B.; White, P. Thermal degradation of polybenzyl. *J. Polym. Sci. Polym. Chem. Ed.* **1973**, *11*, 801–821. [CrossRef]
4. Dreyer, D.; Jarvis, K.; Ferreira, P.; Bielawski, C. Graphite Oxide as a Dehydrative Polymerization Catalyst: A One-Step Synthesis of Carbon-Reinforced Poly(phenylene methylene) Composites. *Macromolecules* **2011**, *44*, 7659–7667. [CrossRef]
5. Hino, M.; Arata, K. Iron oxide as an effective catalyst for the polycondensation of benzyl chloride, the formation of para-substituted polybenzyl. *Chem. Lett.* **1979**, *8*, 1141–1144. [CrossRef]
6. Braendle, A.; Perevedentsev, A.; Cheetham, N.; Stavrinou, P.; Schachner, J.; Mösch-Zanetti, N.; Niederberger, M.; Caseri, W. Homoconjugation in poly(phenylene methylene)s: A case study of non-π-conjugated polymers with unexpected fluorescent properties. *J. Polym. Sci. Part B Polym. Phys.* **2017**, *55*, 707–720. [CrossRef]
7. Muller, P. Glossary of terms used in physical organic chemistry (IUPAC Recommendations 1994). *Pure Appl. Chem.* **1994**, *306*, 1077–1184. [CrossRef]
8. Ferguson, L.; Nnadi, J. Electronic interactions between nonconjugated groups. *J. Chem. Educ.* **1965**, *42*, 529–535. [CrossRef]
9. Rosen, S.L. *Fundamental Principles of Polymeric Materials*; Wiley: New York, NY, USA, 1993.
10. Rahman, M.; Brazel, C.S. The plasticizer market: An assessment of traditional plasticizers and research trends to meet new challenges. *Prog. Polym. Sci.* **2004**, *29*, 1223–1248. [CrossRef]
11. Small, P.A. The diffusion of plasticisers from polyvinyl chloride. *J. Chem. Technol. Biotechnol.* **1947**, *66*, 17–19. [CrossRef]
12. Wypych, G. 13-Plasticizers in various industrial products. In *Handbook of Plasticizers*, 3rd ed.; Elsevier: Toronto, ON, Canada, 2007; pp. 495–605.
13. Immergut, E.H.; Mark, H.F. Principle of Plasticization. In *Plasticization and Plasticizer Processes*; ACS Editorial Library: Washington, DC, USA, 1965; pp. 1–26.
14. Rehberg, C.E.; Fisher, C.H. Properties of Monomeric and Polymeric Alkyl Acrylates and Methacrylates. *Ind. Eng. Chem.* **1948**, *40*, 1429–1433. [CrossRef]
15. Tavares, A.; Schneider, P.H.; Merlo, A.A. 3,5-Disubstituted Isoxazolines as Potential Molecular Kits for Liquid-Crystalline Materials. *Eur. J. Org. Chem.* **2009**, *2009*, 889–897. [CrossRef]

16. Spreti, N.; Brinchi, L.; Germani, R.; Mancini, M.V.; Savelli, G. A New Carrier for Selective Removal of Heavy Metal Ions from Aqueous Solutions through Bulk Liquid Membranes. *Eur. J. Org. Chem.* **2004**, *2004*, 3865–3871. [CrossRef]
17. Zhao, Y. Nanoparticles and Nanoparticles Compositions. U.S. Patent 2013/0101516 A1, 25 April 2013.
18. Gunes, D.; Yagci, Y.; Bicak, N. Synthesis of Soluble Poly(p-phenylene methylene) from Tribenzylborate by Acid-Catalyzed Polymerization. *Macromolecules* **2010**, *43*, 7993–7997. [CrossRef]
19. M'Hiri, T.; Catusse, C.; Catusse, R.; Dubry, J.L.J. Polymerization of benzyl alcohol and its derived compounds with an ion exchange resin. *React. Kinet. Catal. Lett.* **1983**, *22*, 425–428. [CrossRef]
20. Banker, G.S. Film Coating Theory and Practice. *J. Pharm. Sci.* **1966**, *55*, 81–89. [CrossRef] [PubMed]
21. Trueba, M.; Trasatti, S.P. Electrochemical approach to repassivation kinetics of Al alloys: Gaining insight into environmentally assisted cracking. *Corros. Rev.* **2015**, *33*, 373–393. [CrossRef]
22. Comotti, I.M.; Trueba, M.; Trasatti, S.P. The pit transition potential in the repassivation of aluminium alloys. *Surf. Interface Anal.* **2013**, *45*, 1575–1584. [CrossRef]
23. Truba, M.; Trasatti, S.P. Study of Al alloy corrosion in neutral NaCl by the pitting scan technique. *Mater. Chem. Phys.* **2010**, *121*, 523–533. [CrossRef]

applied sciences

MDPI

Article

Corrosion of α-Brass in Solutions Containing Chloride Ions and 3-Mercaptoalkyl-5-amino-1*H*-1,2,4-triazoles

Oleg Kozaderov [1], Khidmet Shikhaliev [1], Chetti Prabhakar [2], Anuj Tripathi [2], Dmitry Shevtsov [1,*], Alexei Kruzhilin [1], Ekaterina Komarova [1], Andrei Potapov [1], Ilya Zartsyn [1] and Yuri Kuznetsov [3]

1 Faculty of Chemistry, Voronezh State University, Universitetskaya pl. 1, 394018 Voronezh, Russia
2 Department of Chemistry, National Institute of Technology, Kurukshetra 136119, Haryana, India
3 Laboratory of Physico-Chemical Fundamentals of Corrosion Inhibition of Metals and Alloys, A.N. Frumkin Institute of Physical Chemistry and Electrochemistry, Russian Academy of Sciences, (IPCE RAS), 31 Leninski Prospect, 119071 Moscow, Russia
* Correspondence: shevtsov@chem.vsu.ru

Received: 14 June 2019; Accepted: 10 July 2019; Published: 15 July 2019

Abstract: The protective effect of 3-mercaptoalkyl derivatives of 5-amino-1*H*-1,2,4-triazole against corrosion of α-brass in a chloride media was studied using polarization curves, electrochemical impedance spectroscopy, and full-scale corrosion testing. The brass electrode remains passive up to the activation potential, which is much higher than in solutions without organic additives, and it increases with the concentration of the inhibitor. The protection degree of all the studied inhibitors reaches its maximum of over 99% in solutions with the concentration of the inhibitor $C_{inh} \geq 0.10$ mM. The protective effect increases with the length of the alkyl chain. All the studied derivatives are effective against atmospheric corrosion of α-brass. A protective film is formed on the brass surface, and it most probably includes oxides as well as complex compounds of zinc and copper with the molecules of the inhibitors. The impedance spectroscopy demonstrated that the presence of the inhibitor results in a decrease in the double-layer capacitance and an increase in the polarization resistance, which proves that the protective film actually forms on the brass surface. The quantum chemical analysis of the optimized molecular structures demonstrates that all the studied inhibitors should have a similar protective effect, which agrees with the experimental results.

Keywords: brass; chloride; triazole derivatives

---

1. Introduction

Due to their high thermal conductivity, workability, and low cost, copper and copper-based alloys are widely used in heat-exchange equipment. Brasses, which are alloys of copper and zinc, are relatively harder and stronger, but are prone to corrosion in aggressive environments. Apart from the uniform attack, the most common forms of brass corrosion are selective leaching (dezincification) [1], stress-corrosion cracking [2,3], and pitting. Selective leaching goes significantly faster in hydrochloric acid solutions, the rate of dezincification increases at higher temperatures [4]. In neutral chloride solutions, selective leaching occurs during the early stages of the corrosion process [5]. Corrosion models [6] demonstrate that, during selective leaching in brasses, zinc anodically dissolves, and as a result, vacancies and copper adatoms appear on the surface of the alloy. By means of surface diffusion, they form a highly developed phase characterised by the increased thermodynamic activity of copper [7]. Consequently, the corrosion of brasses becomes pseudo-selective, as both components are ionized followed by the deposition of copper [8–10]. As opposed to brass corrosion in an acidic environment, when the ionization/deposition results in the formation of a sponge Cu-metal, neutral chloride solutions produce additional multiple surface layers containing oxides of zinc and copper [11–13]. According to [5,14–16], the formation of such layers from ZnO xH_2O, Cu_2O, and CuO passivate the surface of α- and β-brasses in buffer chloride solutions and promotes corrosion resistance. According to [17], the sublayer resulting from dezincification is rich in corrosion-resistant copper, which also contributes to the protective effect. This sublayer, however, appears due to the sparingly soluble multicomponent structure formed when alloys are kept in alkali-chloride solutions for a long time. It is this structure that effectively protects the material from corrosion attacks when the concentration of chloride ions is relatively low, with CuCl providing additional protection [18]. When the chloride concentration reaches a certain critical point, the rate of corrosion increases dramatically due to the formation of soluble $CuCl_2$- and discontinuity of the passive film. Compared to pure copper, brasses are much more prone to pitting, as the multicomponent surface layer containing ZnO is less resistant to dissolution in a corrosive environment [14,15]. In the presence of chloride ions, dezincification and localized attacks on the surface of α-brass take place simultaneously [19], which adds to the corrosive effect.

Authors [1] suggests deoxygenation of solutions, anodic protection, and doping as methods of protection of brasses from selective leaching. These methods, however, have a number of drawbacks. Namely, they require constantly low concentrations of dissolved oxygen, additional sacrificial materials, or changing the composition of the copper-zinc alloy. These obstacles can be avoided by using inhibitors, which is quite a common technique to control corrosion of copper and its alloys in water solutions [20–23]. The protective effect is found in organic compounds containing heteroatoms of nitrogen, sulphur, phosphorus, and oxygen, polar groups, heterocycles, and benzene rings with delocalized π-electrons [24–29].

There are quite a number of substances that produce an anti-corrosion effect on brasses. Thus, quinoline derivatives are effectively adsorbed on α-brass in an acid environment [30], while aminopropanol derivatives do so in neutral chloride solutions [31]. Diphenylamine derivatives form thin protective films [32], and phenol derivatives produce self-organizing layers on copper-zinc alloys [33]. However, the most effective corrosion inhibitors proved to be derivatives of azoles, widely used to control corrosion of copper, the main component of brasses [34–40]. Thus, adding 1,2,3-benzotriazole (BTA), benzimidazole, pyrazole, or their derivatives to a corrosive environment, including a chloride environment, results in the formation of a chemisorptive compound [41] or a multicomponent structure on brass, which, besides copper and zinc oxides, includes polymer systems with ions Cu(I) and Zn(II) [42–51]. This creates a protective barrier for both metal components of the copper-zinc alloy and prevents further oxidation and consequent dezincification and pitting of brass.

BTA is known to be toxic [52–54], so various organic and inorganic compounds may be considered as alternatives. It was found, however, that sodium salt of glycogenic and polyphosphoric acids [55], as well as chromates, molybdates, tetraborate, tungstates [56,57], and natural aragonite [58], whose corrosion protection properties were studied, do not protect brasses as effectively as the derivatives of environmentally friendly 1,2,4-triazole. Thus, the degree of protection of α-brass in hydrochloric acid solutions with 5-amino-1*H*-1,2,4-triazole is up to 96% [59]. The effect is accounted for by the formation of a barrier layer that includes aminotriazole-copper systems and a copper oxide (I).

For neutral chloride solutions, 3-amino-1,2,4-triazole is highly effective, with the Cu-40Zn brass corrosion inhibition effect increasing in the presence of sulphides [60] or if the inhibitor is polymerized [61]. The inhibition effect of azoles derivatives can be improved by introducing additional heteroatoms, such as S, in the structure of the inhibitor. Taking into account that mercapto derivatives of azoles are biodegradable, they seem to be the most effective and environmentally friendly corrosion inhibitors for brasses.

Their effectiveness against corrosion of copper in a neutral chloride environment was demonstrated in experiments with 2-mercaptobenzimidazole [62,63], 2-mercaptobenzoxazole [64], and 2-mercaptobenzothiazole [65]. Reference [66] demonstrates that mono- and poly-molecular chemisorption of 5-alkylmercapto-3-amino-1,2,4-triazoles on an oxidised Cu-electrode passivate copper very effectively. The protective effect of 5-alkylmercapto-3-amino-1,2,4-triazoles in a neutral chloride-borate buffer solution is most probably accounted for by the adsorption of the inhibitor from the solution and the formation of a protective nano-layer of sparingly soluble complex compounds, both on copper [67] and on Cu-20Zn brass [68]. The critical role of the complex formation in the passivation of Cu-20Zn brass in solutions containing various aggressive anions and a 2-mercaptobenzothiazole additive was proved in [69].

The purpose of the paper is to study the corrosion inhibition effect of 3-mercaptoalkyl derivatives of 5-amino-1*H*-1,2,4-triazole on α-brass in chloride environments using polarization curves, electrochemical impedance spectroscopy, full-scale corrosion testing, and quantum chemical simulation.

2. Materials and Methods

Electrochemical measurements were performed at room temperature (~25 °C) on brass (Cu/Zn: 63/37) electrodes in an unstirred borate buffer water solution (pH 7.4) with natural aeration, in the presence of the inhibitor and 10 mM NaCl. A classical electrolytic three-electrode cell with unseparated electrode spaces was used in order to accelerate transient measurements. Derivatives of 5-amino-1*H*-1,2,4-triazole synthesised at Voronezh State University were used. Their names, structures, and solubility in water are given in Table 1.

Table 1. Inhibitor names, structures and solubility in water.

Symbol	Name	Formula	Solubility
A	3-mercaptopropyl-5-amino-1*H*-1,2,4-triazole	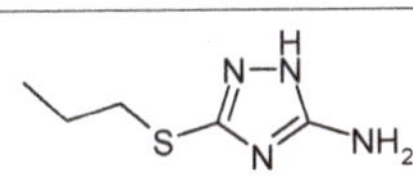	>10 mM
B	3-mercaptobutyl-5-amino-1*H*-1,2,4-triazole	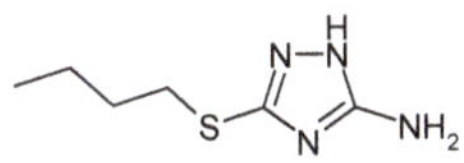	
C	3-mercapto(3-methylbutyl)-5-amino-1*H*-1,2,4-triazole	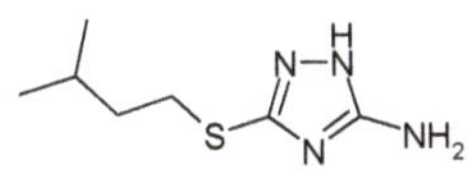	

The saturated silver chloride reference electrode was placed in a separate container connected to the electrolytic cell by an agar-agar based salt bridge and filled with potassium nitrate saturated solution. The auxiliary electrode was a platinum grid. The working brass electrode was polished by K3000 sandpaper, degreased in 96% ethanol, and washed with distilled water. The potentials of the working electrode (E) are given according to the standard hydrogen electrode (SHE) scale. The current density i was calculated by dividing the actual current I by the geometric area of the working electrode (2.25 cm^2).

Electrochemical measurements were performed using IPC-PRO potentiostat. In order to remove the oxide film appearing in air, the working Cu-electrode was polarized at $E = -0.60$ V for 15 min prior to the experiment. The electrode was then held in the solution for about 3 to 5 min until the stable free corrosion potential (E_{corr}) established. The working solution was then mixed with an NaCl solution while stirring, so that the concentration of chloride ions would be $C_{Cl^-} = 10$ mM. The studied inhibitors were also added to the solution in the concentration of $C_{inh} = 0.01$, 0.10, and 1.00 mM. After the new E_{corr} value was established, the I, E polarization curve was registered by scanning the potential towards either the anodic or the cathodic direction at 0.2 mV/s. Pitting potential (E_{pit}) was identified by the rapid increase in the current on the anodic polarization curve. Pits were then visually identified on the surface of the electrode. The measurement error for E_{pit} was below 0.03 V.

The rate of corrosion in current units (i_{corr}) was determined by the polarization resistance technique as summarized by Mansfeld [70,71]. The polarization curve was previously registered by scanning the potential from $E = E_{corr} - 0.03$ V to $E = E_{corr} + 0.03$ V at 0.2 mV/s. The polarization resistance R_p was then calculated as the slope of the polarization curve, replotted in $\Delta E, i$-coordinates ($\Delta E = E - E_{corr}$), at

E_{corr}. The polarization curve was then plotted in coordinates $2.3{\cdot}R_p{\cdot}i - \Delta E$, and the Tafel slopes of the half-reaction on the anode b_a and the cathode b_c were determined by means of the TableCurve software as the approximation parameter of the equation $2.3{\cdot}R_p{\cdot}i = \frac{b_a \cdot b_c}{b_a + b_c}\left[\exp\left(\frac{\Delta E}{b'_a}\right) - \exp\left(\frac{\Delta E}{b'_c}\right)\right]$, which at $\Delta E = 0$ gives i_{corr}. The measurements were performed at least 5 times for each of the studied concentrations of the inhibitor until the reproducible data was obtained. The data was then statistically analysed.

The effectiveness of the inhibition activity of the derivatives of 5-amino-1*H*-1,2,4-triazole was evaluated judging by the protection level:

$$Z_i = \frac{i_{corr,\,0} - i_{corr,\,inh}}{i_{corr,\,0}} \cdot 100\% \quad (1)$$

where $i_{corr,0}$ and $i_{corr,inh}$ are the corrosion current density with, and without, the inhibitor, respectively.

The electrochemical impedance spectroscopy (EIS) was performed using the IPC-PRO potentiostat with a frequency response analyser FRA-2. After E_{corr} was stable for 30 min, the frequency dependent impedance was registered within the range from 0.1 to 5000 Hz with no current. The results were presented as a Nyquist diagram. The analysis of the obtained results, the selection of the equivalent circuit, and determining its component values was carried out using DCS software.

Atmospheric corrosion of brass was accelerated by means of salt spray testing in order to determine the inhibition effect of derivatives of 5-amino-1*H*-1,2,4-triazole for inter-operational protection of brass products. The experiments were carried out on brass plates with the surface area $S = 20 \times 50$ mm^2 and the thickness $d = 1.00$ mm, which were previously polished by K3000 sandpaper and degreased in acetone. The protective inhibitor films were obtained by holding the brass plates in a water solution with the inhibitor for 60 min at 60 °C. The samples were dried and placed in the cabin at 95–100% air humidity. A 5% solution of NaCl (pH 6.5 ÷ 7.2, GOST R 52763-2007) was sprayed into the cabin every hour. The samples were examined 3 times every 24 h in order to register the moment (τ_{corr}) of appearance of the first signs of corrosion attack.

For microscopic analysis of the surface, a Jeol JSM-680LV scanning electron microscope (Japan) was used together with Oxford Instrument INCA 250 X-ray microanalysis module (UK) which determines the chemical composition of the surface layer.

Computational methodology was as follows. All the three molecules are fully optimized using density functional theory (DFT) with B3LYP functional at 6-311 + G (d, p) basis set in the program Gaussian 16 [72]. These optimized structures have no imaginary frequencies and hence are situated at the lowest energy of the potential energy surface. The theoretical calculation also involves determining HOMO and LUMO energies, and the energy gap along with ionization potential, electron affinity, electronegativity (χ), absolute hardness (η), and softness (σ) for all the molecules.

Following the density functional theory, the η is defined as [73]:

$$\eta = \frac{1}{2}\frac{\partial \mu}{\partial N} = \frac{1}{2}\frac{\partial^2 E}{\partial N^2} \quad (2)$$

where μ is the chemical potential, N is the number of electrons, and E is the energy. R. G. Pearson proposed [73] the following operational definition for absolute hardness

$$\eta = \frac{IP - EA}{2} \quad (3)$$

where IP and EA are ionization energy and electron affinity respectively. According to Koopman's theorem, IP and EA calculated from orbital energies are defined as

$$IP = -E_{HOMO},\ EA = -E_{LUMO} \quad (4)$$

E_{HOMO} and E_{LUMO} are the energies of the highest occupied and the lowest unoccupied molecular orbitals. Based on the formula given above, the hardness can be represented as a measure of change resistance of the electron cloud of the chemical system. Softness (σ) is defined as the reciprocal of hardness. This is an important parameter for chemical reactions since most reactive species have higher softness value. Electronegativity (χ) is also defined in terms of energies of HOMO and LUMO as

$$\chi = \frac{IP + EA}{2} \tag{5}$$

3. Results and Discussion

The results of measuring E_{corr} of a brass electrode in a neutral chloride solution in the presence of mercaptoalkyl derivatives of 5-amino-1*H*-1,2,4-triazole are given in Table 2. In the presence of 3-mercaptopropyl-5-amino-1*H*-1,2,4-triazole (inhibitor A) with the concentration of C_{inh} = 0.01 mM the corrosion potential of the electrode decreases slightly. As the concentration of the inhibitor increases up to 0.10–1.00 mM, E_{corr} shifts noticeably to the anodic region.

Table 2. Open circuit potential E_{corr}, polarization resistance R_p, corrosion current density i_{corr}, and degree of protection of brass Z_i, in 10 mM NaCl solutions with various 3-mercaptoalkyl derivatives of 5-amino-1*H*-1,2,4-triazole.

Inhibitor	C_{inh}, mM	E_{corr}, V	R_p, kΩ·cm^2	i_{corr} μA·cm^{-2}	Z_i, %
-	-	0.111	1.51 ± 0.11	26 ± 4	-
A	0.01	0.100	1.73 ± 0.10	38 ± 5	−46.2
	0.10	0.200	382 ± 24	0.06 ± 0.02	99.8
	1.00	0.200	244 ± 18	0.11 ± 0.03	99.6
B	0.01	0.110	0.76 ± 0.09	13 ± 3	50.0
	0.10	0.200	195 ± 13	0.12 ± 0.04	99.5
	1.00	0.250	170 ± 15	0.19 ± 0.06	99.3
C	0.01	0.120	20.9 ± 2.3	4.0 ± 0.7	84.6
	0.10	0.150	143 ± 18	0.22 ± 0.05	99.2
	1.00	0.180	281 ± 24	0.12 ± 0.03	99.5

When 3-mercaptobutyl-5-amino-1*H*-1,2,4-triazole (inhibitor B) is used with the concentration of 0.01 mM, the corrosion potential remains stable. When $C_{inh} \geq 0.10$ mM, the potential increases at least by 0.1 V. 3-mercapto(3-methylbutyl)-5-amino-1*H*-1,2,4-triazole (inhibitor C) added to the chloride solution also results in E_{corr} shifting to the anodic region, with the shift increasing at higher concentrations of the inhibitor. Thus, the analysis of E_{corr} demonstrates that the derivatives of 5-amino-1*H*-1,2,4-triazole can be classified as anodic inhibitors.

The polarization curves of brass in neutral chloride solutions, with the inhibitors under consideration, are given in Figure 1a–c. We can see that when the anodic polarization is low, the current density for all mercaptoalkyl derivatives is lower than in the solution without the inhibitors independent of their concentration. When the concentration of the additive is 0.01 mM (Figure 1a, curve 2), an increase in polarization results in the anodic maximum, while when $C_{inh} \geq 0.10$ mM, the polarization curve has no peaks and the anodic current density does not exceed 1 mA·cm^{-2}. In such solutions, the brass electrode remains passive up to the potential of local activation by chlorides. The potential is much higher than in solutions without the organic additive, and increases at higher concentrations of the inhibitor.

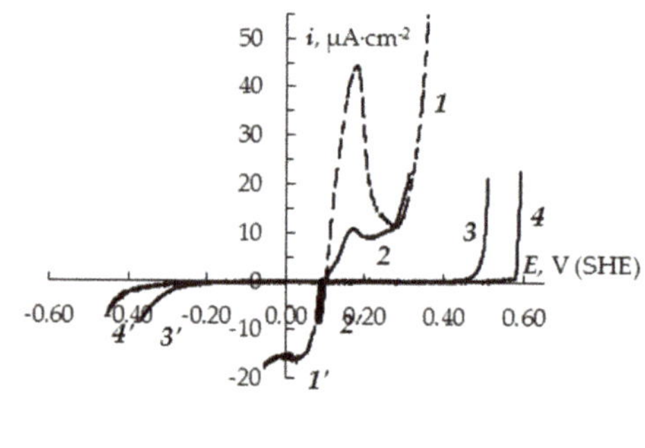

(a)

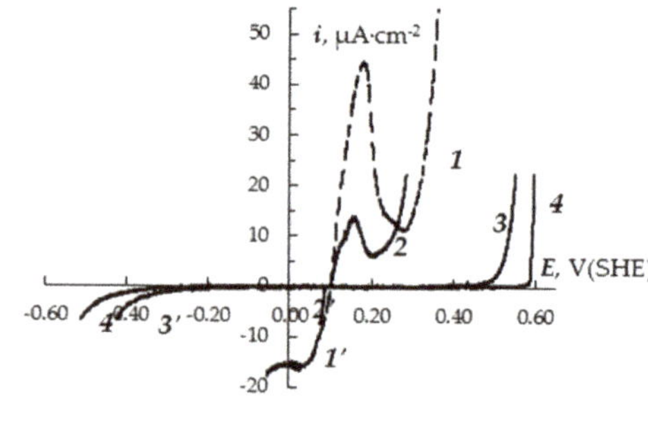

(b)

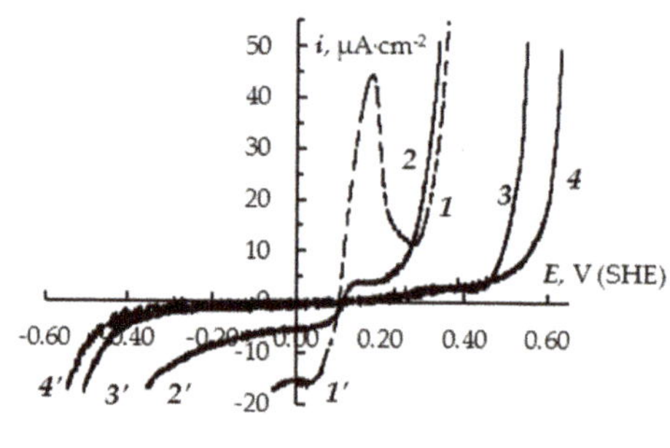

(c)

Figure 1. Anodic (1-4) and cathodic (1′-4′) polarization curves of α-brass in a borate buffer (pH 7.4) with 0.01 M NaCl and 3-mercaptopropyl-5-amino-1*H*-1,2,4-triazole (**a**), 3-mercaptobutyl-5-amino-1*H*-1,2,4-triazole (**b**) and 3-mercapto(3-methylbutyl)-5-amino-1*H*-1,2,4-triazole (**c**) at the following concentrations: 1,1′–0; 2,2′–0.01 mM; 3,3′–0.10 mM; 4,4′–1.00 mM.

The difference between cathodic curves of brasses in solutions with, and without, inhibitors becomes noticeable only when $C_{inh} \geq 0.10$ mM. In fact, cathodic curves obtained without any inhibitors and when their concentration was 0.01 mM are almost identical (Figure 1a–c, curves 1′ and 2′). The higher the concentration of the inhibitor, the lower the current density (Figure 1a–c, curves 3′ and 4′). This means that the cathodic process is slowing down. The region of the rapid growth of the cathodic current density is shifted towards negative values by at least 0.3 V as compared to that of the chloride solution without inhibitors.

Analysis of the polarization resistance R_p, the rate of corrosion, and the protection level (Table 2) demonstrate that the protective effect of all the studied 3-mercaptoalkyl derivatives of 5-amino-1*H*-1,2,4-triazole depends on their concentration and reaches its maximum when $C_{inh} \geq$ 0.10 mM. In the latter case R_p increases by over 100 times and the corrosion rate deteriorates correspondingly. When the concentration of the additive is 0.01 mM, inhibitor A does not demonstrate any protective effect, while inhibitors B and C slow down the rate of corrosion by two, and six times, respectively. Thus, even when the concentration is low, the protective effect of the inhibitor increases as the alkyl chain gets longer.

The same results were obtained during the salt spray testing (Table 3).

Table 3. Registered time of first corrosion attacks during the salt spray testing (τ_{corr}, hours).

Inhibitor	C_{inh}, mM			
	0.0	1.0	5.0	10.0
A		68	98	152
B	3	173	194	212
C		94	117	283

Indeed, the period when the first signs of corrosion attack appear τ_{corr}, if C_{inh} = 10 mM, is shorter for inhibitor A, and longer for inhibitors B and C. With lower C_{inh}, 3-mercaptobutyl-5-amino-1*H*-1,2,4-triazole demonstrates the best protective effect. τ_{corr} increases with higher C_{inh}, and reaches 6–12 days at 10 mM, which proves the effectiveness of 3-mercaptoalkyl derivatives of 5-amino-1*H*-1,2,4-triazole for α-brass in moist atmosphere.

Another proof are the images of the surface of brass samples held in the salt spray cabin (Figure 2). They show that in the presence of the inhibitor the brass electrode remains bright and smooth even if the concentration of the organic additive is minimal. Without the inhibitors, the surface of the electrode darkens, and corrosion spots appear.

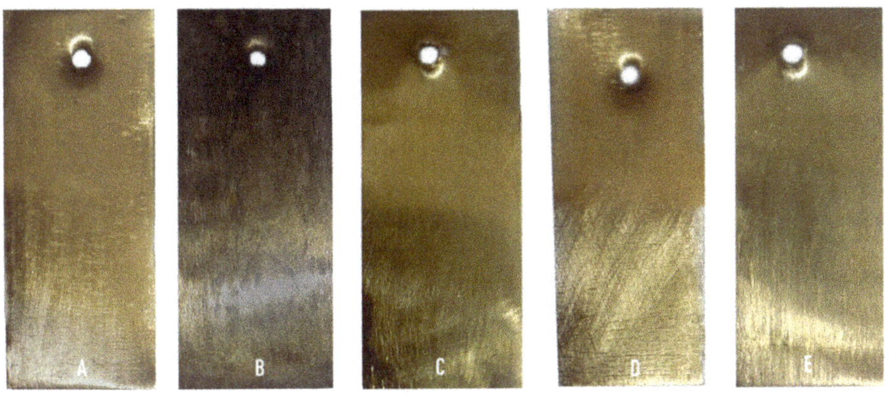

Figure 2. Brass samples before (**A**) and after salt spray corrosion testing without an inhibitor (**B**), and in the presence of 3-mercaptobutyl-5-amino-1*H*-1,2,4-triazole at concentrations of 1 (**C**), 5 (**D**), 10 mM (**E**).

To explain the effect, we carried out scanning electron microscopy and analyzed the micrographs of the brass electrode surface taken before, and after, the potentio-dynamic anodic polarization in a neutral chloride solution (Figure 3). Polarization was performed at intervals from the current-less potential to the activation potential. Figure 3 demonstrates that, without the inhibitors, multiple corrosion spots of about 10 μm appear on the surface. However, when the brass is polarized in the solution with 1 mM of 3-mercapto(3-methylbutyl)-5-amino-1*H*-1,2,4-triazole, no pits can be observed on the surface. The micrographs show only scratches left from polishing the electrode.

The results of the analysis of the elemental composition of the surface layer of the brass electrode before and after anodic polarization are given in Table 4.

It is interesting that anodic polarization does not result in any change in the ratio of the atomic fractions of copper and zinc, which means that there is no dezincification. It is possible that oxides are formed during the anodic oxidation of both components of the alloy. This is also demonstrated by the presence of oxygen. The increase in the oxygen concentration on the brass surface in the presence of the inhibitor is observed because of the formation of a multicomponent passive film, which includes $ZnO{\cdot}xH_2O$, Cu_2O, and CuO oxides [5,14–16], together with sparingly soluble complex compounds of zinc and copper with molecules of the studied inhibitors [67–69].

Figure 3. SEM micrographs of α-brass surface (**a**) before polarization, (**b**) after polarization in borate buffer (pH 7.4) + 10 mM NaCl, (**c**) after polarization in borate buffer (pH 7.4) + 10 mM NaCl + 1.00 mM of 3-mercapto(3-methylbutyl)-5-amino-1*H*-1,2,4-triazole (structure B).

Table 4. Elemental surface composition (at. %) of the brass electrode.

Polarization Mode	**Element**			
	Cu	**Zn**	**O**	**Cu/Zn**
No polarization	61.16	38.84	0.00	61/39
After polarization in borate buffer + 10 mM NaCl	59.04	37.51	3.45	61/39
After polarization in borate buffer + 10 mM NaCl + 1.00 mM of 3-mercapto(3-methylbutyl)-5-amino-1*H*-1,2,4-triazole	57.13	35.47	7.39	62/38

The electrochemical impedance spectroscopy also proves that a protective film forms on brass in neutral chloride solutions with 3-mercaptoalkyl-5-amino-1*H*-1,2,4-triazoles. Figure 4 presents Nyquist diagrams obtained for the chloride-borate solution with (Figure 4a) and without the A, B, and C inhibitors with concentrations of C_{inh} = 0.01–1.00 mM (Figure 4b–d). The hodographs are presented as imperfect semi-circumferences, the imperfection being usually accounted for by the impedance frequency dispersion and the geometric and/or energetic inhomogeneity of the electrode's surface.

We can see that when the organic additives are added to the neutral solution, or when C_{inh} increases, the shape of the impedance diagram does not change, but the radius of the semi-circumference increases significantly. The regression analysis of the impedance spectroscopy data was used to determine the most probable equivalent circuit simulation the processes at the brass/solution interface (Figure 5). Table 5 presents the values of the following circuit elements: R_{Ω} (ohmic resistance of the solution),

C_{DL} (double-layer capacitance), R_p (polarization resistance of the electrochemical reaction), and *BW* (Warburg impedance).

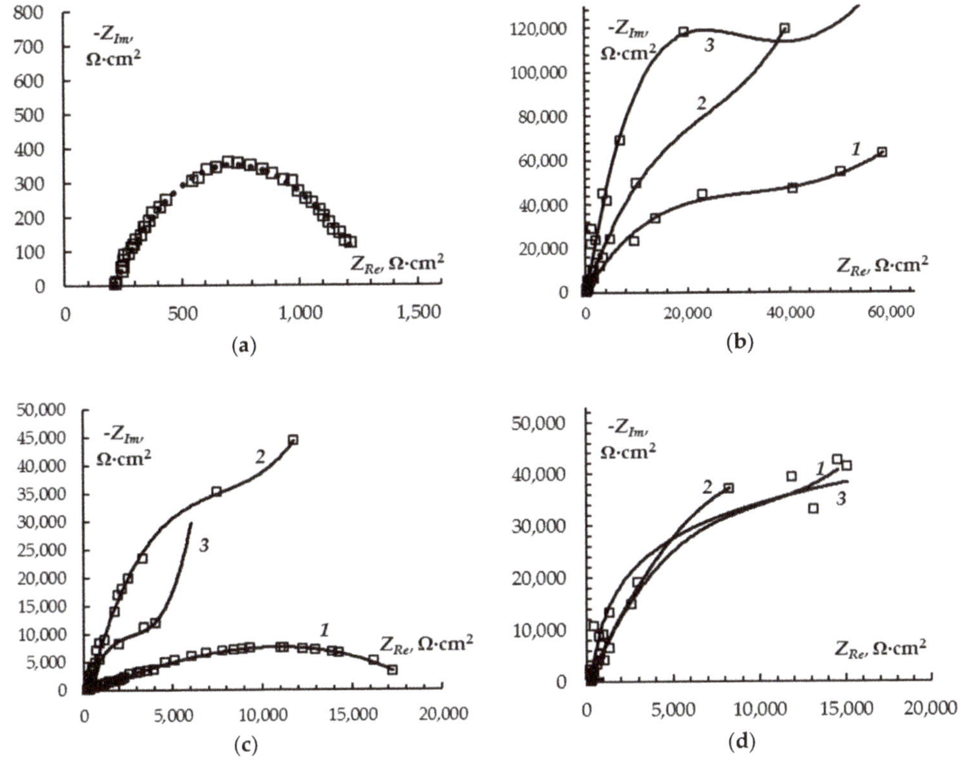

Figure 4. The Nyquist diagrams of brass in borate buffer (pH 7.4) + 10 mM NaCl without an inhibitor (**a**), in the presence of 3-mercaptopropyl-5-amino-1*H*-1,2,4-triazole (**b**), 3-mercaptobutyl-5-amino-1*H*-1,2,4-triazole (**c**) and 3-mercapto(3-methylbutyl)-5-amino-1*H*-1,2,4-triazole (**d**) at concentrations of 0.01 (*1*), 0.10 (*2*), and 1.00 mM (*3*).

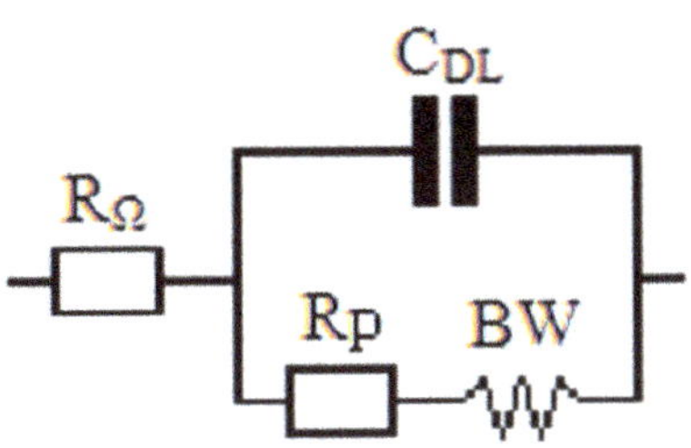

Figure 5. The electric equivalent circuit model used to fit the electrochemical impedance spectroscopy response (R_{Ω}-direct-current resistance, C_{DL}-double-layer capacitance, R_p-polarization resistance of the electrochemical reaction, *BW*-Warburg impedance).

Table 5. Equivalent circuit elements and degree of protection of brass electrode in 10 mM NaCl solutions with various derivatives of 5-amino-1*H*-1,2,4-triazole.

Inhibitor	C_{inh}, mM	R_{Ω}, $\Omega \cdot cm^2$	C_{DL}, $\mu F \cdot cm^{-2}$	R_p^{EIS}, $k\Omega \cdot cm^2$	BW, $k\Omega \cdot cm^2 \cdot s^{-0,5}$		Degree of protection, η_{inh}, %
					BWa	BWb	
-	0.00	240	26.2	0.176	1.28	0.20	-
A	0.01	256	1.57	72	73	1.77	99.76
	0.10	250	1.00	141	298	6.00	99.88
	1.00	258	1.54	473	135	0.26	99.96
B	0.01	200	2.56	0.9	13	0.81	79.86
	0.10	255	1.58	149	151	23.27	99.88
	1.00	244	1.02	35	9336	10^{-6}	99.50
C	0.01	275	0.86	52	239	2.87	99.66
	0.10	275	1.42	158	20605	10^{-6}	99.89
	1.00	275	0.79	111	23526	10^{-6}	99.84

R_p is significantly higher in solutions with an organic additive and increases when C_{inh} is higher, while C_{DL} noticeably decreases. We can thus say that the changes in the hodograph are most probably caused by the decrease in the double-layer capacitance resulting from the adsorption of the inhibitor molecules at the brass/solution interface, on the one hand, and the increase in the polarization resistance resulting from the formation of a protective film on the brass surface, on the other. Further evidence of the formation of the film is that the Warburg impedance grows rapidly in the presence of the inhibitor (Table 5). This means either that the diffusion of the soluble corrosion product on the electrode's surface slows down, or that the barrier properties of the protective film blocking the aggressive elements of the solution increase.

$$\eta_{inh} = \left(1 - \frac{R_{p,0}^{EIS}}{R_{p,inh}^{EIS}}\right) \cdot 100\% \quad (6)$$

The inhibition effectiveness is calculated using the values of charge transfer resistance without ($R_{p,0}^{EIS}$) and with the inhibitor ($R_{p,inh}^{EIS}$), is about 100% (Table 5) for all the studied inhibitors, when $C_{inh} \geq$ 0.10 mM. This agrees well with the results of the regression analysis of the polarization curves carried out using Mansfeld method (Table 2).

The results of the electrochemical studies and corrosion testing were interpreted within the framework of the quantum chemical approach to the inhibition effect assessment. Optimized molecular structures of 3-mercaptoalkyl-5-amino-1*H*-1,2,4-triazoles are given in Figure 6. Table 6 gives the values of the ionization potential, electron affinity, electronegativity, absolute hardness and softness in eV calculated at B3LYP/6-311 + G (d, p), and optimized geometry.

Table 6. Calculated energies of HOMO, LUMO, HOMO LUMO gap (HLG), Ionization Potential (IP), Electron Affinity (EA), Electronegativity (χ), Absolute Hardness (η) and Softness (σ) in eV at B3LYP/6-311 + G (d, p) level.

Molecule	HOMO	LUMO	HLG	IP	EA	χ	η	σ
A	−6.36	−0.61	5.75	6.36	0.61	3.49	2.88	0.35
B	−6.38	−0.61	5.77	6.38	0.61	3.50	2.89	0.34
C	−6.35	−0.62	5.73	6.35	0.62	3.48	2.87	0.35

It is known that the ability of inhibitors to interact with a metal surface increases with the growth of HOMO and decrease in LUMO [74]. The smaller the HOMO LUMO gap (HLG), the higher the reactivity of the inhibitor to the metal surface and the stronger the inhibition effect. In Table 6, of all the three triazole molecules, 3-mercapto(3-methylbutyl)-5-amino-1*H*-1,2,4-triazole i.e., molecule C, has the smallest HLG and hardness value of 5.73 eV, and 2.87 eV, respectively. While, 3-mercaptopropyl-5-amino-1*H*-1,2,4-triazole (molecule A) and 3-mercaptobutyl-5-amino-1*H*-1,2,4-triazole (molecule B) have the same LUMO energy level of 0.61 eV.

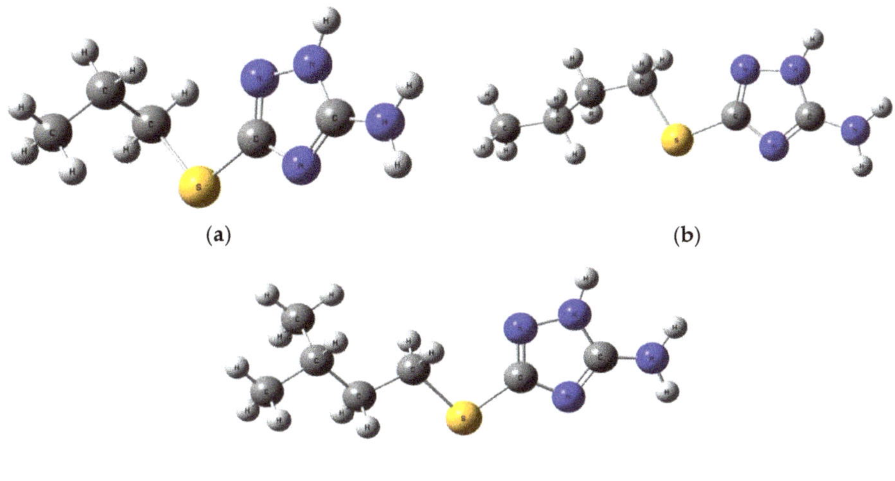

(a) (b) (c)

Figure 6. Optimized structures of the molecules considered for study: 3-mercaptopropyl-5-amino-1*H*-1,2,4-triazole (**a**), 3-mercaptobutyl-5-amino-1*H*-1,2,4-triazole (**b**) and 3-mercapto (3-methylbutyl)-5-amino-1*H*-1,2,4-triazole (**c**).

Destabilization of HOMO level 0.02 eV was observed upon changing the substituent from propyl (*A*) to butyl (*B*). Of the three molecules, 3-mercaptobutyl-5-amino-1*H*-1,2,4-triazole (molecule *B*) has the largest values of HLG, hardness and electronegativity −5.77 eV, 2.89 eV, and 3.50 eV respectively. We can, thus, see that the calculated quantum chemical parameters of all the studied molecules of 3-mercaptoalkyl-5-amino-1*H*-1,2,4-triazoles are very similar. Therefore, their protective effect is also similar, which was proved by testing in neutral chloride solutions when the concentration of the inhibitor was at least 0.10 mM.

This section may be divided by subheadings. It should provide a concise and precise description of the experimental results, their interpretation as well as the experimental conclusions that can be drawn.

4. Conclusions

The study demonstrates that 3-mercaptoalkyl derivatives of 5-amino-1*H*-1,2,4-triazole are effective corrosion inhibitors for α-brass in chloride environment. Their protective effect increases with the length of the alkyl chain and at higher concentrations of the additive. The maximum protection (>99%) is possible at C_{inh} > 0.10 mM, when the inhibitors become comparable in degree of protection. This can be accounted for by the similarity of the quantum chemical parameters of the organic molecules responsible for the inhibition. The inhibition effect results from the formation of a passive film on the brass surface. The film includes oxides of the brass components as well as sparingly soluble complex compounds of zinc and copper with molecules of the considered 3-mercaptoalkyl derivatives of 5-amino-1*H*-1,2,4-triazole. The fact that the protective film actually forms on the brass surface is proved by a decrease in the double-layer capacitance, an increase in the polarization resistance of the brass electrode, and a rapid increase in the Warburg impedance upon the introduction of the inhibitor.

Author Contributions: Conceptualization, I.Z.; Data curation, Y.K.; Investigation, D.S., A.K., E.K. and A.P.; Methodology, A.K., E.K. and A.P.; Project administration, K.S.; Software, C.P. and A.T.; Writing—original draft, O.K., D.S. and A.K.; Writing—review & editing, K.S., C.P., A.T., I.Z. and Y.K.

Funding: The study received financial support from the Ministry of Education and Science of the Russian Federation within the framework of State Contract with universities regarding scientific research in 2017–2019, project No. 4.3633.2017/4.6.

Conflicts of Interest: The authors declare no conflict of interest.

References

1. Marshakov, I.K. Corrosion resistance and dezincing of brasses. *Prot. Met.* **2005**, *41*, 205–210. [CrossRef]
2. Shih, H.C.; Tzou, R.J. Effect of benzotriazole on the stress corrosion cracking and the electrochemical polarization of 70/30 brass in fluoride solutions. *J. Electrochem. Soc.* **1991**, *138*, 958–961. [CrossRef]
3. Nishimura, R.; Yoshida, T. Stress corrosion cracking of Cu-30% Zn alloy in Mattsson's solutions at pH 7.0 and 10.0 using constant load method—A proposal of SCC mechanism. *Corros. Sci.* **2008**, *50*, 1205–1213. [CrossRef]
4. El-Mahdy, G.A. Electrochemical impedance study on brass corrosion in NaCl and $(NH_4)_2SO_4$ solutions during cyclic wet–dry conditions. *J. Appl. Electrochem.* **2005**, *35*, 347–353. [CrossRef]
5. Alfantazi, A.M.; Ahmed, T.M.; Tromans, D. Corrosion behavior of copper alloys in chloride media. *Mater. Des.* **2009**, *30*, 2425–2430. [CrossRef]
6. Martin, H.; Carro, P.; Hernandez Creus, A.; Morales, J.; Fernandez, G.; Esparza, P.; Gonzales, S.; Salvarezza, R.C.; Arvia, A.J. Inerplay of surface diffusion and surface tension in the evolution of solid/liquid interfaces. Dealloying of β-brass in aqueous sodium chloride. *J. Phys. Chem. B.* **2000**, *104*, 8229–8237. [CrossRef]
7. Zarnitsyn, I.D.; Vvedenskyi, A.V.; Marshakov, I.K. Nonequilibrium behavior of the surface-layer in anodic-dissolution of homogeneous alloys. *Russ. J. Electrochem.* **1994**, *30*, 492–512.
8. Polunin, A.V.; Pchelnikov, A.P.; Losev, V.V.; Marshakov, I.K. Electrochemical studies of the kinetics and mechanism of brass dezincification. *Electrochim. Acta* **1982**, *27*, 467–475. [CrossRef]
9. Pchelnikova, A.P.; Sitnikov, A.D.; Marshakov, I.K.; Losev, V.V. A study of the kinetics and mechanism of brass dezincification by radiotracer and electrochemical methods. *Electrochim. Acta* **1981**, *26*, 591–600. [CrossRef]
10. Zhou, P.; Hutchison, M.J.; Erning, J.W.; Scully, J.R.; Ogle, K. An in situ kinetic study of brass dezincification and corrosion. *Electrochim. Acta* **2017**, *229*, 141–154. [CrossRef]
11. Holliday, J.E.; Pickering, H.W.; Holliday, J.E.; Pickering, H.W. A Soft X-Ray Study of the Near Surface Composition of Cu30Zn Alloy during Simultaneous Dissolution of Its Components. *J. Electrochem. Soc.* **1973**, *120*, 470–475. [CrossRef]
12. Valcarce, M.B.; de Sanchez, S.R.; Vazquez, M. Brass dezincification in a tap water bacterial suspension. *Electrochim. Acta* **2006**, *51*, 3736–3742. [CrossRef]
13. Valcarce, M.B.; de Sanchez, S.R.; Vazquez, M. A comparative analysis of copper and brass surface films in contact with tap water. *J. Mater. Sci.* **2006**, *41*, 1999–2007. [CrossRef]
14. Morales, J.; Fernandez, G.T.; Esparza, P.; Gonzalez, S.; Salvarezza, R.C.; Arvia, A.J. A comparative study on the passivation and localized corrosion of α, β, and α+β brass in borate buffer solutions containing sodium chloride—I. Electrochemical data. *Corros. Sci.* **1995**, *37*, 211–229. [CrossRef]
15. Morales, J.; Esparza, P.; Fernandez, G.T.; Esparza, P.; Gonzalez, S.; Salvarezza, R.C.; Arvia, A.J. A comparative study on the passivation and localized corrosion of α-and β-brass in borate buffer solutions containing sodium chloride—II. X-ray photoelectron and Auger electron spectroscopy data. *Corros. Sci.* **1995**, *37*, 231–239. [CrossRef]
16. Warraky, A.E. Early stages of dezincification of α brass immersed in 4% NaCl solution. *Br. Corros. J.* **1997**, *32*, 57–62. [CrossRef]
17. Milošev, I.; Mikić, T.K.; Gaberšček, M. The effect of Cu-rich sub-layer on the increased corrosion resistance of Cu–xZn alloys in chloride containing borate buffer. *Electrochim. Acta* **2006**, *52*, 415–426. [CrossRef]
18. Ismail, K.M.; El-Egamy, S.S.; Abdelfatah, M. Effects of Zn and Pb as alloying elements on the electrochemical behaviour of brass in borate solutions. *J. Appl. Electrochem.* **2001**, *31*, 663–670. [CrossRef]
19. El-Rahman, H.A. Passivation and pitting corrosion of α-brass (Cu/Zn: 63/37) in neutral buffer solutions containing chloride ions. *Mater. Corros.* **1990**, *41*, 635–639. [CrossRef]
20. Wang, Z.Q.; Gong, Y.L.; Jing, C.; Huang, H.J.; Li, H.R.; Zhang, S.T.; Gao, F. Synthesis of dibenzotriazole derivatives bearing alkylene linkers as corrosion inhibitors for copper in sodium chloride solution: A new thought for the design of organic inhibitors. *Corros. Sci.* **2016**, *113*, 64–77. [CrossRef]

21. Ma, F.B.; Li, W.H.; Tian, H.W.; Hou, B.R. The use of a new thiadiazole derivative as a highly efficient and durable copper inhibitor in 3.5% NaCl solution. *Corros. Sci.* **2015**, *10*, 5862–5879.
22. Gong, Y.L.; Wang, Z.Q.; Gao, F.; Zhang, S.T.; Li, H.R. Synthesis of new benzotriazole derivatives containing carbon chains as the corrosion inhibitors for copper in sodium chloride solution. *Ind. Eng. Chem. Res.* **2015**, *54*, 12242–12253. [CrossRef]
23. Finšgar, M. EQCM and XPS analysis of 1,2,4-triazole and 3-amino-1,2,4-triazole as copper corrosion inhibitors in chloride solution. *Corros. Sci.* **2013**, *77*, 350–359. [CrossRef]
24. Chirkunov, A.A.; Kuznetsov, Y.I.; Shikhaliev, K.S.; Agafonkina, M.O.; Andreeva, N.P.; Kazansky, L.P.; Potapov, A.Y. Adsorption of 5-alkyl-3-amino-1, 2, 4-triazoles from aqueous solutions and protection of copper from atmospheric corrosion. *Corros. Sci.* **2018**, *144*, 230–236. [CrossRef]
25. Bi, H.; Burstein, G.T.; Rodriguez, B.B.; Kawaley, G. Some aspects of the role of inhibitors in the corrosion of copper in tap water as observed by cyclic voltammetry. *Corros. Sci.* **2016**, *102*, 510–516. [CrossRef]
26. Qin, T.T.; Li, J.; Luo, H.Q.; Li, M.; Li, N.B. Corrosion inhibition of copper by 2,5-dimercapto-1,3,4-thiadiazole monolayer in acidic solution. *Corros. Sci.* **2011**, *53*, 1072–1078. [CrossRef]
27. Sherif, E.M.; El Shamy, A.M.; Ramla, M.M.; El Nazhawy, A.O.H. 5-(Phenyl)-4H-1,2,4-triazole-3-thiol as a corrosion inhibitor for copper in 3.5% NaCl solutions. *Mater. Chem. Phys.* **2007**, *102*, 231–239. [CrossRef]
28. Otmačić, H.; Stupnišek-Lisac, E. Copper corrosion inhibitors in near neutral media. *Electrochim. Acta* **2003**, *48*, 985–991. [CrossRef]
29. Fan, H.; Li, S.; Zhao, Z.; Wang, H.; Shi, Z.; Zhang, L. Inhibition of brass corrosion in sodium chloride solutions by self-assembled silane films. *Corros. Sci.* **2011**, *53*, 4273–4281. [CrossRef]
30. Fouda, A.S.; Ismael, M.A.; Shahba, R.A.; Kamel, L.A.; El-Nagggar, A.A. Corrosion Inhibition of Copper and α-Brass in 1 M HNO3 Solution using New arylpyrimido [5, 4-c] quinoline-2, 4-dione derivative. *Int. J. Electrochem. Sci.* **2017**, *12*, 3361–3384. [CrossRef]
31. Gao, G.; Liang, C.H. 1, 3-Bis-diethylamino-propan-2-ol as volatile corrosion inhibitor for brass. *Corros. Sci.* **2007**, *49*, 3479–3493. [CrossRef]
32. Keleş, H.; Akça, S. The effect of Variamine Blue B on brass corrosion in NaCl solution. *Arabian J. Chem.* **2019**, *12*, 236–248. [CrossRef]
33. Asan, A.; Kabasakaloglu, M.; Işiklan, M.; Kiliç, Z. Corrosion inhibition of brass in presence of terdentate ligands in chloride solution. *Corros. Sci.* **2005**, *47*, 1534–1544. [CrossRef]
34. Antonijević, M.M.; Petrović, M.B. Copper corrosion inhibitors. A review. *Int. J. Electrochem. Sci.* **2008**, *3*, 1–28.
35. Antonijević, M.M.; Milić, S.M.; Petrović, M.B. Films formed on copper surface in chloride media in the presence of azoles. *Corros. Sci.* **2009**, *51*, 1228–1237. [CrossRef]
36. Kuznetsov, Y.I.; Kazansky, L.P. Physicochemical aspects of metal protection by azoles as corrosion inhibitors. *Russ. Chem. Rev.* **2008**, *77*, 219–232. [CrossRef]
37. Mihajlović, M.B.P.; Antonijević, M.M. Copper Corrosion Inhibitors. Period 2008–2014. A Review. *Int. J. Electrochem. Sci.* **2015**, *10*, 1027–1053.
38. Milošev, I.; Kovačević, N.; Kovač, J.; Kokalj, A. The roles of mercapto, benzene and methyl groups in the corrosion inhibition of imidazoles on copper: I. Experimental characterization. *Corros. Sci.* **2015**, *98*, 107–118. [CrossRef]
39. Finšgar, M.; Petovar, B.; Xhanari, K.; Maver, U. The corrosion inhibition of certain azoles on steel in chloride media: Electrochemistry and surface analysis. *Corros. Sci.* **2016**, *111*, 370–381. [CrossRef]
40. Kovačević, N.; Milošev, I.; Kokalj, A. How relevant is the adsorption bonding of imidazoles and triazoles for their corrosion inhibition of copper? *Corros. Sci.* **2017**, *124*, 25–34. [CrossRef]
41. Allah, A.G.; Badawy, M.W.; Rehan, H.H.; Abou-Romia, M.M. Inhibition of corrosion of α-brass (Cu-Zn, 67/33) in acid chloride solutions by some amino pyrazole derivatives. *J. Appl. Electrochem.* **1989**, *19*, 928–932. [CrossRef]
42. Walker, R. Triazole, benzotriazole, and naphthotriazole as corrosion inhibitors for brass. *Corrosion* **1976**, *32*, 414–417. [CrossRef]
43. Kosec, T.; Milošev, I.; Pihlar, B. Benzotriazole as an inhibitor of brass corrosion in chloride solution. *Appl. Surf. Sci.* **2007**, *253*, 8863–8873. [CrossRef]
44. Kosec, T.; Merl, D.K.; Milošev, I. Impedance and XPS study of benzotriazole films formed on copper, copper–zinc alloys and zinc in chloride solution. *Corros. Sci.* **2008**, *50*, 1987–1997. [CrossRef]

45. Mamaş, S.; Kiyak, T.; Kabasakaloğlu, M.; Koc, A. The effect of benzotriazole on brass corrosion. *Mater. Chem. Phys.* **2005**, *93*, 41–47. [CrossRef]
46. Antonijević, M.M.; Milić, S.M.; Šerbula, S.M.; Bogdanović, G.D. The influence of chloride ions and benzotriazole on the corrosion behavior of Cu37Zn brass in alkaline medium. *Electrochim. Acta* **2005**, *50*, 3693–3701. [CrossRef]
47. Ravichandran, R.; Rajendran, N. Influence of benzotriazole derivatives on the dezincification of 65–35 brass in sodium chloride. *Appl. Surf. Sci.* **2005**, *239*, 182–192. [CrossRef]
48. Obot, I.B.; Edouk, U.M. Benzimidazole: Small planar molecule with diverse anti-corrosion potentials. *J. Mol. Liq.* **2017**, *246*, 66–90. [CrossRef]
49. Ravichandran, R.; Nanjundan, S.; Rajendran, N. Effect of benzotriazole derivatives on the corrosion and dezincification of brass in neutral chloride solution. *J. Appl. Electrochem.* **2004**, *34*, 1171–1176. [CrossRef]
50. Ravichandran, R.; Nanjundan, S.; Rajendran, N. Effect of benzotriazole derivatives on the corrosion of brass in NaCl solutions. *Appl. Surf. Sci.* **2004**, *236*, 241–250. [CrossRef]
51. Ravichandran, R.; Rajendran, N. Electrochemical behaviour of brass in artificial seawater: Effect of organic inhibitors. *Appl. Surf. Sci.* **2005**, *241*, 449–458. [CrossRef]
52. Stupnišek-Lisac, E.; Gazivoda, A.; Madžarac, M. Evaluation of non-toxic corrosion inhibitors for copper in sulphuric acid. *Electrochim. Acta* **2002**, *47*, 4189–4194. [CrossRef]
53. Okafor, P.C.; Ikpi, M.E.; Uwah, I.E.; Ebenso, E.E.; Ekpe, U.J.; Umoren, S.A. Inhibitory action of Phyllanthus amarus extracts on the corrosion of mild steel in acidic media. *Corros. Sci.* **2008**, *50*, 2310–2317. [CrossRef]
54. Fouda, A.S.; Shalabi, K.; Idress, A.A. Ceratonia siliqua extract as a green corrosion inhibitor for copper and brass in nitric acid solutions. *Green Chem. Lett. Rev.* **2015**, *8*, 17–29. [CrossRef]
55. Nagiub, A.; Mansfeld, F. Evaluation of corrosion inhibition of brass in chloride media using EIS and ENA. *Corros. Sci.* **2001**, *43*, 2147–2171. [CrossRef]
56. Muñoz, A.I.; Antón, J.G.; Guiñón, J.L.; Herranz, V.P. Comparison of inorganic inhibitors of copper, nickel and copper–nickels in aqueous lithium bromide solution. *Electrochim. Acta* **2004**, *50*, 957–966. [CrossRef]
57. Xu, Q.J.; Zhou, G.D.; Wang, H.F.; Cai, W.B. Electrochemical studies of polyaspartic acid and sodium tungstate as corrosion inhibitors for brass and Cu30Ni alloy in simulated cooled water solutions. *Anti-Corros. Met. Mater.* **2006**, *53*, 207–211. [CrossRef]
58. Nihorimbere, M.; Kerroum, Y.; Guenbour, A.; Kacimi, M.; Bellaouchou, A.; Touir, R.; Zarrouk, A. Corrosion inhibition of brass in artificial drinking water by mineral compound. *J. Mater. Environ. Sci.* **2016**, *7*, 4121–4128.
59. Abd El-Rahman, H.A. Evaluation of AHT as corrosion inhibitor for α-brass in acid chloride solutions. *Corrosion* **1991**, *47*, 424–428. [CrossRef]
60. Mountassir, Z.; Srhiri, A. Electrochemical behaviour of Cu–40Zn in 3% NaCl solution polluted by sulphides: Effect of aminotriazole. *Corros. Sci.* **2007**, *49*, 1350–1361. [CrossRef]
61. Elbakri, M.; Touir, R.; Touhami, M.E.; Srhiri, A.; Benmessaoud, M. Electrosynthesis of adherent poly (3-amino-1,2,4-triazole) films on brass prepared in nonaqueous solvents. *Corros. Sci.* **2008**, *50*, 1538–1545. [CrossRef]
62. Finšgar, M. 2-Mercaptobenzimidazole as a copper corrosion inhibitor: Part I. Long-term immersion, 3D-profilometry, and electrochemistry. *Corros. Sci.* **2013**, *72*, 82–89. [CrossRef]
63. Finšgar, M. 2-Mercaptobenzimidazole as a copper corrosion inhibitor: Part II. Surface analysis using X-ray photoelectron spectroscopy. *Corros. Sci.* **2013**, *72*, 90–98. [CrossRef]
64. Finšgar, M.; Kek Merl, D. 2-Mercaptobenzoxazole as a copper corrosion inhibitor in chloride solution: Electrochemistry, 3D-profilometry, and XPS surface analysis. *Corros. Sci.* **2013**, *80*, 82–95. [CrossRef]
65. Finšgar, M.; Merl, D.K. An electrochemical, long-term immersion, and XPS study of 2-mercaptobenzothiazole as a copper corrosion inhibitor in chloride solution. *Corros. Sci.* **2014**, *83*, 164–175. [CrossRef]
66. Agafonkina, M.O.; Kuznetsov, Y.I.; Andreeva, N.P.; Shikhaliev, K.S.; Potapov, A.Y. Adsorption and passivation of copper by S-containing heterocyclic compounds in neutral aqueous solutions. *Corros. Mater. Prot.* **2016**, *1*, 29–38.
67. Arkhipushkin, I.A.; Vagramyan, T.A.; Shikhaliev, K.S.; Kazansky, L.P. The study of adsorption of 5-mercaptomethyl-3-amino-1,2,4-triazole on copper in neutral solutions. *Corros. Mater. Prot.* **2016**, *7*, 18–24.
68. Arkhipushkin, I.A.; Shikhaliev, K.S.; Potapov, A.Y.; Sapronova, L.V.; Kazansky, L.P. Inhibition of Brass (80/20) by 5-Mercaptopentyl-3-Amino-1,2,4-Triazole in Neutral Solutions. *Metals* **2017**, *7*, 488. [CrossRef]
69. Kazansky, L.P.; Pronin, Y.E.; Arkhipushkin, I.A. XPS study of adsorption of 2-mercaptobenzothiazole on a brass surface. *Corros. Sci.* **2014**, *89*, 21–29. [CrossRef]

70. Mansfeld, F. Tafel slopes and corrosion rates obtained in the pre-Tafel region of polarization curves. *Corros. Sci.* **2005**, *47*, 3178–3186. [CrossRef]
71. Shih, H.; Mansfeld, F. Software for quantitative analysis of polarization curves. *Comput. Model. Corros. ASTM Int.* **1992**, 174–183. [CrossRef]
72. Frisch, M.J.; Trucks, G.W.; Schlegel, H.B.; Scuseria, G.E.; Robb, M.A.; Cheeseman, J.R.; Scalmani, G.; Barone, V.; Petersson, G.A.; Nakatsuji, H.; et al. *Gaussian 16, Revision B.01*; Gaussian, Inc.: Wallingford CT, UK, 2016.
73. Parr, R.G.; Pearson, R.G. Absolute hardness: Companion parameter to absolute electronegativity. *J. Am. Chem. Soc.* **1983**, *105*, 7512–7516. [CrossRef]
74. Özcan, M.; Dehri, I.; Erbil, M. Organic sulphur-containing compounds as corrosion inhibitors for mild steel in acidic media: Correlation between inhibition efficiency and chemical structure. *Appl. Surf. Sci.* **2004**, *236*, 155–164. [CrossRef]

applied sciences

MDPI

Article

Corrosion Behavior and Conductivity of TiNb and TiNbN Coated Steel for Metallic Bipolar Plates

Kun Shi [1], Xue Li [1], Yang Zhao [1], Wei-Wei Li [1], Shu-Bo Wang [1,*], Xiao-Feng Xie [1,2,*], Li Yao [3], Jens Oluf Jensen [4] and Qing-Feng Li [4]

1 Institute of Nuclear and New Energy Technology, Tsinghua University, Beijing 100084, China; shi-k16@mails.tsinghua.edu.cn (K.S.); lixue13@mails.tsinghua.edu.cn (X.L.); yzhao16@mails.tsinghua.edu.cn (Y.Z.); willar@mail.tsinghua.edu.cn (W.-W.L.)
2 Shanxi Research Institute for Clean Energy, Tsinghua University, Taiyuan 030032, China
3 Advanced Technology Department, SAIC Motor Corporation Limited, Shanghai 201804, China; YaoLi01@saicmotor.com
4 Department of Energy Conversion and Storage, Technical University of Denmark, 2800 Lyngby, Denmark; jojen@dtu.dk (J.O.J.); qfli@dtu.dk (Q.-F.L.)
* Correspondence: wangshubo@tsinghua.edu.cn (S.-B.W.); xiexf@tsinghua.edu.cn (X.-F.X.)

Received: 10 March 2019; Accepted: 30 May 2019; Published: 24 June 2019

Abstract: To improve corrosion resistance and electronic conductivity of bipolar plates for proton exchange membrane fuel cells (PEMFC), coatings of TiNb and TiNbN on 316L stainless steel (SS) were prepared by magnetron sputtering. X-ray diffraction measurements confirmed the existence of metallic nitrides in the TiNbN coating. Scanning electron microscope tests showed that the deposited coatings provided smooth surfaces. Further electrochemical measurements indicated that the corrosion resistance of the TiNb coating was significantly higher than that of the substrate. At 0.19 V vs MSE, the long-term stabilized current density of TiNb/316L SS was lower than 1 $\mu A \cdot cm^{-2}$. The interfacial contact resistance values between coating and carbon paper suggested that TiNb and TiNbN films had better contact conductivity than the 316L SS substrate. In conclusion, TiNb coated 316L SS metallic bipolar plate material is a promising option for PEMFC.

Keywords: coating; metallic bipolar plate; PEMFC; TiNb; TiNbN

1. Introduction

The proton exchange membrane fuel cell (PEMFC) is a kind of energy conversion device, which transforms chemical energy into electricity. Owing to its high power-density and zero-emission features, PEMFC has been considered as one of the most prospective power suppliers for vehicles, portable devices and distributed generation [1]. PEMFC is composed of a proton exchange membrane, catalysts, gas diffusion layers, and bipolar plates. The lifetime of PEMFC depends on the durability of individual components. One of the reasons for failure is corrosion of the bipolar plates. This may not only destroy the bipolar plate itself, it also produces ions, which can be detrimental to both membrane [2] and catalyst. Therefore, the corrosion resistance of the bipolar plates is one of the decisive factors influencing the lifetime of a PEMFC. The American Department of Energy (DOE) has established a series of required properties for bipolar plates and one crucial target is to improve corrosion resistance and conductivity [3,4]. Earlier studies concentrated on graphite materials and graphite-based composite bipolar plates [5–7], considering the high corrosion resistance of graphite. However, graphite is brittle and in order to prevent breaking, it is usually machined into quite thick bipolar plate, which increases weight and manufacturing costs of the PEMFC stack [8]. Along with the development of metal forming and welding techniques, metallic bipolar plates have become increasingly relevant for PEMFC due to the good mechanical strength. This is especially for automotive stacks where weight and volume are critical parameters. In order to reduce cost, the plates should preferably be manufactured as coated inexpensive substrates [9–11]. Austenitic stainless steel (SS), especially 316L, is prone to passivation and can be a suitable bipolar plate material [12,13]. As for coatings, noble metals like Au have been investigated widely [14,15], but although most noble metallic bipolar plates established great corrosion resistance and conductivity, high prices have confined their applications. Compared with noble metals, the low-cost transition metals and transition metal nitrides [16] have showed good performances in PEMFC surroundings attracting significant interest. The niobized 304 SS was reported to a decrease of alloy's corrosion current density, meeting the DOE demand (<1 $\mu A \cdot cm^{-2}$) [17]. In acid solution, the corrosion resistance of Ti still needs to be further improved [13]. Metal nitrides such as TiN and NbN have been deposited on different substrates establishing low interfacial contact resistance (ICR). However, after a transient polarization test, the TiN coated sample achieved higher current density than the base SS substrate material [18]. Similarly, the NbN coating's corrosion resistance at high potential also needs to be improved [19]. Considering the embarrassing situation that it is hard to obtain high conductivity and great corrosion resistance at the same time, the idea of mixing Ti, Nb, TiN and NbN properly to combine the best of each of the advantages is proposed. The alloyed TiNb and TiNbN have acceptable prices when applied as films. Additionally, a TiNbN coating can be expected to have attractive mechanical property and has been used in biomedical or mechanical areas [20,21]. TiNb alloys have also been studied earlier for their unique shape memory effect [22]. Aukland et al. [23] evaluated the chemical durability and surface resistances of Ti alloys with 3 at.% Nb and confirmed that the oxidized Ti-3Nb alloys had lower resistances than Ti oxides. However, to the best of our knowledge, there is still a lack of adequate literature investigating the corrosion behavior and conductivity of TiNb or TiNbN coated SS with low Ti content under simulated PEMFC working conditions. As for the deposition method, physical vapor deposition (PVD) provides firm, uniform, and dense coatings compared with electroplating [24]. Among all the PVD ways, magnetron sputtering is used extensively in industry due to its high stability and limited contamination. During sputtering, influenced by electrical and magnetic fields, argon ions in the vacuum chamber bombard the targets causing atoms or clusters to separate from the targets and finally be deposited on the substrate. Usually, coating particles form firm bonds with the base material.

In this study, TiNb and TiNbN coatings with low Ti content were deposited on 316L SS by magnetron sputtering to explore their performances in simulated PEMFC cathode environments. The microstructures, morphologies, and chemical compositions of the coatings were characterized by X-ray diffraction (XRD), scanning electron microscopy (SEM), energy dispersive spectrometry (EDS), and X-ray photoelectron spectroscopy (XPS). Electrochemical experiments and ICR tests were conducted to evaluate the corrosion resistance and conductivity of coated and uncoated 316L SS.

2. Materials and Methods

2.1. Preparation of Coatings

The TiNb and TiNbN coatings were prepared by magnetron co-sputtering and reactive sputtering with a Process Equipment (KJLC, Kurt Company, US) using a radio frequency (RF) Ti target and a direct current (DC) Nb target. The substrate was 316L SS with a thickness of 0.2 mm. Before sputtering, the test specimens were ground with 800, 1500, and 2000 mesh sandpaper, polished with diamond polishing agents and ultrasonically cleaned in ethanol and acetone to remove the solid particles and grease. High purity (99.9999%) argon and nitrogen were used as working gases. The chamber was depressurized to below 7 mPa. During deposition, the substrate temperature was 523.15 K and the working pressure was in 0.5–0.8 Pa. To improve the adhesion between substrate and coating, a thin titanium layer was deposited at 7.7 $W \cdot cm^{-2}$ for 10 minutes followed by transition layers with different atomic ratios of Ti and Nb. Finally, the target powers of Ti and Nb were regulated to 4.4 $W \cdot cm^{-2}$ and held for 2 h to obtain the TiNb coating. The deposition parameters of the TiNbN coating were similar to that of TiNb with lower temperature (473.15 K). The gas flow volumetric ratio of reactive gas (nitrogen) and inert gas (argon) was set as 1:5. The thicknesses of the TiNb and the TiNbN coatings were about 660 and 2080 nm, respectively, according to coatings' deposition rates.

2.2. Characterization of Coatings

The structures of the TiNb and TiNbN coatings were determined by XRD (D8 Advance, Bruker, GER), using Cu Kα radiation in glancing angle mode at 1°. The scan speed was 2°·min^{-1} with a step size of 0.02°. The surface morphologies of the coatings were characterized by SEM (Merlin, Zeiss Company, GER) equipped with EDS. To evaluate the chemical composition of the coatings, XPS (EscaLab 250XI, Thermo Fisher Scientific, UK) technique was used. The corrosion resistance of the bipolar plate material was tested by electrochemical methods including potentiodynamic polarization, electrochemical impedance spectroscopy (EIS) and potentiostatic polarization in 0.5 mol·L^{-1} H_2SO_4 solutions saturated with O_2 at room temperature, using an electrochemical workstation (CHI760E, Shanghai Chenhua Instruments Limited, CN). The sample size was 20 by 20 mm for the tests. The testing container was a three-electrode cell. Working electrodes were SS samples, sealed by circular rings with a hole of 1 cm^2 exposure area. To avoid the possible addition of adverse reactive ions from other electrodes, a saturated mercurous sulfate electrode (MSE) and a platinum foil served as reference electrode and counter electrode, respectively. Prior to tests, samples were immersed in electrolyte solutions for 30 min to get stable open circuit potential (OCP). The scan rate of potentiodynamic polarization was set as 2 mV·s^{-1}. EIS was carried out at both OCP and 0.19 V vs MSE, within a 0.01–10^5 Hz frequency range and with a 10 mV potential amplitude. Besides electrochemical evaluation, ICR measurements were performed. Two pieces of Toray conductive carbon papers were sandwiched between SS sample (30 by 30 mm) and two Au coated copper plates like Wang's method [25]. An electrical current of 2A was applied via the copper plates. The ICR value between the coating and carbon paper was calculated from the voltage drop and the current with the resistances between the other contact interfaces measured by the similar method subtracted.

3. Results and Discussion

3.1. Structural Characterization

To avoid the possible effects from rough SS substrates, TiNb and TiNbN surface films for structural characterization were deposited on well-polished silicon wafers. The XRD patterns are shown in Figure 1.

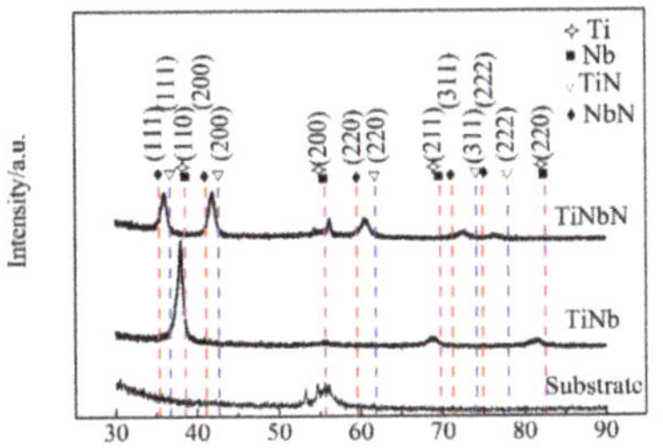

Figure 1. X-ray diffraction (XRD) patterns for TiNb and TiNbN coatings on silicon wafers. The hatched lines indicate peak positions of selected materials. The peaks of cubic Ti and Nb are showed by coinciding lines.

The vertical hatched lines represent peaks recorded in PDF cards No. 38-1155 (cubic NbN), No. 38-1420 (cubic TiN), No. 34-0370 (cubic Nb), and No. 44-1288 (cubic Ti). The diffraction peaks of cubic Nb and Ti are very close and consequently indicated by the same lines. The TiNb film follows the same patterns suggesting that the alloy has the same cubic structure. It is reasonable to assume a solid solution of Ti and Nb. The angles are shifted to lower values indicating some stress. For the TiNbN coating, the Bragg angles of diffraction peaks are between those of TiN and NbN, which denotes that TiN and NbN are likewise a solid solution [26]. Moreover, diffraction peaks arising from (111), (200), (220), (311), and (222) plane reflections are clearly identified, confirming the face centered cubic (fcc) crystalline structure.

3.2. *Surface Morphology*

The SEM images of 316L SS, TiNbN/316L SS and TiNb/316L SS before and after potentiostatic polarization corrosion tests are shown in Figure 2.

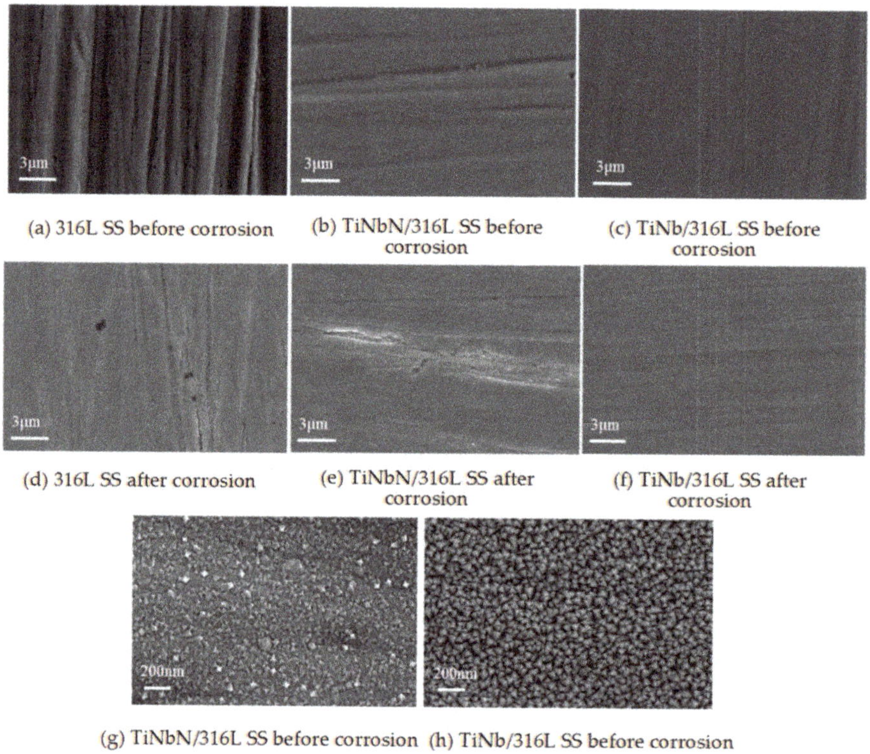

Figure 2. Scanning electron microscopy (SEM) images of the surface of 316L stainless steel (SS), TiNbN/316L SS and TiNb/316L SS.

In Figure 2a, the polished SS surface is uneven due to grinding and polishing, and a number of holes are randomly distributed on it. In Figure 2b, the TiNbN surface is much smoother than that of the SS. However, small holes still exist, and at the bottom of the surface valley, some large particles appear. The micrograph of the TiNb coating shown in Figure 2c is smooth and nearly no obvious holes can be seen. Obviously, the deposition of TiNbN or TiNb has smoothened the SS samples. This may also limit the formation of micro corrosion cells. The morphologies of uncoated and coated 316L SS after potentiostatic polarization tests are shown in Figure 2d–f. After corrosion, 316L SS shows several holes (Figure 2d) which are larger than those observed on 316L SS before corrosion. In Figure 2e, localized corrosion is apparent especially for the valley and defect areas on the TiNbN coating while the surface of TiNb/316L SS in Figure 2f shows few large or deep corrosion pits. This is relevant with the coating defects formed on TiNbN before corrosion. For the hole defect, the electrolyte solution diffusion at the bottom of it is slow. The concentrations of O_2 and ions inside the hole differ from that in solution, resulting in the different electrochemical activities of grains. Due to the lack of O_2 and the dense passive film, the inner region of the hole becomes the anode of a corrosion battery with a smaller area and a higher corrosion current density, compared with other surface zones. Thus the localized corrosion is induced and more defects are formed. Images in Figure 2g,h under higher magnification indicate that the grain size of sputtering deposited TiNbN varies greatly but with a smaller average value compared with TiNb which might be attributed to the incorporation of N. For the TiNbN coating, the distinction in grain size could result in an inhomogeneous distribution of grain boundaries and finally contributing to the non-uniformity of passive layer [27]. Additionally, the SEM results show the particle agglomeration phenomenon of TiNbN is more serious than of TiNb. And the agglomeration may cause the differences in coating's surface structure and composition by which the localized corrosion of TiNbN can be induced.

In order to characterize the coating composition, EDS was applied and results are shown in Figure 3. Since the coating is thin, the detected images may include intensity peaks from some elements in the substrate, the substrate oxidation layer and adsorbed substances on sample surface. In the film of TiNbN/SS, the atomic ratio of Ti, Nb and N is about 7:22:10 (with the usual reservations regarding quantifications of light elements). For the coating of TiNb/SS, the atomic ratio of Ti and Nb is about 1:4. The EDS results illustrate under the same sputtering power mentioned above, the deposition speed of Nb is faster than that of Ti. Possible reason for this phenomenon is that Nb is a DC power target, and Ti is powered by a RF electric source. For the same material, the sputtering rate of the RF target is lower than that of the DC target at the same set power.

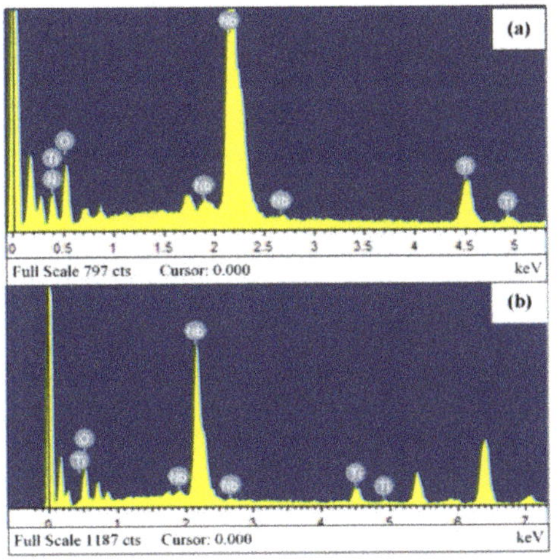

Figure 3. Energy dispersive spectrometry (EDS) results of (**a**) TiNbN/316L SS and (**b**) TiNb/316L SS.

3.3. X-ray Photoelectron Spectroscopy Analysis

The chemical states of Ti and Nb in the surface layers of the TiNb and TiNbN films were examined by XPS. The original Nb 3d and Ti 2p spectra are shown in Figures 4 and 5. After fitting all the spectra with backgrounds subtracted, Nb 3d and Ti 2p peaks were analyzed according to the NIST XPS Database [28], relative literatures and possible reactions in testing environment. In Figure 4a, it's deduced that the highest doublet peaks are related to $3d_{5/2}$ and $3d_{3/2}$ of Nb in Nb_2O_5 and the lower peaks represent metal Nb [29–32]. From the Ti 2p fitted spectra in Figure 4b, two peaks are observed corresponding to Ti $2p_{3/2}$ and Ti $2p_{1/2}$ in TiO_2, respectively [33,34]. The testing results reveal that the surface layer of the TiNb coating is primarily composed of Nb_2O_5 and TiO_2, which means the TiNb coating can be oxidized spontaneously in air and form a stable metallic oxide film. Figure 5 shows XPS spectra of Nb 3d and Ti 2p of the TiNbN coating. The fitted Nb 3d spectra depicted in Figure 5a indicate that the major peaks representing Nb $3d_{5/2}$ and Nb $3d_{3/2}$ can be ascribed to the formation of Nb_2O_5, NbN, and NbO [31,32,35]. The Ti 2p fitted spectra in Figure 5b reveal the existence of TiN [36]. But the two largest peaks stem from Ti $2p_{3/2}$ and Ti $2p_{1/2}$ in TiO_2 [37,38]. From the fitted results, it is obvious that under the sputtering conditions mentioned above, TiN and NbN were synthesized. Due to the exposure of the coating to air, stable oxides of Ti and Nb were also formed in the outer layer of TiNbN coating. Since the main surface composition difference between TiNb and TiNbN coatings lies in TiN and NbN, metallic nitrides could play an important role in exhibiting some different properties for TiNb/316L SS and TiNbN/316L SS.

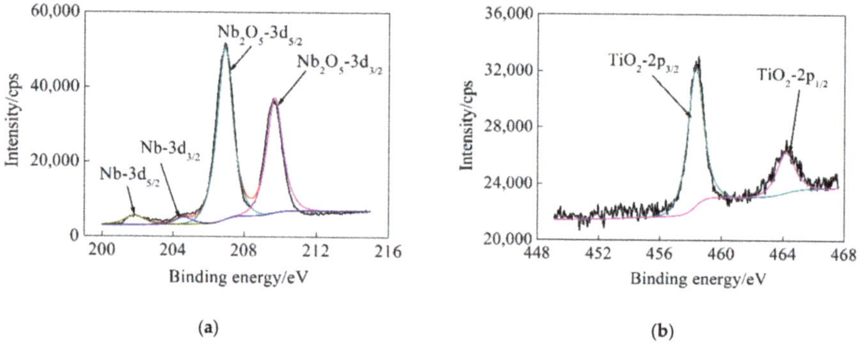

Figure 4. X-ray photoelectron spectroscopy (XPS) spectra of (**a**) Nb 3d and (**b**) Ti 2p (original-black lines, fitted-red lines) measured on TiNb/316L SS.

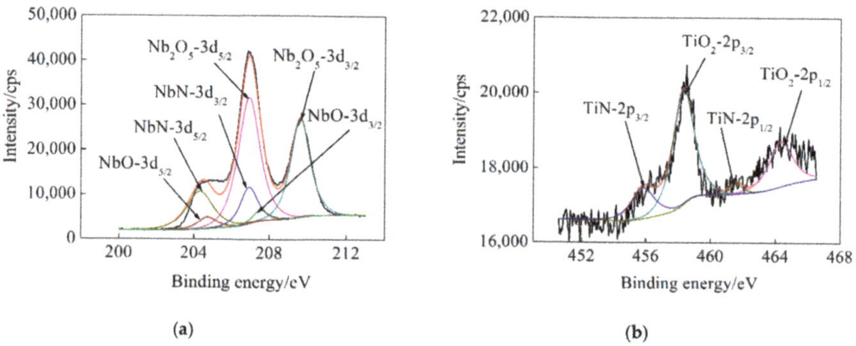

Figure 5. X-ray photoelectron spectroscopy (XPS) spectra of (**a**) Nb 3d and (**b**) Ti 2p (original-black lines, fitted-red lines) measured on TiNbN/316L SS.

3.4. Corrosion Behavior

The corrosion behaviors of 316L SS, TiNb/316L SS and TiNbN/316L SS were characterized electrochemically by potentiodynamic polarization, EIS, and potentiostatic polarization. Figure 6 displays potentiodynamic polarization curves of bare and coated 316L SS. From the curves, corrosion potentials (ϕ_{corr}) are obtained. The ϕ_{corr} is related to coating's thermodynamic stability, and the higher ϕ_{corr} value normally means better corrosion resistance from the aspect of thermodynamics [3]. In the tests, the ϕ_{corr} of 316L SS is −0.761 V vs MSE, whereas TiNbN/316L SS has a more positive value of −0.744 V. Compared with the substrate, the TiNb film increases the ϕ_{corr} by 138 mV. When potentials are swept from negative to positive, all three samples show obvious passivation areas. In the passivation area, as the potential value increases, the current density remains nearly constant. At 0.19 V vs MSE, all the samples are passivated and 316L SS exhibits a current density of 17.3 $\mu A \cdot cm^{-2}$, while it is 2.28 $\mu A \cdot cm^{-2}$ and 44.6 $\mu A \cdot cm^{-2}$ for TiNb/316L SS and TiNbN/316L

SS, respectively. The results indicate that the TiNb film has high corrosion resistance at 0.19 V vs MSE. This is owing to the formation of corrosion resistant passive film on TiNb coating obtained from anodic polarization process. Moreover, it is noteworthy that although the TiNbN film is thicker, TiNb alloy establishes better corrosion resistance. This phenomenon is mainly associated with the surface condition and composition of TiNbN film.

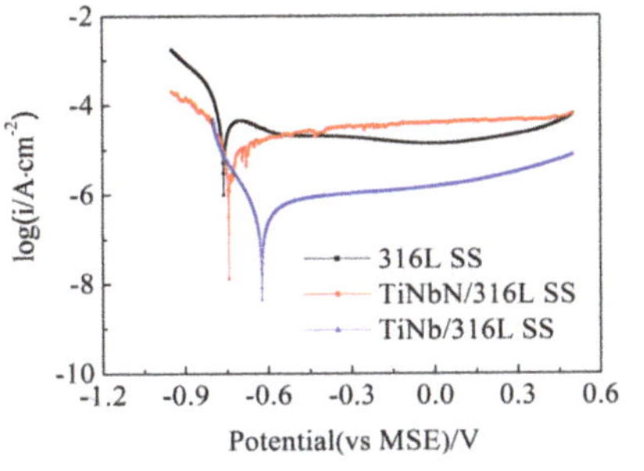

Figure 6. Potentiodynamic polarization curves of 316L SS, TiNb/316L SS, and TiNbN/316L SS.

EIS was performed and Figure 7 shows Nyquist impedance spectra of coated and uncoated 316L SS measured at OCP and 0.19 V vs MSE. In Figure 7a, influenced by surface defects, TiNbN/316L SS shows two time constants at OCP which reflect the impedance information from coating and substrate [12,39]. In the Nyquist impedance spectra of TiNbN/316L SS at 0.19V, TiNb/316L SS, and 316L SS, incomplete semicircular arcs appear. The depressed semicircular shape of SS is attributed to the roughness of sample's surface. To fit the impedance spectrum of TiNbN/316L SS measured at OCP, an equivalent circuit with two constant phase elements (CPE), shown in Figure 8a [40], is adopted. The CPE is represented by Q. Under ideal conditions, a CPE corresponds to a capacitor [41]. R_s is the resistance of electrolyte solution between working electrode and reference electrode. R_f and Q_f are coating's resistance and capacitance, respectively, and R_{ct} is the charge transfer resistance, while Q_1 represents the double layer capacitance [42]. The other spectra in Figure 7a,b are fitted by the equivalent circuit in Figure 8b [43]. Table 1 lists the fitted values of R_{ct}. The higher R_{ct} suggests the enhanced corrosion resistance [44,45].

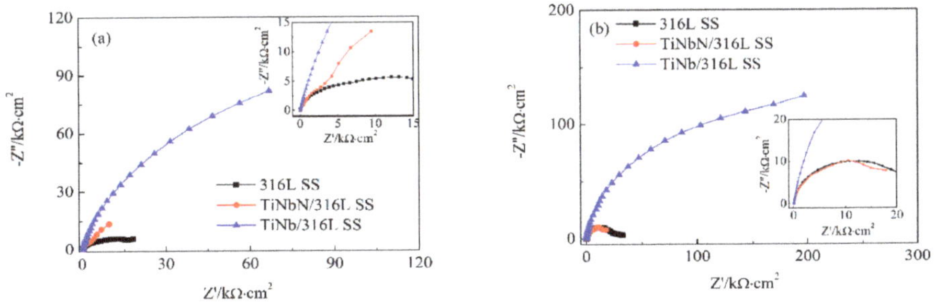

Figure 7. Nyquist impedance spectra of 316L SS, TiNb/316L SS and TiNbN/316L SS @ (**a**) open circuit potential (OCP), (**b**) 0.19 V vs mercurous sulfate electrode (MSE).

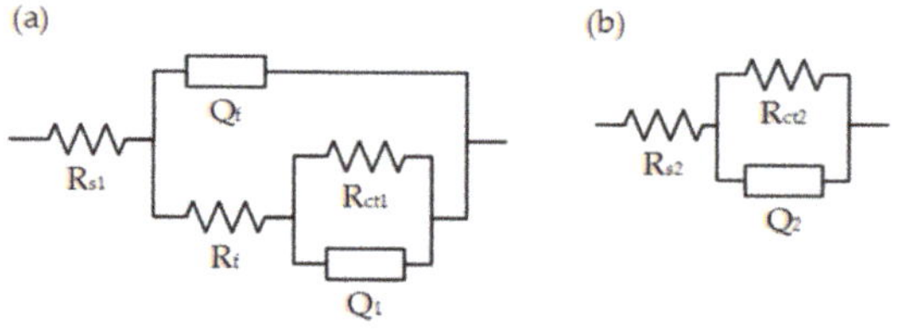

Figure 8. Equivalent circuits used for the Nyquist impedance spectra of (**a**) TiNbN/316L SS @ OCP; (**b**) TiNbN/316L SS @ 0.19 V vs MSE, 316L SS, and TiNb/316L SS.

Table 1. R_{ct} values of 316L SS, TiNb/316L SS and TiNbN/316L SS at different potentials.

Materials	316L SS	TiNb/316L SS	TiNbN/316L SS
R_{ct} @ OCP/ kΩ·cm^2	16.2	198	21.13
R_{ct} @ 0.19 V vs MSE/ kΩ·cm^2	26.74	229	23.52

Comparing R_{ct} values of different materials in Table 1, it is observed that the charge transfer impedance of TiNb coated 316L SS is much higher than for the uncoated substrate. Part of reasons for this phenomenon is that the surface layer of TiNb could be oxidized into TiO_2 and Nb_2O_5 spontaneously in air according to XPS measurements, providing a high corrosion resistance. At 0.19 V vs MSE, the high potential and the continuous supply of oxygen could passivate TiNb and lead to the formation of dense passive film which is corrosion resistant and could act as a barrier to inhibit the bulk solution from approaching the sub-layer directly. In comparison with TiNb, the R_{ct} value of the TiNbN film is much smaller, either at OCP or 0.19 V vs MSE. Two major factors may contribute to this phenomenon. From a morphology point of view, the particle clusters and pores observed on the TiNbN coating can cause localized corrosion. The other reason is that the bonds between nitrogen and metallic atoms may impede the formation of more homogeneous and corrosion resistant passive layer. According to the above analysis, from the EIS and potentiodynamic polarization test results draw one can the conclusion that TiNb has good corrosion resistance while TiNbN and bare 316L SS are likely to dissolve in the aggressive cathode environment.

To evaluate the stability of the materials, potentiostatic polarization experiments were conducted at 0.19 V vs MSE. The results are shown in Figure 9. In the beginning, the current density of each coated 316L SS sample declines while that of 316L SS first increases sharply, then decreases gradually. After 2000 s, 316L SS experiences an obvious fluctuation due to the surface states changes affected by electrochemical dissolution, but throughout this period, its average corrosion rate is increasing. As for TiNbN/316L SS, the time needed for total passivation is about 3000 s, and its stabilized current density is approximately 6–7 μA·cm^{-2}. Although TiNbN/316L SS and 316L SS show comparable current density values and variation trends at the initial stage of tests, the TiNbN coating is much more stable than bare 316L SS, judging from the smoothness of the curves. That is because the coating deposited by magnetron sputtering is uniform and has less defects than the substrate. The current density of TiNb/316L SS decays rapidly first and then stabilizes at 0.9–1 μA·cm^{-2}. This demonstrates that the TiNb coating can be easily passivated and keep a low corrosion current density for a relatively long period, influenced by the high potential and the oxygen saturated acid solution, which shows the good corrosion resistance and stability of the TiNb coating. From the potentiostatic polarization tests above, it is assumed that TiNb/316L SS has the potential to work as a PEMFC cathode flow plate material for a long time.

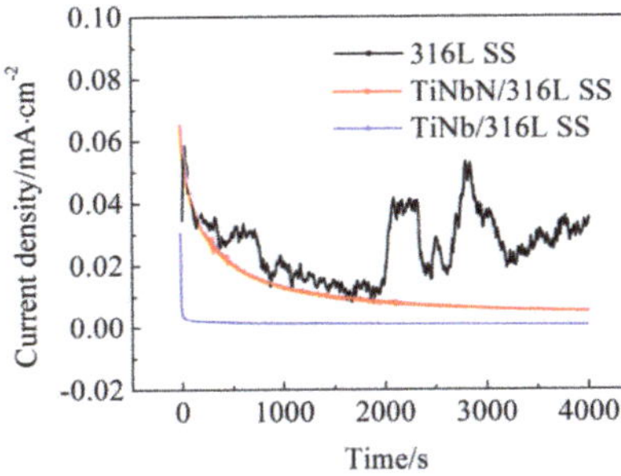

Figure 9. Potentiostatic polarization curves of 316L SS, TiNb/316L SS, and TiNbN/316L SS.

The electrochemical testing shows that both TiNb and TiNbN coatings can form steady passive films at 0.19 V vs MSE, and the long-term corrosion current density of TiNb/316L SS is lower than 1 μA· cm^{-2}, illustrating the high corrosion resistance of TiNb coating. Compared with TiNb/316L SS, TiNbN/316L SS is of poorer corrosion resistance.

3.5. *Interfacial Contact Resistance*

The conductivity of bipolar plates has great impact on the PEMFC's working efficiency [46]. Normally, the bulk resistance of metallic bipolar plates can be ignored since the ICR between the bipolar plate and the carbon paper is much greater. In Figure 10, the ICR of the material with a carbon paper is plotted as a function of the compression force.

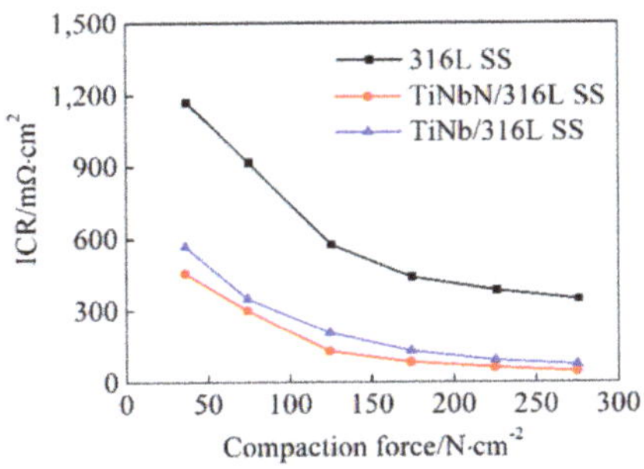

Figure 10. Interfacial contact resistances (ICR) between 316L SS, TiNb/316L SS, TiNbN/316L SS and carbon paper as a function of compaction force.

In the low-force region, as compaction force raises, the contact area between carbon paper and sample increases leading to the decrease of ICR, and when the force is high enough, the ICR is dominated by the surface composition and nature [46,47]. At the same force, the contact resistance of 316L SS, TiNb/316L SS, and TiNbN/316L SS decreases in turn, and their conductivity increases sequentially. I.e., both the TiNb and the TiNbN coatings provide lower ICR than 316L SS's, but the effect of TiNbN is more remarkable. At 276 N·cm^{-2}, TiNbN coating's ICR is approximately 48 mΩ·cm^2, and the ICR value of TiNb/316L SS is nearly 1/5 of that of 316L SS. One of the reasons behind this is that both coatings could reduce the roughness of 316L SS substrate leading to an increased contact area. For the TiNbN coating, the dopant of TiN and NbN in its surface layer improves conductivity. In comparison with TiO_2 and Nb_2O_5, TiN [48] and NbN have higher conductivity. Overall, the TiNbN coating is of relatively high electrical conductivity, and the TiNb film reduces the substrate's ICR. However, considering the moderate corrosion resistance of TiNbN, TiNb/316L SS is more promising for use as bipolar plate. Further studies are relevant assessing a combined film with good conductivity from TiNbN and great corrosion resistance from TiNb.

4. Conclusions

TiNb and TiNbN films with low Ti content were coated on 316L SS by magnetron sputtering with a view to application for bipolar plates in PEMFC. The electrochemical testing indicates that TiNb coating has the best corrosion resistance in cathode simulation electrolyte solutions. At 0.19 V vs MSE, the stable corrosion current density of TiNb/316L SS is lower than 1 $\mu A \cdot cm^{-2}$. The ICR measurements show that the contact conductivity of TiNb/316L SS is higher than that of substrate. Compared with 316L SS and the TiNb coating, TiNbN film shows moderate corrosion resistance but the lowest ICR. According to XRD analysis, the main component of TiNbN is a solid solution of cubic NbN and TiN. The morphologies of TiNb and TiNbN characterized by SEM reveal that both coatings provide smooth surfaces. It can be concluded that the TiNb/316L SS is a kind of promising bipolar plate material with high corrosion resistance. Nevertheless, the conductivity of TiNb/316L SS still needs to be improved. In the future, multilayer coatings consisting of TiNb and TiNbN with optimized atomic ratios will be studied with aim to improve material's corrosion resistance and conductivity further.

Author Contributions: Conceptualization, K.S.; methodology, K.S. and X.-F.X.; resources, X.L., Y.Z. and W.-W.L.; writing—original draft preparation, K.S.; writing—review and editing, X.-F.X., J.O.J., S.-B.W. and L.Y.; project administration, X.-F.X., J.O.J. and Q.-F.L.

Funding: This research was funded by National Key R&D Program of China (2016YFE0102700), Major Science and Technology Projects of Shanxi Province (20181101006) and Innovation Fund Denmark by the project Wind2H (5185-00025B).

Conflicts of Interest: The authors declare no conflict of interest.

References

1. Wang, Y.; Chen, K.S.; Mishler, J.; Cho, S.C.; Adroher, X.C. A review of polymer electrolyte membrane fuel cells: Technology, applications, and needs on fundamental research. *Appl. Energy* **2011**, *88*, 981–1007. [CrossRef]
2. Antunes, R.A.; Oliveira, M.C.L.; Ett, G.; Ett, V. Corrosion of metal bipolar plates for PEM fuel cells: A review. *Int. J. Hydrog. Energy* **2010**, *35*, 3632–3647. [CrossRef]
3. Wang, L.; Tao, Y.; Zhang, Z.; Wang, Y.; Feng, Q.; Wang, H.; Li, H. Molybdenum carbide coated 316L stainless steel for bipolar plates of proton exchange membrane fuel cells. *Int. J. Hydrog. Energy* **2019**, *44*, 4940–4950. [CrossRef]
4. Kim, M.; Lim, J.W.; Lee, D.G. Electrical contact resistance between anode and cathode bipolar plates with respect to surface conditions. *Compos. Struct.* **2018**, *189*, 79–86. [CrossRef]
5. Adloo, A.; Sadeghi, M.; Masoomi, M.; Pazhooh, H.N. High performance polymeric bipolar plate based on polypropylene/graphite/graphene/nano-carbon black composites for PEM fuel cells. *Renew. Energy* **2016**, *99*, 867–874. [CrossRef]
6. Dhakate, S.R.; Sharma, S.; Chauhan, N.; Seth, R.K.; Mathur, R.B. CNTs nanostructuring effect on the properties of graphite composite bipolar plate. *Int. J. Hydrog. Energy* **2010**, *35*, 4195–4200. [CrossRef]
7. Li, W.; Jing, S.; Wang, S.; Wang, C.; Xie, X. Experimental investigation of expanded graphite/phenolic resin composite bipolar plate. *Int. J. Hydrog. Energy* **2016**, *41*, 16240–16246. [CrossRef]
8. Taherian, R. A review of composite and metallic bipolar plates in proton exchange membrane fuel cell: Materials, fabrication, and material selection. *J. Power Sources* **2014**, *265*, 370–390. [CrossRef]
9. Yu, H.; Yang, L.; Zhu, L.; Jian, X.; Wang, Z.; Jiang, L. Anticorrosion properties of Ta-coated 316L stainless steel as bipolar plate material in proton exchange membrane fuel cells. *J. Power Sources* **2009**, *191*, 495–500. [CrossRef]
10. Mawdsley, J.R.; Carter, J.D.; Wang, X.; Niyogi, S.; Fan, C.Q.; Koc, R.; Osterhout, G. Composite-coated aluminum bipolar plates for PEM fuel cells. *J. Power Sources* **2013**, *231*, 106–112. [CrossRef]
11. Bi, F.; Peng, L.; Yi, P.; Lai, X. Multilayered Zr–C/a-C film on stainless steel 316L as bipolar plates for proton exchange membrane fuel cells. *J. Power Sources* **2016**, *314*, 58–65. [CrossRef]
12. Mani, S.P.; Rajendran, N. Corrosion and interfacial contact resistance behavior of electrochemically nitrided 316L SS bipolar plates for proton exchange membrane fuel cells. *Energy* **2017**, *133*, 1050–1062. [CrossRef]
13. Asri, N.F.; Husaini, T.; Sulong, A.B.; Majlan, E.H.; Daud, W.R.W. Coating of stainless steel and titanium bipolar plates for anticorrosion in PEMFC: A review. *Int. J. Hydrog. Energy* **2017**, *42*, 9135–9148. [CrossRef]

14. Wind, J.; Spah, R.; Kaiser, W.; Bohm, G. Metallic bipolar plates for PEM fuel cells. *J. Power Sources* **2002**, *105*, 256–260. [CrossRef]
15. Tsai, S.Y.; Bai, C.Y.; Lin, C.H.; Shi, G.N.; Hou, K.H.; Liu, Y.M.; Ger, M.D. The characteristics and performance of electroless nickel and immersion Au plated aluminum alloy bipolar plates in polymer electrolyte membrane fuel cells. *J. Power Sources* **2012**, *214*, 51–58. [CrossRef]
16. Wang, C.; Wang, S.; Peng, L.; Zhang, J.; Shao, Z.; Huang, J.; Sun, C.; Ouyang, M.; He, X. Recent progress on the key materials and components for proton exchange membrane fuel cells in vehicle applications. *Energies* **2016**, *9*, 603. [CrossRef]
17. Wang, L.; Sun, J.; Li, P.; Jing, B.; Li, S.; Wen, Z.; Ji, S. Niobized AISI 304 stainless steel bipolar plate for proton exchange membrane fuel cell. *J. Power Sources* **2012**, *208*, 397–403. [CrossRef]
18. Zhang, D.; Duan, L.; Guo, L.; Tuan, W.H. Corrosion behavior of TiN-coated stainless steel as bipolar plate for proton exchange membrane fuel cell. *Int. J. Hydrog. Energy* **2010**, *35*, 3721–3726. [CrossRef]
19. Cha, B.C.; You, Y.Z.; Hong, S.T.; Kim, J.H.; Kim, D.W.; Lee, B.S.; Kim, S.K. Nitride films as protective layers for metallic bipolar plates of polymer electrolyte membrane fuel cell stacks. *Int. J. Hydrog. Energy* **2011**, *36*, 4565–4572. [CrossRef]
20. Serro, A.P.; Completo, C.; Colaço, R.; dos Santos, F.; da Silva, C.L.; Cabral, J.M.S.; Araújo, H.; Pires, E.; Saramago, B. A comparative study of titanium nitrides, TiN, TiNbN and TiCN, as coatings for biomedical applications. *Surf. Coat. Technol.* **2009**, *203*, 3701–3707. [CrossRef]
21. Gispert, M.P.; Serro, A.P.; Colaço, R.; Pires, E.; Saramago, B. Wear of ceramic coated metal-on-metal bearings used for hip replacement. *Wear* **2007**, *263*, 1060–1065. [CrossRef]
22. Saud, S.N.; Hosseinian, R.; Bakhsheshi-Rad, H.R.; Yaghoubidoust, F.; Iqbal, N.; Hamzah, E.; Ooi, C.H.R. Corrosion and bioactivity performance of graphene oxide coating on Ti-Nb shape memory alloys in simulated body fluid. *Mater. Sci. Eng. C Mater. Biol. Appl.* **2016**, *68*, 687–694. [CrossRef] [PubMed]
23. Aukland, N.; Boudina, A.; Eddy, D.S.; Mantese, J.V.; Thompson, M.P.; Wang, S.S. Alloys that form conductive and passivating oxides for proton exchange membrane fuel cell bipolar plates. *J. Mater. Res.* **2004**, *19*, 1723–1729. [CrossRef]
24. Yi, P.; Zhu, L.; Dong, C.; Xiao, K. Corrosion and interfacial contact resistance of 316L stainless steel coated with magnetron sputtered ZrN and TiN in the simulated cathodic environment of a proton-exchange membrane fuel cell. *Surf. Coat. Technol.* **2019**, *363*, 198–202. [CrossRef]
25. Wang, H.; Sweikart, M.A.; Turner, J.A. Stainless steel as bipolar plate material for polymer electrolyte membrane fuel cells. *J. Power Sources* **2003**, *115*, 243–251. [CrossRef]
26. Xu, J.; Huang, H.J.; Li, Z.Y.; Xu, S.; Tao, H.; Munroe, P.; Xie, Z.H. Corrosion behavior of a ZrCN coated Ti alloy with potential application as a bipolar plate for proton exchange membrane fuel cell. *J. Alloys Compd.* **2016**, *663*, 718–730. [CrossRef]
27. Gollapudi, S. Grain size distribution effects on the corrosion behaviour of materials. *Corros. Sci.* **2012**, *62*, 90–94. [CrossRef]
28. National Institute of Standards and Technology. NIST X-ray Photoelectron Spectroscopy Database, NIST Standard Reference Database Number 20. Available online: https://srdata.nist.gov/xps/ (accessed on 15 April 2019).
29. Bahl, M.K. ESCA studies of some niobium compounds. *J. Phys. Chem. Solids* **1975**, *36*, 485–491. [CrossRef]
30. Nyholm, R.; Martensson, N. Core level binding energies for the elements Zr-Te (Z = 40–52). *J. Phys. C Solid St. Phys.* **1980**, *13*, L279–L284. [CrossRef]
31. Nefedov, V.I.; Salyn, Y.V.; Chertkov, A.A.; Padurets, L.N. X-ray electron study of electron-density distribution in hydrides of transition-elements. *Zh. Neorg. Khim.* **1974**, *19*, 1443–1445.
32. Latta, E.E.; Ronay, M. Catalytic-oxidation of niobium. *Phys. Rev. Lett.* **1984**, *53*, 948–951. [CrossRef]
33. Blasco, T.; Camblor, M.A.; Corma, A.; Perez-Pariente, J. The state of Ti in titanoaluminosilicates isomorphous with zeolite β. *J. Am. Chem. Soc.* **1993**, *115*, 11806–11813. [CrossRef]
34. Sanjinés, R.; Tang, H.; Berger, H.; Gozzo, F.; Margaritondo, G.; Lévy, F. Electronic structure of anatase TiO_2 oxide. *J. Appl. Phys.* **1994**, *75*, 2945–2951. [CrossRef]
35. Zhang, L.; You, L.; Ying, L.; Peng, W.; Wang, Z. Characterization of surface oxidation layers on ultrathin NbTiN films. *Phys. C Supercond. Appl.* **2018**, *545*, 1–4. [CrossRef]
36. Shulga, Y.M.; Troitskii, V.N.; Aivazov, M.I.; Borodko, Y.G. X-ray photoelectron-spectra of scandium, titanium, vanadium and chromium mononitrides. *Zh. Neorg. Khim.* **1976**, *21*, 2621–2624.
37. Netterfield, R.P.; Martin, P.J.; Pacey, C.G.; Sainty, W.G.; McKenzie, D.R.; Auchterlonie, G. Ion-assisted deposition of mixed TiO_2-SiO_2 films. *J. Appl. Phys.* **1989**, *66*, 1805–1809. [CrossRef]

38. Fierro, J.L.G.; Arrua, L.A.; Nieto, J.M.L.; Kremenic, G. Surface properties of co-precipitated V-Ti-O catalysts and their relation to the selective oxidation of isobutene. *Appl. Catal.* **1988**, *37*, 323–338. [CrossRef]
39. Jeon, W.S.; Kim, J.G.; Kim, Y.J.; Han, J.G. Electrochemical properties of TiN coatings on 316L stainless steel separator for polymer electrolyte membrane fuel cell. *Thin Solid Films* **2008**, *516*, 3669–3672. [CrossRef]
40. Cao, M.; Liu, L.; Yu, Z.; Fan, L.; Li, Y.; Wang, F. Electrochemical corrosion behavior of 2A02 Al alloy under an accelerated simulation marine atmospheric environment. *J. Mater. Sci. Technol.* **2019**, *35*, 651–659. [CrossRef]
41. Kartsonakis, I.A.; Stanciu, S.G.; Matei, A.A.; Karaxi, E.K.; Hristu, R.; Karantonis, A.; Charitidis, C.A. Evaluation of the protective ability of typical corrosion inhibitors for magnesium alloys towards the Mg ZK30 variant. *Corros. Sci.* **2015**, *100*, 194–208. [CrossRef]
42. Nam, N.D.; Jo, D.S.; Kim, J.G.; Yoon, D.H. Corrosion protection of CrN/TiN multi-coating for bipolar plate of polymer electrolyte membrane fuel cell. *Thin Solid Films* **2011**, *519*, 6787–6791. [CrossRef]
43. Li, M.; Luo, S.; Zeng, C.; Shen, J.; Lin, H.; Cao, C. Corrosion behavior of TiN coated type 316 stainless steel in simulated PEMFC environments. *Corros. Sci.* **2004**, *46*, 1369–1380. [CrossRef]
44. Kartsonakis, I.A.; Stanciu, S.G.; Matei, A.A.; Hristu, R.; Karantonis, A.; Charitidis, C.A. A comparative study of corrosion inhibitors on hot-dip galvanized steel. *Corros. Sci.* **2016**, *112*, 289–307. [CrossRef]
45. Shen, S.; Zuo, Y.; Zhao, X. The effects of 8-hydroxyquinoline on corrosion performance of a Mg-rich coating on AZ91D magnesium alloy. *Corros. Sci.* **2013**, *76*, 275–283. [CrossRef]
46. Xu, J.; Li, Z.Y.; Xu, S.; Munroe, P.; Xie, Z.H. A nanocrystalline zirconium carbide coating as a functional corrosion-resistant barrier for polymer electrolyte membrane fuel cell application. *J. Power Sources* **2015**, *297*, 359–369. [CrossRef]
47. Feng, K.; Li, Z.; Sun, H.; Yu, L.; Cai, X.; Wu, Y.; Chu, P.K. C/CrN multilayer coating for polymer electrolyte membrane fuel cell metallic bipolar plates. *J. Power Sources* **2013**, *222*, 351–358. [CrossRef]
48. Lee, E.K.; Kim, J.K.; Kim, T.J.; Song, H.; Kim, J.H.; Park, S.A.; Jeong, T.G.; Yun, S.W.; Lee, J.; Goo, J.; et al. Enhanced corrosion resistance and fuel cell performance of Al1050 bipolar plate coated with TiN/Ti double layer. *Energy Convers. Manag.* **2013**, *75*, 727–733. [CrossRef]

applied sciences

MDPI

Article

The Influence of Replacing Aggregates and Cement by LFS on the Corrosion of Steel Reinforcements

María Isabel Prieto [1,*], María de las Nieves González [2,*], Ángel Rodríguez [3] and Alfonso Cobo [1]

1 Tecnología de la Edificación, Universidad Politécnica de Madrid, 28040 Madrid, Spain; alfonso.cobo@upm.es
2 Construcciones Arquitectónicas y su Control, Universidad Politécnica de Madrid, 28040 Madrid, Spain
3 Construcciones Arquitectónicas e Ingeniería de la Construcción y del Terreno, Universidad de Burgos, 09001 Burgos, Spain; arsaizmc@ubu.es
* Correspondence: mariaisabel.prieto@upm.es (M.I.P.); mariadelasnieves.gonzalez@upm.es (M.d.l.N.G.)

Received: 21 January 2019; Accepted: 13 February 2019; Published: 17 February 2019

Abstract: The aim of this study is to investigate the corrosion behavior of steel reinforcement embedded in mortar specimens in which both the aggregate and cement are partially replaced by ladle furnace slag (LFS) and different percentages of chloride ions by weight of cement are introduced into the mix at the time of kneading. The corrosion behavior was studied by using electrochemical techniques in order to evaluate the corrosion rate and the symptoms produced in steels of specimens with and without slag LFS. From the analysis of the results, it is concluded that the use of LFS in a partial replacement of aggregate and cement in mortar specimens does not compromise the behavior of the mortar with regard to corrosion of the steel reinforcement; consequently, partial replacement by LFS is fully feasible from this standpoint.

Keywords: corrosion; reinforcements; concrete; slag; LFS

1. Introduction

Due to its versatility and low cost, reinforced concrete is the most frequently used structural material in the field of building and engineering; it reaches annual volumes of 10 km^3/year and causes major environmental problems. The manufacture of Portland cement produces between 5% and 8% of global CO_2 emissions due to human activity [1–3].

The regulations [4–7] specify the importance of durability in reinforced concrete and take as a determining factor the corrosion of embedded reinforcement steel.

Corrosion of embedded steel reinforcement may occur through carbonation of the concrete cover, the presence of chlorides, or the combination of both. These are the factors which trigger the corrosion process and transfer the reinforcement from the passive state, in which the corrosion rates are barely significant, to the active state, in which corrosion affects the durability of concrete structures.

Besides these triggering factors, the continuous presence of oxygen and humidity is necessary as these are the controlling factors for the rate of the corrosion process. In practice, the presence of chlorides in concrete is a determining factor in the onset of reinforcement corrosion. In general, a threshold level of 0.4% of chloride ions by weight of cement is considered, as proposed by RILEM Committee 60-CSC [8].

Moreover, the particular feature of steel as an indefinitely recyclable material has meant that the use of scrap iron has become increasingly widespread in steel production at a global level. World crude steel production reached 1689.4 million tons (Mt) for the year 2017 [9]. Steel production in Spain has been increasing over the years, reaching 14.5 million tons in 2017 [10]. This provides major environmental benefits as a ton of steel manufactured in electric arc furnaces (EAF) consumes only 9–12 GJ/tcs, with a consequent reduction in CO_2 emissions, though the continuous increase in steel production has been the cause of a significant increase in environmental problems over the years [11,12].

EAF steel production generates approximately 120 to 180 kg of black slag and 20 to 30 kg of white slag for each ton of steel produced [13]. The volume of ladle furnace slag (LFS) generated in the European Union was 2.0 million tons in 2016 [14], while in Spain, with 70.5% of the steel produced by EAF, the volume of white slags was between 0.20 and 0.31 million tons in 2017 [10].

Among the most immediate applications of the LFS are the manufacture of clinker, the partial substitution of aggregates and/or cement in mortars and concretes, and the stabilization of soils and pavements [15,16]. The characteristics of the slag that can be used in concrete depend on the different treatments they have received [17].

There is some research on the characterization of ladle furnace slags and their hydration properties. These studies show that the LFSs are a dusty material and are composed mainly of calcium oxide (50%), silicon (15%), aluminum (12%), and magnesium (9%). Its density is of the order of 2.7 g/cm^3 and associated important volumetric expansions are due to the high proportion of free lime. If this expansion is controlled, it does not imply any inconvenience for its use in construction materials [18–20]. There has also been some research that studied the behavior of mortars or concretes in which the aggregate and/or the cement was partially replaced by LFS and that indicate that its application is viable, improving the properties of mortars and sustainability both in traditional concretes and in self-compacting concretes [21–27].

Taking into account the previous premises, it is observed that although the corrosion process of steel bars embedded in concrete has been studied in various research projects [28–30], no studies have analyzed this process when ladle furnace slag is introduced into the concrete instead of some of its components. In this work, corrosion behavior was tested using electrochemical techniques in order to evaluate the corrosion rate and the effects produced in the reinforcements of the specimens with and without LFS. These specimens were initially subjected to a natural corrosion process and then to accelerated corrosion; this allowed us to determine the influence range of the LFS slag.

2. Experimental Investigation

2.1. Materials

The materials employed in the manufacturing of the mortar specimens were Portland cement CEM I/42.5 R [31] (Table 1), silica sand specified as "siliceous rolled aggregate 0/4" (Table 2), urban drinking water, retardant SikaTard, calcium chloride ($CaCl_2$), and ladle furnace slag (Table 3).

Table 1. Main properties of Portland cement.

Chemical (%)		Physical	
SO_3	3.4	Specific surface area (Blaine)	414 m^2/kg
Cl^-	0.01	Density	3.15 g/cm^3
Calcination loss	1.72	Le Chatelier expansion	0 mm
Insoluble residue	0.4	Initial setting time	108 min
		Final setting time	160 min

Table 2. Characterization of the sand used in the specimens.

Aggregate	0.78%
Sand equivalent	78
Real density	2.619 g/cm^3
Normal absorption coefficient	15%
Saturated surface dry density	2.630 g/cm^3
Clay lumps	0.01%
Low specific weight particles	0.00%
Coefficient of type of course	0.26%
Soft particles	0.93%
S, S03, Cl^-	0%

Table 3. Physical properties and chemical composition of ladle furnace slag (LFS).

Density	2.83 g/cm^3
Specific surface	2064 cm^2/g
Chlorides	Absence
Total sulfur, expressed as sulfate ions	<1%
Clay lumps	Absence
Organic material	Absence
CaO	56%
SiO_2	17%
Al_2O_3	11%
MgO	10%
Others (Fe_2O_3+MnO+TiO_2+SO_3+Na_2O+K_2O)	6%

Taking into account the results obtained in other research work [25], the following guidelines for dosages were considered:

- A ratio of cement/sand/water by weight of 1:6:w, with w relating to the necessary quantity of water to achieve a slump of 175 $\pm$ 10 mm;

- A compressive mechanical strength at 28 days of at least 7.5 N/mm^2;

- 30% of cement and 25% of sand substituted for ladle furnace slag in the LFS specimens;

- 0.5% by weight of cement of retardant was used.

Two series of mortar test specimens were manufactured: one used as a standard test specimen (MCC) and another in which part of the cement and sand was replaced by LFS (MCE) (Table 4). During kneading, chloride ions were added in the form of calcium chloride, corresponding to 0%, 0.4%, 0.8%, 1.2%, and 2% by weight of cement.

Table 4. Dosages of the samples with and without LFS.

Label	Cement (g)	Sand (g)	Water (g)	LFS (g)	Retardant (g)	$CaCl_2$ (g)	Ion Cl^- (%)
MCC-0.0	99.6	597.5	102.9	—	0.50	0.00	0.0
MCC-0.4	99.6	597.5	102.9	—	0.50	0.80	0.4
MCC-0.8	99.6	597.5	102.9	—	0.50	1.60	0.8
MCC-1.2	99.6	597.5	102.9	—	0.50	2.40	1.2
MCC-2.0	99.6	597.5	102.9	—	0.50	3.99	2.0
MCE-0.0	68.6	441.1	113.9	176.4	0.35	0.00	0.0
MCE-0.4	68.6	441.1	113.9	176.4	0.35	0.55	0.4
MCE-0.8	68.6	441.1	113.9	176.4	0.35	1.10	0.8
MCE-1.2	68.6	441.1	113.9	176.4	0.35	1.65	1.2
MCE-2.0	68.6	441.1	113.9	176.4	0.35	2.75	2.0

2.2. Experimental Setup

This experimental work involved the preparation of prismatic mortar test specimens measuring 6 × 8 × 2 cm^3, in which 3 parallel steel rebars with a diameter of 6 mm were embedded (Figure 1). The study is a continuation of research into LFS as a partial replacement for aggregate and cement in mortars [32,33] from the viewpoint of its behavior with regard to rebar corrosion and the symptom this produces with the time.

All the steel bars were measured, numbered, and weighed before their introduction into the mold. The steel–concrete–atmosphere interface was protected by adhesive tape over a length of 5 cm in order to avoid the possibility of localized corrosion attacks by differential aeration, leaving an effective length of 6 cm contact between the steel and the concrete. The mortar was kneaded in a mechanical mixer for 90 s. The mold was filled in two pourings, coinciding with the arrangement of the steel bars. Thereafter, test specimens were placed in a humidity chamber for 24 h. After this period, the specimens were demolded and placed again in the humidity chamber for 28 days, and then they were left to dry normally. With the aim of studying the corrosion of the bars by using electrochemical techniques, the specimens were once again placed in a humidity chamber for 550 days, at a temperature of 20 °C and a relative humidity 95%, associating each measurement of corrosion potential and corrosion rate with the degree of moistness of the mortar.

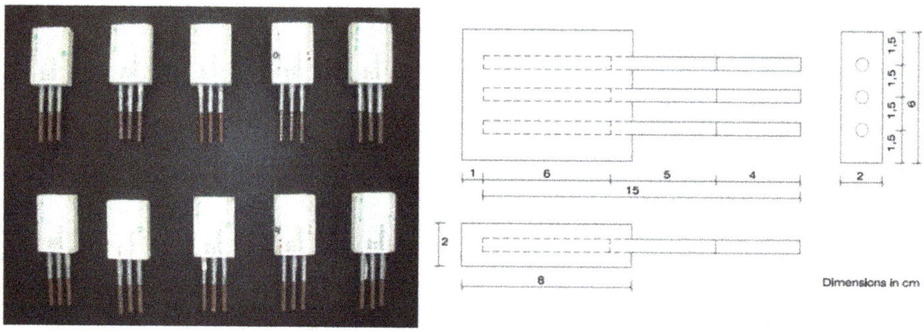

Figure 1. Geometrical characteristics of the specimens and specimens corresponding to the MCC dosages.

To highlight the symptoms that rebar corrosion produces in the mortar test specimens with mixed-in chlorides, the corrosion process was further accelerated by connecting the central bar of the standard test specimens as well as the test specimens containing LFS, both having mixed-in chloride ions percentages of 1.2% and 2%, to a direct current source. Once connected, a constant anodic current of 1.3 mA, corresponding to 10 $\mu A/cm^2$, was passed through each central bar for 282 days. During the periodic measurements of the voltage needed to maintain this preset current and observation of the symptoms, the degree of moistness of the test specimens, with measurements taken before and after their moistening on each occasion, was again associated with the stabilized potential.

2.3. Measurement Techniques

Electrochemical measurements were performed with an AUTOLAB/PGSTAT302N potentiostat (Version 4.9 AUTO83745, ECO CHEMIE, Utrech, Netherland, 2008) [34]. The central steel bar was used as the working electrode while the parallel bars on the sides were used as the counter electrodes. For the electrochemical measurements a silver/silver chloride reference electrode was used (SSCE; +0.222 V SHE). During the measurements a damp flannel was placed on the mortar surface in order to improve electrolytic contact between the mortar and reference electrode (Figure 2).

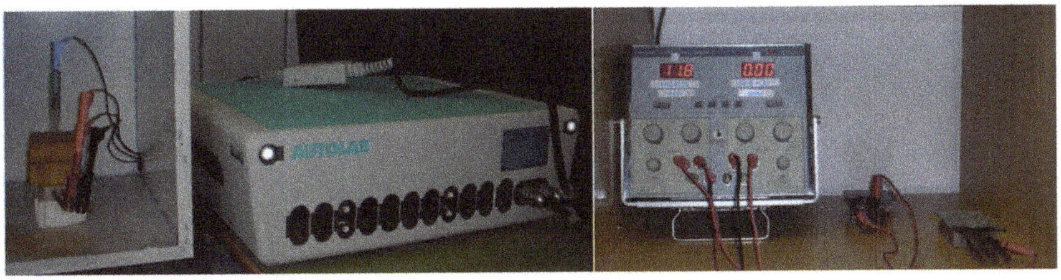

Figure 2. Setup for carrying out electrochemical measurements using an AUTOLAB/PGSTAT302N potentiostat and setup of the accelerated corrosion process using a POWER SUPPLY FAC-662B POWER SUPPLY FAC-662B (PROMAX, Barcelona, Spain, 1998).

All the electrochemical measurements were associated with the humidity content of the test specimens, which were gradually moistening and were kept in a humidity chamber.

The use of the potentiostat enabled the evolution of the corrosion to be studied by means of the polarization curve, producing data on the open circuit corrosion potential (E_{corr}), and through the polarization resistance on the corrosion rate (i_{corr}). Since midway through the 20th century the polarization resistance technique has become a common technique to study corrosion [35].

The first evaluation of rebar behavior with regard to corrosion was made by measuring their corrosion potentials, which are useful for a qualitative determination of whether the steel bars are in an active or a passive state. The data obtained were interpreted according to the ASTM C 876 standard [36], from which, taking into account the reference electrode used, it was concluded that $E_{corr} < -231$ mV would indicate a 90% probability that corrosion exists in the active state, a potential in the region -231 mV $< E_{corr} < -91$ mV would indicate uncertainty, and $E_{corr} > -91$ mV would indicate a probability of 10% that corrosion exists in the active state.

The following step was to interpret the results for the rate or intensity of corrosion as this information enables us to quantitatively interpret the active or passive state of the bars.

According to different research [37], $i_{corr} < 0.1$ μA/cm^2 indicates a passive state, 0.1 μA/cm^2 < $i_{corr} < 0.5$ μA/cm^2 is equivalent to a low level of corrosion, 0.5 μA/cm^2 < i_{corr} < 1 μA/cm^2 to a high level of corrosion, and $i_{corr} > 1$ μA/cm^2 to a very high level of corrosion.

The central bar was connected to a POWER SUPPLY FAC-662B (PROMAX, Barcelona, Spain, 1998) to accelerate the corrosion process in both the standard test specimens and the test specimens with LFS, for specimens containing 1.2% and 2% of mixed-in chloride ions by weight of cement, making the embedded central steel bar act as an anode. On the upper face of the test specimens was placed a damp flannel with a lead sheet on top; this acted as a continuous counter electrode [38]. The voltage necessary to maintain the intensity of the preset current (1.3 mA) was then measured, moistening the test specimens frequently, as the potential varies substantially with the humidity content (Figure 2).

3. Results and Analysis

Figure 3a shows the evolution of the corrosion potential (E_{corr}) in standard test specimens while Figure 3b provides the data from test specimens with slag for different chloride ion percentages. The corrosion potentials in the standard test specimens for chloride ion percentages equal to or below 0.4% are clearly less negative than those in the test specimens with chloride ion percentages above the limit set by the standards of different countries, which show potentials suggesting the active state. In the test specimens with LFS, regardless of the chloride ion percentages possessed by the test specimens, the corrosion potentials correspond to the active state. In all test specimens the potential becomes more negative over time and, consequently, with their humidity content, as well as with the chloride ion percentages of the test specimens.

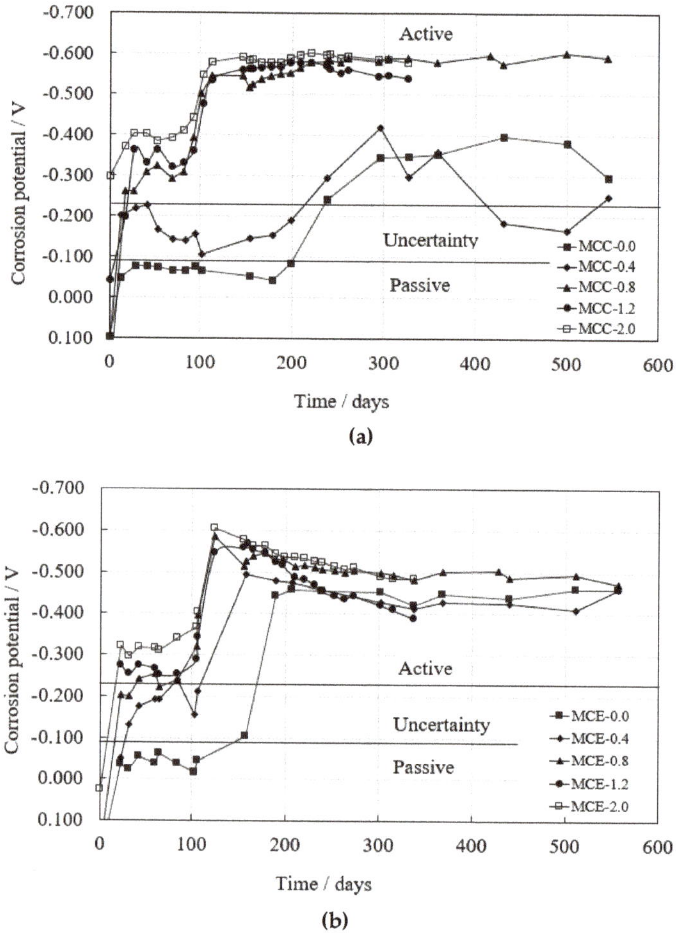

Figure 3. Evolution of E_{corr} over time in standard test specimens (**a**) and in LFS specimens (**b**) at different chloride ion percentages.

The corrosion rate increases over time and, therefore, with the humidity content, together with the chloride ion percentage introduced at the time of kneading. In the test specimens with chloride ion percentages within the limit of the EHE "Instrucción de Hormigón Estructural" Instruction (Figure 4a) (0.4% of chloride ions by weight of cement), the corrosion rates are characteristic of the passive state as the maximum values are around 0.1 $\mu A/cm^2$, regardless of whether the test specimens contain LFS, with data corresponding to 545 days of exposure in the humidity chamber. In the test specimens with chloride ion percentages above 0.4% by weight of cement (Figure 4b), the corrosion rates are characteristic of the active state, with values above 1 $\mu A/cm^2$. In the test specimens with 0.8% chloride ions, with data obtained for up to 545 days of exposure, the corrosion rates are slightly higher in the standard test specimens than in the test specimens with LFS. In the test specimens with chloride ion percentages of 1.2% and 2%, data are held on the corrosion rate for up to 326 days of exposure in the moist chamber, as from that moment they underwent an accelerated corrosion process; the results are discussed below. In these test specimens, the corrosion rates increased with the chloride percentage contained in the test specimens, which were slightly higher in the standard test specimens than in the LFS test specimens.

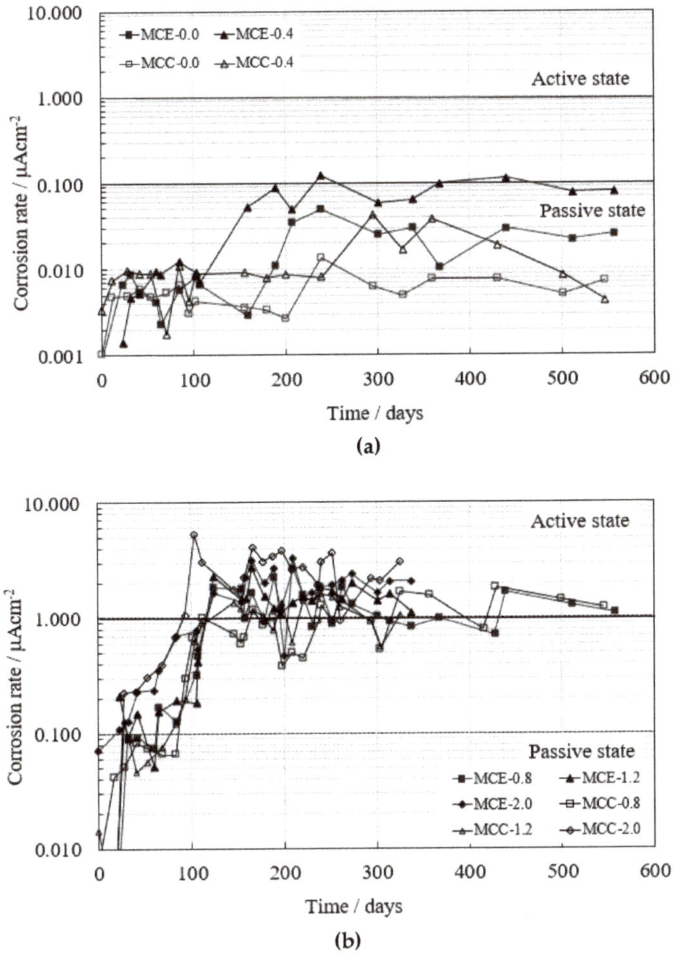

Figure 4. Evolution of i_{corr} over time in test specimens without chlorides and with 0.4% chloride ions by weight of cement (**a**) and in test specimens with chloride ion percentages of 0.8%, 1.2% and 2% (**b**).

Figure 5 shows the evolution of the voltage needed to maintain the preset current at a level of 1.3 mA. The voltage increases over time and with the chloride ion percentage introduced at the time of kneading. Moreover, standard specimens need higher voltage to maintain the preset current than do the specimens with slag LFS.

Figures 6 and 7 show the symptoms produced by the corrosion of the steel bars subjected to an accelerated corrosion process by an impressed anodic current for chloride ion percentages of 1.2% and 2.0% by weight of cement. Each image shows the number of days that had elapsed since the start of the process of natural corrosion, the number of days subjected to accelerated corrosion, and the potential and the intensity of the current passed through the rebar.

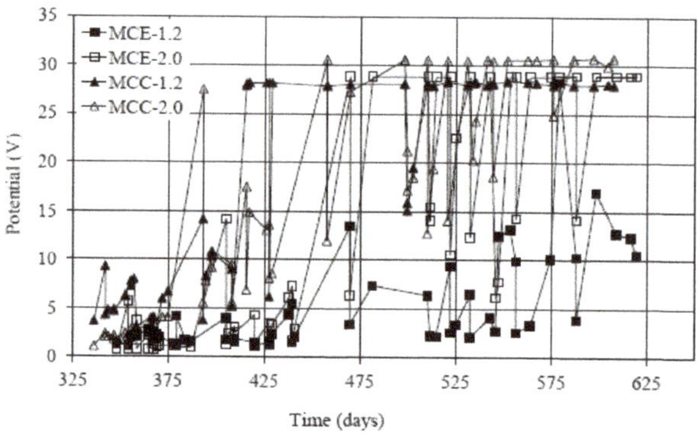

Figure 5. Evolution of the potential (V) necessary to maintain a constant current of 1.3 mA in each rebar.

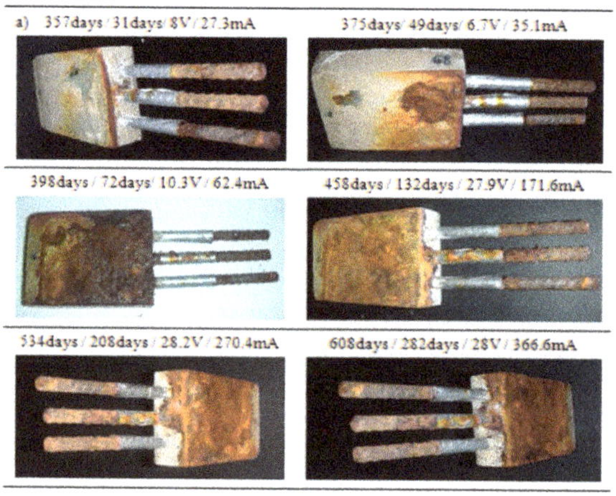

(a)

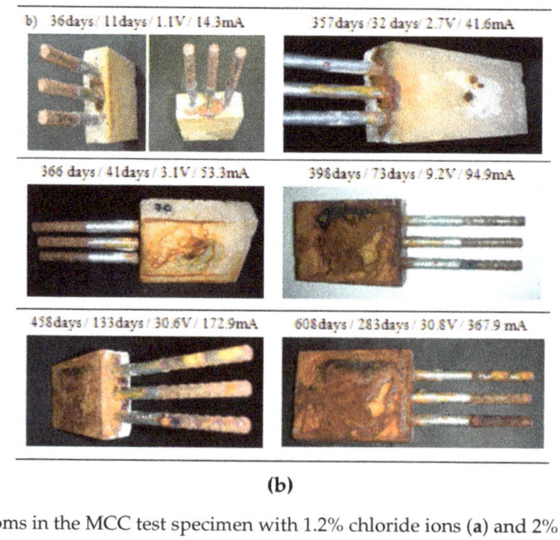

(b)

Figure 6. Symptoms in the MCC test specimen with 1.2% chloride ions (**a**) and 2% chloride ions (**b**).

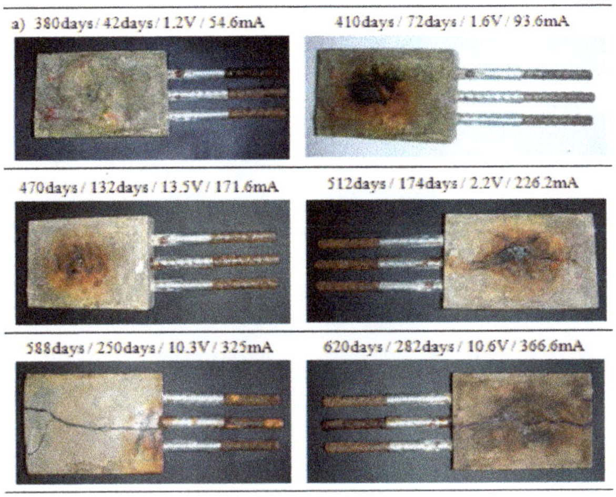

(a)

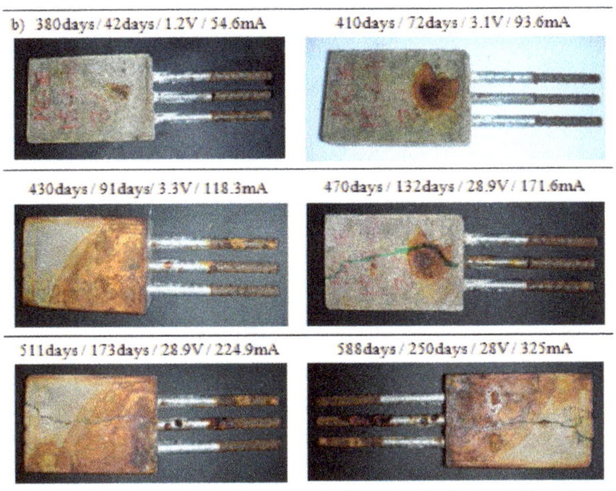

(b)

Figure 7. Symptoms in the MCE test specimen with 1.2% chloride ions (**a**) and 2% chloride ions (**b**).

In the standard test specimen with a chloride ion percentage of 1.2%, it was observed that the first rust stains began to appear 31 days after the start of the accelerated corrosion, appearing both on the edge of the test specimen and on the lower face, gradually increasing up to 72 days when the stain extended over the entire face of the test specimen, while on the upper face there were no signs of corrosion. The standard test specimen showed no signs of cracking after the central bar had been connected to the current for 282 days, an externally applied electrical charge of 366.6 mA having passed through the central bar (Figure 6a).

In the final test specimen which was subjected to the accelerated corrosion process, corresponding to the other standard test specimen with 2% chloride ions, rust stains began to appear on the edge of the test specimen 11 days after the start of the process, with stains appearing on the lower face of the test specimen after 32 days. From that moment, the rust stains became more uniform on the lower face of the test specimen after 73 days with an electrical charge passed through of 94.9 mA. The test specimen continued to show signs of corrosion over time in the form of increasingly marked stains, but it presented no cracks after 283 days of accelerated corrosion with an electrical charge of 367.9 mA having passed through the bar (Figure 6b).

In the test specimens with LFS and with 1.2% chloride ions by weight of cement, rust stains began to appear in the upper part of the test specimen coinciding with the position of the bar which was undergoing accelerated corrosion. This initial stain appeared approximately 40 days following the commencement of the accelerated corrosion process and continued to enlarge its surface area and importance with the progressive passage of the current. After 174 days of current, the first cracks appeared, on both the upper and lower faces of the test specimen and even on its edge, coinciding with the central bar. The cracks increased in length and width over time, forking at the end of the test specimen 250 days after the commencement of accelerated corrosion, an overall electrical charge of 325 mA having passed through the bar. The crack width after 282 days, once a charge of 366.6 mA had passed through the bar, was 1 mm (Figure 7a).

The first rust stain in the test specimen with 2% chloride ions and with LFS appeared 42 days after the commencement of accelerated corrosion. The first stain, on the upper face of the test specimen, increased in size over time, leaving an appearance of generalized stains on the test specimen. After 132 days since the commencement of the accelerated corrosion process, the first cracks appeared, coinciding with the central bar and appearing both on the upper and lower faces of the test specimen. The width of the cracks gradually increased; after 250 days and after passing a 325 mA electrical charge through the central bar, numerous cracks in a mesh were formed on the rear of the test specimen, which caused the mortar to disintegrate (Figure 7b).

4. Discussion

Both the qualitative results obtained through the measurement of corrosion potential and the quantitative results deduced from the corrosion rates (Figures 3 and 4) show that for chloride ion percentages of 0.4%, the bars are in a passive state but with higher corrosion rates in test specimens with LFS than in the standard test specimens. For test specimens with chloride ion percentages above 0.4%, the bars demonstrate very high states of corrosion both in standard test specimens and in those with LFS, being slightly larger in the former.

The symptoms produced by corrosion show as growing rust stains appearing over time and, therefore, with the increase in current in the standard test specimens, appearing first in the test specimens with 2% chloride ions. The stains began to coincide with the bar through which the external current passed, only to extend gradually along the edge and the lower face of the test specimens, but without cracking 608 days after the start of the experiment and having undergone 283 days of accelerated corrosion.

In the test specimens with LFS, the rust stains began to form later than in the standard test specimens, on the upper face of the test specimen, coinciding with its central bar. In the test specimens with 1.2% chloride ions, 174 days after the commencement of accelerated corrosion, cracks appeared on the lower face and on the edge of the test specimens; these increased in width and length over time until 282 days, when the upper face cracked. In the test specimens with 2% chloride ions by weight of cement, the cracks began 132 days after the commencement of the accelerated corrosion on both the upper and lower faces and on the edge of the test specimens, coinciding with the central bar of the test specimen. The passage of time and the increase in current increased the cracks, with a mesh of cracks appearing in the lower part of the test specimen after 250 days.

Breaking of the test specimens enabled observation of the penetration of corrosive products on the mortar pores in the test specimens with 2% chloride ions; this showed corrosion in the form of pitting produced by the presence of chlorides, beginning with the loss of corrugation in the bar and continuing with the loss of sections in a localized manner (Figure 8).

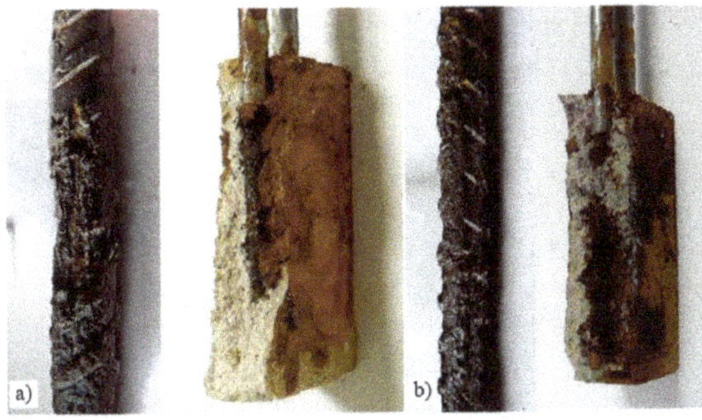

Figure 8. Corrosion products in the test specimens without (**a**) and with LFS (**b**), with 2% chloride ions by weight of cement.

Figure 9 shows the corrosion rates in standard test specimens and those with LFS for different chloride ion percentages within 546 days after the experiment onset. The results show that in test specimens with chloride ion percentages of 0.4%, the corrosion rates were slightly higher in the slag test specimens than in the standard ones. By contrast, for chloride ion percentages higher than the EHE Instruction Limit, the corrosion rates were similar and even higher in standard test specimens.

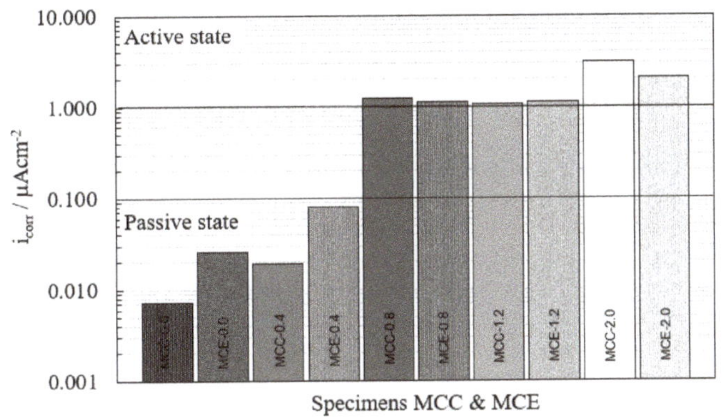

Figure 9. i_{corr} values of test specimens attacked by chlorides, in standard test specimens (MCC) and test specimens with LFS (MCE) after 546 days.

5. Conclusions

Corrosion potentials both in standard test specimens and in test specimens with LFS became more negative as the humidity of the test specimens and the percentage of chloride ions increased, reaching values which indicate a probability of corrosion which is over 90% for chloride ion percentages above 0.4% by weight of cement.

The corrosion rates increased with the humidity content of the test specimens and with the chloride ion percentage introduced at the time of kneading, both in standard test specimens and in those with LFS, reaching values characteristic of the active state for chloride ion percentages above 0.4% by weight of cement.

The standard test specimens offered greater resistance to the passage of an electrical current than the LFS test specimens. The voltage necessary to maintain a constant current fell drastically on increasing the moisture content, increasing progressively as the degree of moistness of the test specimens decreased.

The symptoms produced in standard test specimens with chloride ion percentages of 1.2% and 2.0% by weight of cement subjected to an accelerated corrosion process consisted of rust stains, initially coinciding with the central bar, which progressively lengthened until they occupied the lower face and the edge of the test specimens.

The symptoms produced in test specimens with LFS with chloride ion percentages of 1.2% and 2.0% by weight of cement that underwent the accelerated corrosion process began with the appearance of rust stains that coincided with the central bar of the test specimens and spread over the upper and lower faces and the edge of the test specimens. The pressure exerted by the corrosive products caused progressive cracking of the test specimens, becoming more evident in the test specimens with a greater chloride ion percentage.

The presence of chlorides in the test specimens produced localized pitting in the steel rebar, both in the standard test specimens and in the test specimens with LFS.

From the study on the corrosion behavior through electrochemical techniques and the study of the symptoms produced in the accelerated corrosion process, it may be concluded that in mortars with the presence of mixed-in chlorides, the incorporation of LFS in the mortar at the time of kneading does not negatively affect the corrosive process of the rebar.

Author Contributions: M.I.P.: experimental tests, formal analysis, investigation, and writing—original draft preparation; A.C.: conceptualization, methodology, project administration, supervision, and discussion of the results; A.R.: resources, validation, and writing—review; M.d.l.N.G.: methodology, investigation, and writing—review and editing.

Funding: This research received no external funding.

Conflicts of Interest: The authors declare no conflict of interest.

References

1. Mehta, P.K. Bringing the concrete industry into a new era of sustainable development. In Proceedings of the Symposium on Advances in Concrete Science and Technology, Fifth CANMET/ACI International Conference Proceedings, Rome, Italy, 7 October 1997; pp. 49–67.
2. Gartner, E.M. Potential improvements in cement sustainability. In Proceedings of the 31st Cement and Concrete Science Conference Novel Developments and Innovation in Cementitious Materials, London, UK, 12 September 2011; pp. 12–13.
3. Flatt, R.J.; Roussel, N.; Cheesema, C.R. Concrete: An ECO material that needs to be improved. *J. Eur. Ceram. Soc.* **2012**, *32*, 2787–2798. [CrossRef]
4. EHE. *Instrucción de Hormigón Estructural*; Ministerio de Fomento: Madrid, Spain, 2008.
5. BSI. *BS EN 1992-1-1:2004: Eurocode 2: Design of Concrete Structures. General Rules and Rules for Buildings*; BSI: London, UK, 2004.
6. ACI. *ACI-318S-08: Building Code Requirements for Structural Concrete (ACI-318-08) and Commentary*; American Concrete Institute: Farmington Hills, MI, USA, 2008.
7. CEB-FIB. *Model Code 1990*; Comité Euro-International du Béton: Lausanne, Switzerland, 1993.
8. RILEM (The International Union of Testing and Research Laboratories for Materials and Structures). *Corrosion of Steel in Concrete. Report of the Technical Committee 60-CSC*; Schiessl, P., Ed.; Chapman and Hall: London, UK, 1988.
9. World Steel Association 2017. Available online: https://www.worldsteel.org/ (accessed on 4 February 2019).
10. CEDEX 2017. Catálogo de Residuos Utilizables en Construcción. Available online: www.cedexmateriales.es (accessed on 4 February 2019).
11. Yellishetty, M.; Anjith, P.G.; Gavin, M.; Mudd, A. Tharumarajah Environmental life-cycle comparisons of steel production and recycling: Sustainability issues, problems and prospects. *Environ. Sci. Policy* **2011**, *14*, 650–663. [CrossRef]
12. Motz, H.; Geiseler, J. Products of steel slags: An opportunity to save natural resources. *Waste Manag.* **2001**, *21*, 285–293. [CrossRef]
13. Oliver, S. Las escorias siderúrgicas: Áridos diseñados para el presente y el futuro. *Zuncho* **2012**, *31*, 13–23.

14. Euroslag 2016. Available online: https://www.euroslag.com/products/statistics/statistics-2016/ (accessed on 4 February 2019).
15. Vilaplana, A.S.; Ferreira, V.J.; López-Sabirón, A.M.; Aranda-Usón, A.; Lausín-González, C.; Berganza-Conde, C.; Ferreira, G. Utilization of Ladle Furnace slag from a steelwork for laboratory scale production of Portland cement. *Constr. Build. Mater.* **2015**, *94*, 837–843. [CrossRef]
16. Iacobescu, R.I.; Angelopoulos, G.N.; Jones, P.T.; Blanpain, B.; Pontikes, Y. Ladle metallurgy stainless steel slag as a raw material in Ordinary Portland Cement production: A possibility for industrial symbiosis. *J. Clean. Prod.* **2016**, *112*, 872–881. [CrossRef]
17. Thomas, C.; Rosales, J.; Polanco, J.A.; Agrela, F. *New Trends in Eco-Efficient and Recycled Concrete*; Woodhead Publishing Series in Civil and Structural Engineering; Brito, J., Agrela, F., Eds.; Woodhead Publishing: Duxford, UK, 2019; pp. 169–190.
18. Losáñez, M. Aprovechamiento Integral de Escorias Blancas y Negras de Acería Eléctrica en Construcción y Obra Civil. Ph.D. Thesis, Universidad del País Vasco, Bilbao, Spain, 2005.
19. Herrero, T. Estudio del Efecto de la Hidratación de la Escoria Blanca de Acería de HEA: Aplicación en Pastas y Morteros de Cemento. Ph.D. Thesis, Universidad del País Vasco, Bilbao, Spain, 2015.
20. Choi, S.; Kim, J.; Han, D.; Kim, J. Hydration properties of ladle furnace slag powder rapidly cooled by air. *Constr. Build. Mater.* **2016**, *113*, 682–690. [CrossRef]
21. Herrero, T.; Vegas, I.J.; Santamaría, A.; San-José, J.T.; Skaf, M. Effect of high-alumina ladle furnace slag as cement substitution in masonry mortars. *Constr. Build. Mater.* **2016**, *123*, 404–413. [CrossRef]
22. Setién, J.; Hernández, D.; González, J.J. Characterization of ladle furnace basic slag for use as a construction material. *Constr. Build. Mater.* **2009**, *23*, 1788–1794. [CrossRef]
23. Sideris, K.K.; Tassos, C.; Chatzopoulos, A.; Manita, P. Mechanical characteristics and durability of self compacting concretes produced with ladle furnace slag. *Constr. Build. Mater.* **2012**, *170*, 660–667. [CrossRef]
24. Sideris, K.K.; Tassos, C.; Chatzopoulos, A. Production of Durable Self-compacting Concrete Using Ladle Furnace Slag (LFS) as Filler Material. *Procedia Eng.* **2015**, *108*, 592–597. [CrossRef]
25. Rodríguez, A.; Manso, J.M.; Aragón, A.; González, J.J. Strength and workability of masonry mortars manufactured with ladle furnace slag. *Resour. Conserv. Recy.* **2009**, *53*, 645–651. [CrossRef]
26. Manso, J.M.; Rodríguez, A.; Aragón, A.; González, J.J. The durability of masonry mortars made with ladle furnace slag. *Constr. Build. Mater.* **2011**, *25*, 3508–3519. [CrossRef]
27. Ortega-López, V.; De la Fuente, J.A.; Skaf, M.; Santamaría, A.; Aragón, A.; Manso, J.M. Performance of steel-making slag concrete reinforced with fibers. In Proceedings of the MATEC Web of Conferences, Sharjah, United Arab Emirates, 18–20 April 2017; Volume 120.
28. Vedalakshmi, R.; Palaniswamy, N. Analysis of the electrochemical phenomenon at the rebar–concrete interface using the electrochemical impedance spectroscopic technique. *Mag. Concr. Res.* **2010**, *62*, 177–189. [CrossRef]
29. Darmawan, M.S. Pitting corrosion model for reinforced concrete structures in a chloride environment. *Mag. Concr. Res.* **2010**, *62*, 91–101. [CrossRef]
30. Cobo, A. *Corrosión de Armaduras en Estructuras de Hormigón Armado: Causas y Procedimientos de Rehabilitación*; Fundación Escuela de la Edificación: Madrid, España, 2001; pp. 23–27.
31. *RC-16. Instrucción Para la Recepción de Cementos. Con Comentarios de los Miembros de la Comisión Permanente del Cemento*; Secretaría Técnica: Madrid, Spain, 2016.
32. Prieto, M.I.; Cobo, A.; Rodríguez, A.; Calderón, V. Corrosion behavior of reinforcement bars embedded in mortar specimens containing ladle furnace slag in partial substitution of aggregate and cement. *Constr. Build. Mater.* **2013**, *38*, 188–194. [CrossRef]
33. Rodríguez, A.; Gutiérrez-González, S.; Prieto, M.I.; Cobo, A.; Calderón, V. Analysis of long-term corrosion behavior in mortars containing recycled ladle furnace slag using computerized tomography and SEM. *Mater. Corros.* **2015**, *66*, 199–205. [CrossRef]
34. So, H.S.; Millard, S.G. On-Site Measurements on Corrosion Rate of Steel in Reinforced Concrete. *ACI Mater. J.* **2007**, *104*, 638–642.
35. Stern, M. Surface area relationships in polarization and corrosion. *Corrosion* **1958**, *14*, 43–46. [CrossRef]
36. ASTM. *ASTM Standards ASTM C 876: Standard Test Method for Half-Cell Potentials of Uncoated Reinforcing Steel in Concrete*; ASTM: West Conshohocken, PA, USA, 1991.

37. Network, D. *Manual de Inspección, Evaluación y Diagnóstico de Corrosión en Estructuras de Hormigón Armado;* CYTED Programme: Río de Janeiro, Brazil, 1997.
38. Halmen, C.; Trejo, D. Accelerating Standard Test Method for Assessing Corrosion of Steel in Concrete. *ACI Mater. J.* **2012**, *109*, 421–430.

applied sciences

MDPI

Review

Advances and Challenges of Corrosion and Topology Detection of Grounding Grid

Zhihong Fu [1,*], Xiujuan Wang [1,*], Qian Wang [2], Xiaobin Xu [1], Nengyi Fu [3] and Shanqiang Qin [1]

1. School of Electrical Engineering, State Key Laboratory of Power Transmission Equipment and System Security, Chongqing University, Chongqing 400044, China; xuxiaobin2017@cqu.edu.cn (X.X.); shqiangqin@cqu.edu.cn (S.Q.)
2. Electric Power Research Institute of Chongqing, Chongqing 401120, China; wangq239@163.com
3. Depatrment of Geophysics, Colorado School of Mines, Golden, CO 80401, USA; nengyifu@mymail.mines.edu

* Correspondence: fuzhihong@cqu.edu.cn (Z.F.); 20111101002@cqu.edu.cn (X.W.); Tel.: +86-130-6235-2738 (Z.F.)

Received: 11 April 2019; Accepted: 30 May 2019; Published: 3 June 2019

Abstract: The grounding device plays performs the role of releasing a lightning current and a fault current in the power system, and the corrosion of the conductor will cause damage to the grounding body, which threatens the safe operation of the power system. The grounding grid corrosion detection technology and equipment guarantee the safe operation of the power system. This paper discusses the research status of grounding corrosion and topological detection in detail and introduces the basic principles, research difficulties and existing problems of the methods such as the electric network method, electromagnetic field method, electrochemical method, ultrasonic detection method and electromagnetic imaging method. The methods of electromagnetic imaging and time difference positioning proposed in recent years have been also discussed in detail. The paper points out that the application of grounding grid corrosion detection distance engineering still faces great challenges and that multi-disciplinary, multi-information fusion, new sensing technology, big data platforms and intelligent computing will be the trends to follow in research on grounding grid fault, corrosion detection and life prediction.

Keywords: grounding grid; metal corrosion; topology detection; corrosion detection; nondestructive testing

1. Introduction

The grounding engineering of power systems mainly includes substation grounding, tower grounding, and a high-voltage direct current grounding rod. A major purpose of grounding engineering is to ensure surface point stability and to reinforce the safety of personnel and equipment in the event of a lightning striking or massive amount of current discharged into the ground due to the failure of a power system. However, due to the fact that corrosion will inevitably occur in power systems at the subsurface level, this has become a potential safety hazard of grounding systems. Stray current corrosion, bacterial corrosion, and electrochemical corrosion are the main causes of grounding body corrosion [1,2]. Therefore, the diagnose of grounding body corrosion is one of the major challenges in the evaluation of grounding system conditions [3]. In addition, the detection of subsurface power system involves grounding network topology and grounding impedance detection. Grounding impedance detection technology and equipment are fairly developed [4]. Due to the high difficulty of the grounding body corrosion and topology detection, more advanced technology and equipment are required for such operations.

Research concerning grounding body corrosion detection technology has been conducted for decades, yet it remains difficult to put theories into practice at a large scale. Copper and flat steel are commonly used as grounding conductors. Although copper has stronger corrosion resistance than flat steel, they both have been buried underground for a long time which can be up to more than 30 or 40 years. Thus, their possibility of being corroded cannot be ignored. In China, flat steel is mainly used as a grounding material, but it can be easily corroded. The research of corrosion detection is extensive. Corrosion detection technology of a substation grounding grid mainly includes the network method [5–26], electromagnetic field method [27–35], and electrochemical method [3,36–40]. In recent years, emerging technologies, such as ultrasonic testing [41,42], electromagnetic pulse time difference location [43] and electromagnetic imaging [44–49], have been considered for corrosion detection. Ultrahigh-voltage DC transmission has developed intensively in recent decades, and DC grounding corrosion is a severe challenge. When single pole earth return is used, the carbon steel corrodes rapidly. Thus, grounding corrosion has become the most difficult problem to solve in DC transmission projects. However, the depth, structure, and operation of grounding rods are quite different from those of grounding grids of substation. To our knowledge, no study has been conducted on the diagnosis of grounding rod corrosion. With the rapid development of new energy power generation, the corrosion of grounding bodies caused by a DC stray current in photovoltaic power plants is a new problem that is rarely discussed at present. Charalambos et al. [50] discussed a method of carrying out evaluations for grounding body corrosion caused by DC stray current in photovoltaic power plants. Because the grounding body is buried underground, it is concealed, and the degree of corrosion is difficult to detect directly. At present, evaluating grounding body corrosion mainly depends on site excavation, which is extremely low cost-efficient. It is also restricted by operating conditions. Therefore, the trenchless diagnosis and evaluation technology of grounding body corrosion is demanded.

Recently, grounding grid topology diagnosis technology has come to play an increasingly significant role. In grounding network fault detection and state assessment, accurate drawings of the grounding network are needed. However, in practical projects, old drawings of grounding networks with long a service life are missing or defective due to reconstruction and expansion. Intertwined old and new grids result in a large deviation between the drawing and the actual layout of grounding network, which contributes to the difficulty of fault diagnosis and state assessment of a grounding network. For the grounding project of newly built substations, effective detection methods to judge whether the grounding grid is constructed strictly according to the design drawings are not well-developed. For the fault diagnosis methods of grounding grids such as network method, determining the topological structure of grounding grids in order to establish the diagnosis equation is necessary. Therefore, a considerable amount of attention has been paid to the diagnosis technology of grounding grid topology. Accordingly, the electromagnetic field method, transient electromagnetic method (TEM), and some image processing technologies have been developed in recent years.

2. Corrosion Detection Method of Substation Grounding Grid

According to the detection principle, grounding grid corrosion diagnosis methods can be generalized as the electrical network method, electromagnetic field method, electrochemical method, ultrasonic detection method, and electromagnetic imaging method.

2.1. Electric Network Method

The corrosion or fracture of nodes can be obtained by taking the resistance of ground network conductor as the fault parameter, combining the diagnosis theory with the network theory, adopting the non-linear optimization method, and comparing with the original data. The electric network method can be divided into port resistance method, sensitivity analysis method, and improved sensitivity analysis method.

2.1.1. Port Resistance Method

The basic principle of the port resistance method based on Telegen's theorem is described in [5,6]. It was first proposed by Huang Yong and He Xingbai of Chongqing University in the early 1990s. The principle of the port resistance method is shown in Figure 1. Grounding grids are usually formed by the interconnection of horizontal equalizing conductors. They are buried underground. Single grids are usually rectangular, and the materials are ordinary galvanized flat steel, round steel, or flat copper. Ignoring the influence of soil and other factors, the grounding conductor is regarded as an equivalent of resistors, and the points connecting the down-line and the grid are assumed to be located at the nodes of the grid without any deviation. The grounding conductor network after the completion of grounding grid laying is set to N. According to the structural drawing, and the size and material of the laying conductor, the complete conductor resistance values of each section can be obtained. After using the grounding grid for a long time, the resistance parameters of the grounding conductor are changed due to corrosion. At this time, the network is set to N'. The network N and N' have the same topological structure and can be applied to Tellegen's theorem. By injecting current I_0 between accessible nodes i and j of network N and N' and measuring the port voltage U_{ij}, we can calculate the port resistance R_{ij} and $R\prime_{ij}$ of network N and N' and then obtain the change of port resistance ΔR_{ij}. If the network N and N' have m group accessible nodes, the injected nodes can be replaced in turn, and the m group port resistance can be obtained. According to Kirchhoff's current law (KCL), Kirchhoff's voltage law (KVL), and Tellegen's theorem, the diagnostic equation of the electric network method can be deduced.

$$\begin{cases} \Delta R_{ij\{1} = \sum\limits_{k=1}^{b} \Delta R_k I_{k\{1} I'_{k\{1} / I_0^2 \\ \Delta R_{ij\{2} = \sum\limits_{k=1}^{b} \Delta R_k I_{k\{2} I'_{k\{2} / I_0^2 \\ \cdots \\ \Delta R_{ij\{m} = \sum\limits_{k=1}^{b} \Delta R_k I_{k\{m} I'_{k\{m} / I_0^2 \end{cases} \quad (1)$$

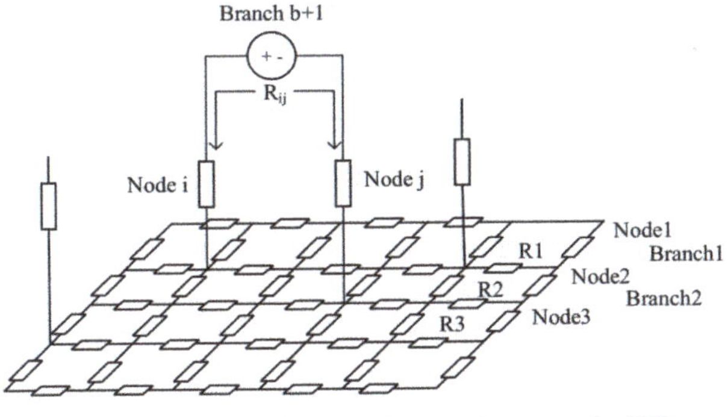

Figure 1. Schematic diagram of port resistance method [5].

In Formula (1), I_0 is the injection current of two accessible nodes, and ΔR_{ij} is the change of port resistance measured by injection current between two accessible nodes. I_k is based on the current of each branch of network N at the injection of I_0. All three quantities are known. ΔR_k is the variation of branch resistance before and after network corrosion, $I\prime_k$ is the branch current of network N'; thus, ΔR_k and $I\prime_k$ are unknown.

Formula (1) is a non-linear system of equations, and the number of equations (m) is usually less than the number of branches (b). The formula belongs to the undetermined system of equations and cannot be solved directly. Generally, an iterative algorithm is implemented to solve this equation.

The effectiveness of port resistance method depends on the accuracy of port resistance measurements. The application of this method is restricted by instrumental error from measuring the small resistance value of grounding conductor and the influence of down-line resistance, contact resistance, and cable resistance.

2.1.2. Sensitivity Analysis Method

Chen [7–9] proposed a sensitivity analysis method to diagnose grounding grid corrosion. In this method, conductor resistance increment of the grounding grid is taken as the fault parameter, and the fault diagnosis equation is established by sensitivity analysis method. The principle of minimum energy and optimization technology are introduced into the solution of the fault diagnosis equation.

For grounding grids with b branches and r accessible nodes, a fault diagnosis Equation (2) is established.

$$\Delta U_r = U_{rb} \cdot X \tag{2}$$

where U_{rb} is the sensitivity matrix and U_{ij} represents the effect of branch j conductor resistance on the potential of measuring node i. ΔU_r is the variation of node voltage before and after corrosion, which can be obtained by measurement. X is a b-dimensional column vector, x_j represents the multiplier of the conductor resistance increment in the j-th segment, which is to be determined. For the actual grounding grid, the fault diagnosis equation is undetermined because the number of accessible nodes m is always less than the number of branches b. The sensitivity matrix is related to branch resistance; that is, the sensitivity matrix is not constant, so Formula (2) is a non-linear equation and is established only when resistance slightly changes.

$$\text{Min P} = \sum_{i=1}^{b} I_j^{*2} R_j^* \tag{3}$$

The application of optimization theory to the solution of the undetermined equation can solve the problem of calculating the actual resistance of conductors in each segment of the grounding grid.

2.1.3. Improved Sensitivity Analysis Method

Because the sensitivity analysis method only makes use of a very limited number of equations and does not take non-linearity into consideration, the resistance of each conductor segment deviates greatly from the actual value. Liu [10] and Luo [11] have improved the sensitivity analysis method. They rotated the excitation position of the current source in the accessible node and measured the voltage of nodes at many locations when each excitation occurs. They made full use of the limited accessible nodes and greatly expanded the number of equations.

The diagnostic equation of the electric network method is a set of undetermined or overdetermined non-linear equations, and the optimization algorithm is a research hotspot. The simplex method [7,8], TABU search algorithm [12], support vector regression [13], differential evolution algorithm [14], constrained total least squares algorithm [15], genetic algorithm [16], and dynamic chaotic PSO [17] have all been applied. Yang [18] proposed Tikhonov regularization on the basis of the least square method in order to reduce the ill-conditioning of non-linear equations, and used the L-curve method to identify grounding grid faults.

Fault diagnosis for large-scale networks requires not only large size matrix operations but measurement errors and mutual-interference. Tearing and block testing methods [19–21] have been developed. In the partial fault diagnosis of grounding grid, only the relevant parts can be tested, which can reduce the testing workload and effectively improve the fault diagnosis efficiency.

Liu [22] and Luo [23] analyzed the branch corrosion measurability of the grounding grid. They eliminated inaccessible nodes by network transformation and carried out inversions to determine the measurability of each branch gradually. The testability of fault diagnosis of the grounding grid is predictable, which has guiding significance for optimizing the test scheme and making scientific use of the diagnosis results.

The substation site is complex, and the practical application of the network method involves many obstacles. The influence of down-line offset [24], portal frame [25], and cable trench [26] is a difficult problem to be solved urgently in the power network method; thus, further research is necessary.

The electric network method has been put forward for nearly 30 years, nevertheless, the testing equipment and implementation methods are not sufficiently practical, as the result, their popularization are obstructed. The network method needs to be developed in more practical manners to determine the topological structure of the grounding network. Establishing an accurate grounding network model is difficult because of irregular construction, lack of drawings, inconsistency of construction and drawings, incomplete grounding network, and connection between the new grounding network and the old one. Moreover, the selection principle of down-line is difficult to determine, and the grounding conductor resistance is very small. In order to suppress strong electromagnetic interference, high power supply is required. There are factors such as the deviation of the positions of the accessible nodes, measuring the resistance of the wire, contact resistance, cable trench and surrounding equipment, which will affect the measurement results. In addition, it has large measurement workload, difficult data management, and low detection efficiency. Hence, operators should have a high technical threshold and limited practicability.

2.2. Surface Electromagnetic Field Method

The surface electromagnetic field method carries out data acquisition from the surface magnetic field, analyzes surface magnetic field characteristics and determines the distribution and fault of grounding grid conductors. Two kinds of grounding grid methods are present: injected and non-injected.

The non-injected type uses the magnetic field induced by the ground current in the ground network to directly measure and analyze its characteristics for fault diagnosis. Gomes et al. [27] designed the surface electromagnetic field detection system and developed a fault diagnosis program on the basis of the finite difference method. This method can test the surface potential, test the distribution of grounding grid, and evaluate the unbalanced current of transformer and the corrosion degree of grounding grid. The non-injected surface electromagnetic field method requires the presence of grid current from power station, which is simple and convenient to set up. However, this method is subject to the change of in situ current and therefore is less used for research and applications.

The injected surface electromagnetic field method was proposed by Dawalibi [28] in 1986 to measure the electromagnetic field distribution on the ground surface which is induced by current injected into the down-line of grounding grids. By comparing data with the theoretical model, local electromagnetic field changes were obtained to determine the locations of grounding grid faults. Dawalibi analyzed the impact of injection mode (location and number of injection points), frequency, and network structure on the diagnosis. Since then, many scholars have studied various forms of injected current, namely DC, alternating current (AC), square wave, and HF. Zhang et al. [29] applied high-frequency AC current injection (up to 1 MHz) to diagnose the grounding grid breakpoint on the basis of the leakage current. The higher the injection current frequency, the better the diagnostic resolution. However, developing a 1 MHz high-frequency current source is difficult. The influence of displacement current should also be considered. Cui and Liu adopted a 379 Hz AC injection current to determine the corrosion degree of ground grid by measuring the drop in magnetic field amplitude [30]. A low-frequency current source reduces the difficulty of power supply design. He [31] injected 50 Hz and 10 A current into high-speed railway grounding systems and realized the corrosion diagnosis of the grounding conductor through the sur face magnetic field. On the basis of the geophysical frequency-domain detection and dual-frequency induced polarization method, the dispersion characteristics of soil and proposed injecting double-frequency square wave current could be realized [32]. The square wave current has broadband characteristics, and some frequency points may amplify the corrosion information of the grounding grid to improve the detection effect of the surface magnetic field method. Qamar [33,34] used the differential operation to amplify the characteristic signal (the third-order differential of the surface magnetic field), locate the grounding grid conductor with peak value, realize grounding network topology detection, and determine the flat steel fracture on the basis of the missing third-order differential peak of the surface magnetic field. However, the differential operation will amplify the noise, which is not practical under the presence of strong electromagnetic interference in the substation. In order to reduce measurement workload, Yang [35] measured the magnetic induction intensity of some measuring points above the grounding grid of substation, established the inverse problem of the grounding grid magnetic field, solved the inverse problem using the reconstruction method based on magnetic detection electrical impedance tomography (MDEIT) method, reconstructed the magnetic field model above the grounding grid, and realized the corrosion fault diagnosis of the grounding grid.

The electromagnetic field method is much simpler than the electric network method in terms of working mode and fault diagnosis characteristic analysis. Moreover, the operability is significantly improved, but the method still has many problems: The noise from the substation is very large, which will affect the judgment accuracy. Moreover, the injection current position is difficult to normalize as it is easily affected by the injection down-line. In addition, the layout of the down-line has not yet been thoroughly investigated. Given the soil dispersion characteristics, the optimal injection frequency and waveform shape are not clear. Moreover, research on the electromagnetic field distribution in the case of the uneven distribution of grounding conductors and complex soil resistivity is still underdeveloped, and the transformer, cable trench, bracket, and other electrical equipment can be easily influenced.

2.3. Electrochemical Method

The electrochemical method determines the corrosion degree or rate of metals by measuring the electrochemical characteristics of the corrosion system between metals and soil, including potentiometry, electrochemical noise, linear polarization, AC impedance, and electrostatic method. The electrochemical method has successfully been applied to detect corrosion of pipelines, concrete reinforcing bars, and metal components. However, the corrosion detection of groundwork has only begun in recent years. The detection targets of the electric network and surface electromagnetic field methods are the corrosion degree of the grounding grid. The electrochemical method mainly detects the corrosion rate of metals. Therefore, it can be used to evaluate the state of the grounding grid and predict its service life span. The electrochemical method has been widely used in the study of corrosion resistance of grounding conductor materials [3,36,37]; by contrast, the corrosion state detection in grounding grid is seldom used. Han et al. [38] proposed a constant current step method to detect the corrosion of the grounding grid and obtained the polarization resistance value by analyzing the charging curve to reflect the corrosion of the grounding grid at the test location. Yang and Peng [39] used the quasi-steady-state measurement of step voltage excitation to obtain the corrosion rate of the ground conductor.

The physical and chemical significance of electrochemical measurement method is clear and it reveals the nature of corrosion. From this point of view, it is a research direction with development prospects. However, electrochemical method is not yet mature and the problems with it focus on:

1. Reliability of corrosion rate detection. This problem has three causes: (1) The influence of AC and DC dispersion. (2) The conductor dispersion is not uniform. The dispersion is affected by the location of the injection point and the distribution of the geoelectric field, and the dispersion of conductors in each segment is extremely uneven. (3) Sensor installation brings influences to the original soil environment. The electrochemical method can only detect the corrosion rate of a test point, and the measurement results are discrete in time and location.
2. Sensor current limiting problem. As a whole, the ground grid is a working electrode. The polarization current is not confined to the measured conductor. The uncertainty of the polarized ground network area makes the corrosion rate impossible to calculate. Han et al. [38] designed a small hole current-limiting device for corrosion detection of the grounding grid. The basic

principle of this device is to use the electric field to guide the focusing and directional flow of the polarization current in soil, so as to achieve current limiting. This method has been applied in concrete corrosion monitoring [40].

3. Strong electromagnetic noise. Strong electromagnetic interference is present in power plants, substations and transmission lines, ground grid dispersion, and stray current effects. There are methods such as wavelet denoising, but the polarization current is on the order of μA. Compared with ground grid and stray current, it is far less than an order of magnitude, and the signal feasibility is not high.
4. Poor accuracy of corrosion assessment. The electrochemical method usually obtains the corrosion rate by measuring the corrosion current and then estimates the corrosion degree by integral. It is an indirect measurement method. The integral calculation produces cumulative error.

In addition, the electrochemical method for the grounding grid is a microscopic measurement. The measurement can only measure the corrosion rate of the measurement point, but it is impossible to judge the location of the breakpoint and cannot applied in the case of the reinforced concrete ground.

2.4. Time Difference Location Method

Time difference location is a newly proposed method to determine the defect location of grounding grids. This method involves ultrasonic detection and electromagnetic pulse time difference location (D-TDOA).

The ultrasonic detection method utilizes the uniform velocity propagation of ultrasonic waves to determine the reflection characteristics of defects, determines the defect location by calculating the time difference of the received reflected wave, estimates the defect magnitude according to the reflected wave amplitude, and detects the non-electromagnetic parameter by ultrasonic method [41,42]. This method is not sensitive to strong electromagnetic interference in power systems, which is its main merit. The overall response of the whole network is measured by the electric network method, but the transmission speed of the guided wave is slow. Early signal response conductor near-end corrosion information can eliminate the influence of near-grid conductor. However, the underground network is deeply buried. In addition, the directional problem caused by the mesh structure, influence of the flat steel solder joint, down-line and electrical equipment connection points, and influence of soil non-uniformity are difficult to solve for the ultrasonic guided wave detection method. A time difference location method (D-TDOA) was proposed to inject the Gauss pulse current and detect pulse transmission time with a broadband measurement sensor (1 kHz–1 GHz) [43]. The fault location accuracy is 10 cm. This method has not been tested on site. Moreover, the output power is a huge problem because the excitation source frequency is very high.

2.5. Electromagnetic Imaging

In recent years, the research achievements in geophysical exploration, biomedical imaging, and other fields have brought vitality to the diagnosis of grounding grid corrosion, forming a branch of the electromagnetic imaging method.

TEM is a geophysical method. It establishes a pulsed magnetic field underground by sending current through a loop. During a field intermission, the secondary eddy current field is observed by a receiving coil, and the longitudinal resistivity profile is formed by inversion. TEM has the characteristics of strong penetration, high resolution, small volume effect, and non-contact measurement. It is mainly used in the field of geophysical exploration. Therefore, the author proposed the transient electromagnetic method for grounding grids (GG-TEM method) [44–46]. The basic principle of GG-TEM is shown in Figure 2. The transmitter outputs bipolar pulse current and excites the magnetic field through the coil. At the moment of a magnetic field turn-off, the secondary eddy current field is excited at the ground grid, and the eddy current is confined within the grounded grid and attenuated. The receiving coil measures the secondary eddy current magnetic field in the ground network intermittently in one pulse. The magnitude and attenuation rate of the secondary eddy current field are determined by the thickness of grounding grid conductor, electromagnetic characteristics of the material, and corrosion degree. The apparent resistivity imaging method is used to obtain the sectional map of the grounding grid. The corrosion degree of the grounding body can be analyzed by the apparent resistivity section characteristics.

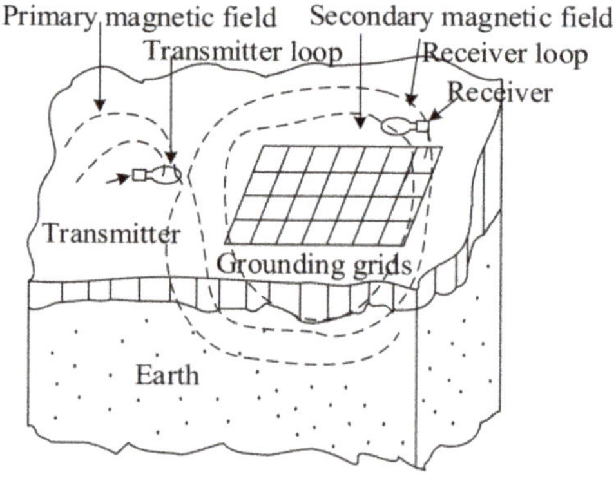

Figure 2. A Schematic field setup of transient electromagnetic method [46].

In 2016, we conducted experiments in the Wuhan Nanrui Grounding Network Experimental Field and achieved fairly ideal detection results (Figures 3–5). The grounding grid, grid spacing, trench width, and depth were24 m × 24 m, 4 m, 0.4 m, and 0.6 m, respectively. The degree of corrosion was simulated by the thickness of the flat steel. The thinner the flat steel, the higher the corrosion degree. Four zones (60 × 6 mm (A), 40 × 4 mm (B), 20 × 3 mm (C), and 40 × 4 mm flat steel (D)) were set, with four fracture ports (numbered 1, 2, 3, and 4).

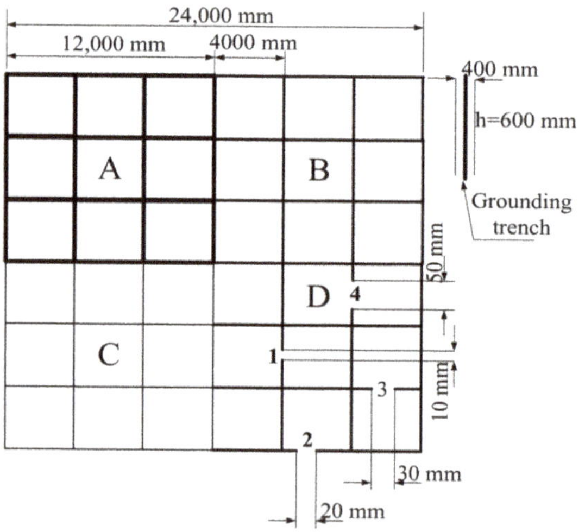

Figure 3. Wuhan Nanrui grounding grid experimental site.

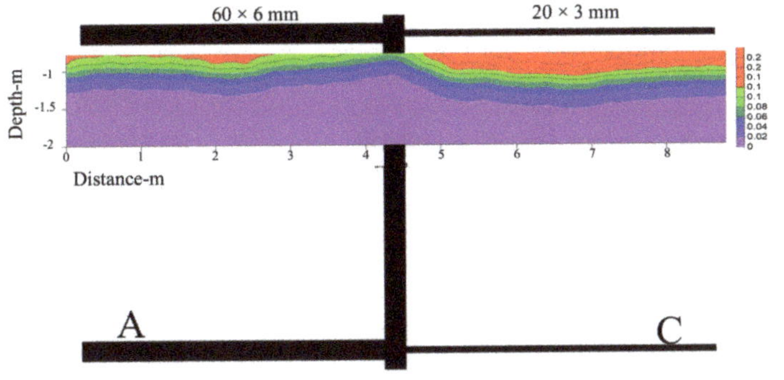

Figure 4. Apparent resistivity section of flat steel.

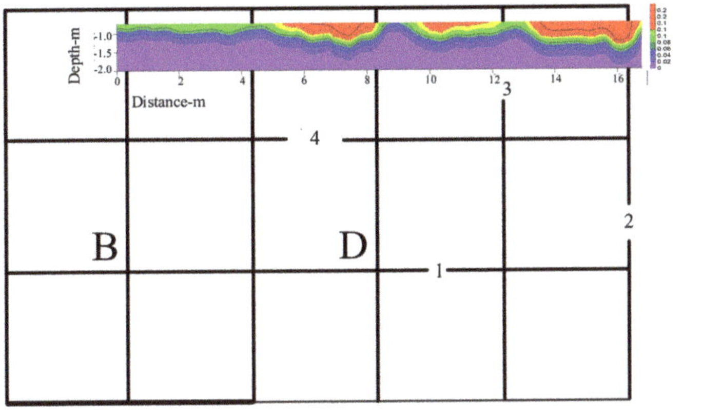

Figure 5. Apparent resistivity section with breakpoints.

Two lines were arranged to detect the imaging characteristics of the thickness and breakpoint of flat steel. In Figure 4, the measuring line passes along the flat steel through areas of different thickness (60 × 6 mm and 20 × 3 mm). As shown in the figure, the high-resistivity layer is thin, and the resistivity is low in the thick conductor section. In the thin conductor section, the high-resistivity layer is thicker, and the resistivity is higher. The electrical resistivity imaging characteristics coincide with the thickness of flat steel. Breakpoint detection: The line passes through the BD area, which has the same flat steel thickness (40 × 4 mm), and the line passes through the breakpoint grids of 4, 1, and 2. Figure 5 shows that the apparent grid resistivity of the three grids where the breakpoints are located is significantly higher than that of the intact grid. Moreover, the thickness of the high-resistivity layer is obvious, and the resistivity of the intact grid is low.

Compared with the surface electromagnetic field method, GG-TEM can solve the problem of current injection point of electromagnetic field method, normalization of injection point, primary field influence problem of electromagnetic induction method, and contact resistance problem and realize contactless measurement.

Yang et al. [47] proposed the grounding net imaging method of endogenous EIT. The 16-channel device mode and imaging principle of biomedical electrical impedance tomography were adopted. The amount of information was greatly increased, and the construction efficiency and diagnostic accuracy were improved. Wu [48] and Wang [49] proposed the radar imaging method, where the ultrawideband emission signal propagates through the soil medium to reach the grounding body and the echo signal is used to obtain the physical state of the grounding body.

In recent years, the electromagnetic imaging method has become a hotspot in the research of grounding grid corrosion diagnosis. However, this method is limited by problems such as reinforced bar shielding of concrete pavement, cable trench, power equipment impact, and large calculation workload.

3. Grounding Network Topology Detection Method

The current injection method and magnetic field excitation method are the main detection methods of grounding grid topology structure.

3.1. Current Injection Method

The current injection method injects a sinusoidal current by selecting a suitable injection node, measures the magnetic field on the surface, and analyzes the magnetic field distribution characteristics to realize the grounding conductor positioning.

Yang [30] designed and developed a different frequency current supply system, mobile weak magnetic induction intensity measurement system, and corresponding software system. Qamar [33] proposed a grounding network topology detection method on the basis of the differential method and established a magnetic field shape function for describing the direction of the ground surface parallel or perpendicular to the grounding grid. This method uses the principal peak characteristic of the even derivative of horizontal component shape function and odd derivative of vertical component shape function to filter and differentiate the measured data of a square magnetic field in the grounding grid after injecting a current, so as to determine the topological structure of the grounding grid. However, the differential method is very sensitive to noise and measurement error. The detection path has a good effect perpendicular to the conductor direction, and the positioning result may have a large error when a large deviation from the vertical direction exists [30,33]. Fu et al. [51] proposed a grounding network topology detection method on the basis of wavelet edge detection technology. Wavelet edge detection effectively strips the magnetic gradient characteristics of each conductor segment from the background field, which greatly reduces the influence of magnetic field superposition on positioning accuracy. At the same time, the strong anti-interference ability of the wavelet transform suppresses the error introduced in the measurement and magnetic field fitting process and ensures the method's stability.

3.2. Magnetic Field Excitation Method

The magnetic field excitation method excites the grounded flat steel through the coil to generate the eddy current field (the secondary field), and the topology of the grounding network is obtained by detecting the eddy current field. The transient electromagnetic method [46] measures the net secondary field, and it has high detection accuracy. The topography of the grounding grid can be visually obtained using the apparent resistivity imaging technology, and the image is intuitive, but it has a large amount of calculation. In addition, the topology of the grounding grid can be obtained by analyzing the amplitude characteristic of the magnetic field which generated by the sinusoidal current. The method is simple in measurement and small in calculation, but the background field is included in the measurement result of the method, so the detection accuracy is relatively low.

4. Challenges and Opportunities

Although a variety of grounding grid corrosion and topological detection methods have been proposed, these methods are still immature, and commercial equipment and technical regulations are lacking. The following are challenges from a technical perspective:

1. Strong electromagnetic interference

The electromagnetic environment of the substation is very complicated, and the power frequency interference magnetic field of the substation operating state can reach more than 10 μT. The high-frequency transient interference caused by high-voltage switching action; large power frequency current from ground power equipment, such as transformers, knife gates, transformers, and gate brackets; power grounding current; harmonic interference; and various secondary and communication equipment will affect the normal operation of the detection instrument by means of spatial and conduction coupling.

The strong electromagnetic interference poses a serious threat to the electrical network method, surface electromagnetic field method, electrochemical method, and electromagnetic imaging method. By using an active source, the excitation source power is increased, and the signal-to-noise ratio is improved. Changing the frequency of the excitation current signal and using a frequency-selective amplification technique, such as notch, amplification, and filtering, can suppress the power frequency and avoid the main interference frequency.

2. Strong metal interference

The substation has a large number of metal facilities, and the connection of several important facilities to the grounding network brings great difficulties to the detection of the grounding grid. For example, a large number of underground metal pipes, cables, leads and cement roads with steel structures will have a great impact on the distribution of surface magnetic fields. This makes it difficult to distinguish the grounding grid information.

3. Corrosion detection of down-line

The grounding grid down conductor is a communication channel between the power equipment and the grounding network. The stray current density is large near the surface layer of the soil, and because of the direct contact of the air, it is susceptible to changes in air and soil humidity, so that the corrosion rate of the part of the grounding conductor is higher than that of the horizontal conductor of the grounding grid. Therefore, the down-line underground part is easily corroded. Moreover, diagnosing down-line corrosion is very difficult, and studies on the corrosion diagnosis of the down-line are lacking.

4. Small target body of flat steel in grounding grid

The common cross-sectional area of the grounding grid conductor is very small compared to other geophysical targets. The cross-sectional area of the grounding grid is much smaller than the length of the conductor, which brings great difficulty to the software simulation. For the electromagnetic induction method, because the conductor cross section is small, the eddy current response intensity and the influence range are small; for the ground penetrating radar, the reflection surface is too small, which brings great difficulty for detection.

5. Multi-information fusion

The electric network method, surface electromagnetic field method, time difference positioning method, and electromagnetic imaging method consider the physical characteristics of the grounding grid corrosion. However, these methods do not reveal the nature of corrosion and can only solve the problem of corrosion degree detection. The electrochemical method can detect the corrosion rate but not the degree of corrosion. Multi-information fusion [52], combined with multiple methods, is expected to simultaneously achieve corrosion degree and corrosion rate detection. Therefore, tapping the potential of existing methods, implementing two or more corrosion detection methods in one device, solving the problem of simultaneous detection of corrosion degree and corrosion rate, and improving the practicality of the detection method are future development directions.

6. Life prediction of grounding grid

With the development of power grid technology and the improvement of voltage levels, the safe operating life of grounding grids has begun to receive attention. It is expected to realize the life prediction of the grounding grid by detecting the corrosion degree of the grounding grid conductor, the corrosion rate, and the parameters of soil aeration, water content, salt content, pH value, soil resistivity, oxidation-reduction potential and soil temperature. Multidisciplinary intersection, multi-information fusion, new sensing technology, big data platform [53] and intelligent computing will certainly promote the maturity and application of grounding grid detection technology and life prediction. These bring new opportunities for research in the field of grounding grid testing.

Author Contributions: Formal analysis, Z.F. and X.W; Methodology, X.W. and X.X.; Project administration, Z.F., Q.W. and S.Q.; Software, S.Q.; Supervision, Q.W.; Validation, N.F.; Writing—original draft, Z.F. and X.W.; Writing—review & editing, X.X. and N.F.

Funding: This research was funded by National Key R&D Program of China, grant number 2017YFC0601804, and National Natural Science Foundation of China, grant number 1777017.

Acknowledgments: This work was supported in part by the National Key R&D Program of China (No. 2017YFC0601804) and the National Natural Science Foundation of China (No. 51777017).

Conflicts of Interest: The authors declare no conflict of interest.

References

1. Schaefer, L.P. Electrical grounding systems and corrosion. *Trans. Am. Inst. Electr. Eng. Part II Appl. Ind.* **1955**, *74*, 75–83. [CrossRef]
2. Rajan, S.; Venugopalan, S.I. Corrosion and grounding systems. *IEEE Ind. Appl.* **1977**, *IA-13*, 297–306. [CrossRef]
3. Lawson, V.R. Problems and detection of line anchor and substation ground grid corrosion. *IEEE Trans. Ind. Appl.* **1988**, *24*, 25–32. [CrossRef]
4. Tong, X.F.; Zheng, Z.H.; Tan, B.; Lu, H.Y.; Lan, L.; Wen, X.S. Corrosion Rate Simulation and Influence Factors of a Vertical DC Grounding Electrode. *IEEE Access* **2018**, *6*, 57230–57238. [CrossRef]
5. Hu, J.; Zeng, R.; He, J.; Sun, W.M.; Yao, J.X. Novel Method of Corrosion Diagnosis for Grounding Grid. In Proceedings of the International Conference on Power System Technology, Perth, WA, Australia, 4–7 December 2000; pp. 1365–1370.
6. Qu, Y.J.; Xu, W.W.; Li, S.L.; Liu, H. Corrosion fault diagnosis for substation's grounding grid based on micro method. In Proceedings of the 2017 China International Electrical and Energy Conference, Beijing, China, 25–27 October 2017; pp. 658–662.
7. Xiao, X.H.; Chen, X.L. Analysis of theory and method about the corrosion as well as the broken point of the grounding grid. *J. Chongqing Univ.* **2001**, *24*, 72–75.
8. Zhang, X.L.; Chen, X.L. The technique of the optimization applied in the grounding grid's failure diagnosis of the power plant and substation. *High Volt. Eng.* **2000**, *26*, 64–66.
9. Xu, G.; Zhu, Z.H.; Yan, P.B. Optimization Algorithm of Corrosion Diagnosis for Grounding Grid. In Proceedings of the 2nd International Conference on Mechanical and Electronics Engineering (ICMEE 2010), Kyoto, Japan, 1–3 August 2010; pp. 42–46.
10. Liu, J.; Wang, J.X.; Wang, S. An improved fault diagnosis algorithm for grounding grids and evaluation of test schemes. *J. Electr. Eng. China* **2005**, *25*, 71–77.
11. Liu, L.Q.; Luo, X.J.; Tao, N.; Wang, K. Grounding Grid Corrosion Diagnosis Based on Large Change Sensitivity. In Proceedings of the Fifth International Conference on Information Assurance and Security, Xi'an, China, 18–20 August 2009; pp. 71–74.
12. Liu, J.; Wang, J.X.; Wang, S. A Corrosion Diagnosis Approach for Grounding Grids Based on TABU Search Algorithm. In Proceedings of the Fifth International Conference on Machine Learning and Cybernetics, Dalian, China, 13–16 August 2006; pp. 1088–1091.
13. Du, J.Y.; Li, N.; Yan, A.J. A Grounding Grid Model of Corrosion Factors Based on Clustering and SVR. *J. Inf. Comput. Sci.* **2013**, *10*, 5243–5251. [CrossRef]
14. Huang, Y.H.; Hu, L.H.; Cao, L.N.; Xia, H.F. Research of Grounding Grid Fault diagnosis based on differential evolution algorithm. In Proceedings of the International Conference on Power System Technology (POWERCON 2014), Chengdu, China, 20–22 October 2014; pp. 1497–1502.
15. Zhang, Y.J.; Luo, X.J.; Niu, T.; Liu, L.Q. A new fault diagnosis model for grounding grid based on constrained total least squares algorithm. *J. Xi'an Jiaotong Univ.* **2010**, *44*, 110–115.
16. Liu, Y.G.; Shang, L.L.; Leng, D.; Mi, H.W.; Ma, J.P. Corrosion diagnosis of grounding grid based on continuum genetic algorithm. *High Volt. Technol.* **2016**, *42*, 1503–1510.

17. Li, T.Y.; Zheng, Y.H.; Li, L.X.; Yao, G.; Xu, Q.S.; Wang, Y.H. Grounding grids corrosion blocking diagnosis algorithm based on dynamic chaotic PSO. *Power Syst. Prot. Control* **2011**, *39*, 78–83.
18. Fan, Y.; Wang, Y.A.; Dong, M.L.; Kou, X.K.; Yao, D.G.; Li, X.; Gao, B.; Irfan, U. A cycle voltage measurement method and application in grounding grids fault location. *Energies* **2017**, *10*, 1929.
19. Liu, J.; Ni, Y.F.; Wang, S.Q.; Li, Z.H.; Wang, J.X.; Wang, S. A Novel Approach of Grounding Grid Corrosion Diagnosis. In Proceedings of the IEEE International Conference on Industrial Technology, Chengdu, China, 21–24 April 2008; pp. 1–5.
20. Liu, Y.G.; Xiao, L.S.; Tian, J.H. Optimized Corrosion Diagnosis of Large-Scale Grounding Grid. In Proceedings of the Power & Energy Society General Meeting, Providence, RI, USA, 25–29 July 2010; pp. 1–6.
21. Zhou, B.; Peng, M.F.; Huang, Q.X.; Jing, J.; Shen, M.E. Grounding grid fault diagnosis based on node tearing and chemical reaction optimization algorithm. *Electr. Power Autom. Equip.* **2017**, *37*, 163–168.
22. Wang, S.Q.; Liu, J.; Wang, S.; Li, Z.Z. Grounding Grid Corrosion Diagnosis and Uncertainly Analysis of Branches. *J. Comput.* **2010**, *5*, 1289–1295. [CrossRef]
23. Liu, L.Q.; Luo, X.J.; Niu, T. Creditability analysis of grounding grid corrosion diagnosis. *Int. J. Appl. Electromagn. Mech.* **2010**, *33*, 1591–1597. [CrossRef]
24. Liu, J.; Ni, Y.F.; Lu, W.; Wang, S.; Li, Z.Z. Influence of Touchable Nodes Deviation on Grounding Grids Corrosion Diagnosis and Its Correction. *High Volt. Eng.* **2008**, *34*, 2349–2354.
25. Liu, Y.G.; Leng, D.; Xiao, L.S.; Tian, J.H. Analysis of Influence of Substation Frame on Corrosion Diagnosis of Grounding Grid. *J. Chongqing Univ.* **2013**, *36*, 92–109.
26. Liu, Y.G.; Cheng, W.J. Influence of grounding conductor in cable trench of substation on corrosion diagnosis. *High Volt. Technol.* **2014**, *40*, 505–512.
27. Gomes, L.V.; de Macedo, E.C.T.; Albuquerque, T.C.; Guedes, E.C.; Junior, G.V.A.; de Castro, M.S.; Freire, R.C.S. Embedded System to Grounding Grid Diagnosis of Energized Substations. In Proceedings of the Instrumentation and Measurement Technology Conference, Graz, Austria, 13–16 May 2012; pp. 796–800.
28. Dawalibi, F. Electromagnetic fields generated by overhead and buried short conductors: Part 2 Ground Networks. *IEEE Trans. Power Deliv.* **1986**, *1*, 112–119. [CrossRef]
29. Zhang, B.; Zhao, Z.B.; Cui, X. Diagnosis of breaks in substaiton's grounding grid by using the electromagnetic method. *IEEE Trans Magn.* **2002**, *38*, 473–476. [CrossRef]
30. Liu, Y.; Cui, X. Research and application of the system on measuring magnetic field in complicated EMI environment in substation. In Proceedings of the IEEE International Symposium on Electromagnetic Compatibility, Qingdao, China, 23–26 October 2007; pp. 146–149.
31. He, Z.Y.; Hu, H.T.; Huang, W.; Pan, Y.L.; Yang, J.W. A Method of Defect Diagnosis for Integrated Grounding System in High-Speed Railway. *IEEE Trans. Ind. Appl.* **2015**, *51*, 5139–5148. [CrossRef]
32. Zhang, P.H.; He, J.J.; Zhang, D.D.; Wu, L.M. A Fault Diagnosis Method for substation grounding grid based on the square-wave frequency domain model. *Metrol. Meas. Syst.* **2012**, *19*, 63–72. [CrossRef]
33. Qamar, A.; Umair, M.; Fan, Y.; Uzairm, M.; Kaleem, Z. Derivative method-based orientation detection of substation grounding grid. *Energies* **2018**, *11*, 1873. [CrossRef]
34. Qamar, A.; Shah, N.; Kaleem, Z.; Uddin, Z.; Orakzi, F. Breakpoint diagnosis of substation grounding grid using derivative method. *Prog. Electromagn. Res.* **2017**, *57*, 73–80. [CrossRef]
35. Liu, K.; Yang, F.; Zhang, S.Y.; Zhu, L.W.; Hu, J.Y.; Wang, X.Y.; Ullah, I. Research on Grounding Grids Imaging Reconstruction Based on Magnetic Detection Electrical Impedance Tomography. *IEEE Trans. Magn.* **2018**, *54*, 1–4.
36. Shao, Y.P.; Mu, M.M.; Zhang, B.; Nie, K.B.; Liao, Q.Q. Corrosion Behavior of Copper-Clad Steel Bars with Unclad Two-End Faces for Grounding Grids in the Red Clay Soil. *J. Mater. Eng. Perform.* **2017**, *26*, 1751–1757. [CrossRef]
37. Li, J.; Su, H.; Feng, C.; Xie, D.M.; Li, L.; Li, X.Y.; Meng, H.M. Corrosion behavior of low-carbon Cr micro-alloyed steel for grounding grids in simulated acidic soil. *J. Iron Steel Res. Int.* **2018**, *25*, 755–766. [CrossRef]
38. Han, L.; Song, S.Z.; Zhang, X.L.; Li, Y.L.; Qiu, Z.G. Portable grounding grid corrosion electrochemical detection system and its application. *Corros. Sci. Prot. Technol.* **2009**, *21*, 337–340.
39. Yang, T.; Peng, M.F.; Hong, H.T. A Practical Method for Detecting the Status of Grounding Grids. In Proceedings of the 2nd International Conference on Industrial Mechatronics and Automation, Wuhan, China, 30–31 May 2010; pp. 270–273.

40. Law, D.W.; Millard, S.G.; Bungey, J.H. Linear polarization resistance measurements using a potentiostatically controlled guard ring. *NDT E Int.* **2000**, *33*, 15–21. [CrossRef]
41. Jiao, J.P.; Zhong, X.; He, C.F.; Wu, B. Experiments on non-destructive testing of grounding grids using SH0. *Insight Non-Destr. Test. Cond. Monit.* **2012**, *54*, 375–379.
42. Jiang, Y.M.; Chen, W.G.; Huang, Y.; Chen, H.G. The Application of Ultrasonic Guided Wave in Grounding Grid Corrosion Diagnosis. *Int. J. Comput. Electr. Eng.* **2013**, *5*, 313–316. [CrossRef]
43. Rodrigues, N.R.; de Oliveira, R.M.; Carvalho, L.F.; de Almeida, J.F.S. A Method Based on High Frequency Electromagnetic Transients for Fault Location on Grounding Grids. In Proceedings of the International Microwave and Optoelectronics Conference, Rio de Janeiro, Brazil, 4–7 August 2013; pp. 1–5.
44. Yu, C.G.; Fu, Z.H.; Wang, Q.; Tai, H.M.; Qin, S.Q. A Novel Method for Fault Diagnosis of Grounding Grids. *IEEE Trans. Ind. Appl.* **2015**, *51*, 5182–5188. [CrossRef]
45. Yu, C.G.; Fu, Z.H.; Hou, X.Z.; Heng, M.T.; Su, X.F. Break Point Diagnosis of Grounding Grids Using Transient Electromagnetic Apparent Resistivity Imaging. *IEEE Trans. Power Deliv.* **2015**, *30*, 2485–2491. [CrossRef]
46. Yu, C.G.; Fu, Z.H.; Wu, G.L.; Zhou, L.Y.; Zhu, X.G.; Bao, M.H. Configuration detection of substation grounding grid using transient electromagnetic method. *IEEE Trans. Ind. Electron.* **2017**, *64*, 6475–6483. [CrossRef]
47. Li, X.; Yang, F.; Ming, J.; Jadoon, A.; Han, S. Imaging the corrosion in grounding grid branch with inner-source electrical impedance tomography. *Energies* **2018**, *11*, 1739. [CrossRef]
48. Wu, X.; Han, X. A RADAR detection system with random noise transmitting wave for power grounding grid. *Front. Artif. Intell. Appl.* **2017**, *299*, 410–415.
49. Wang, Z.M.; Guo, Y.L.; Fu, Q.; Pan, L. A Detection Radar for Grounding Grid. In Proceedings of the 2nd IEEE Advanced Information Management, Communicates, Electronic and Automation Control Conference, Xi'an, China, 25–27 May 2018; pp. 932–935.
50. Charalambous, C.A.; Demetriou, A.; Kokkinos, N.D. Impact of Photovoltaic-Oriented DC Stray Current Corrosion on Large-Scale Solar Farms' Grounding and Third-Party Infrastructure: Modeling and Assessment. *IEEE Trans. Ind. Appl.* **2015**, *51*, 5421–5430. [CrossRef]
51. Fu, Z.H.; Song, S.Y.; Wang, X.J.; Li, J.Q.; Tai, H.M. Imaging the Topology of Grounding Grids Based on Wavelet Edge Detection. *IEEE Trans. Magn.* **2018**, *54*, 1–8. [CrossRef]
52. Chen, L.; Peng, M.F.; Miao, W.H. Application of Information Fusion in Grounding Network Breakpoint Diagnosis. *J. Power Syst. Autom.* **2010**, *22*, 25–30.
53. Du, J.Y.; Wang, L.C. Research on corrosion monitoring system of grounding grid based on cloud platform. In Proceedings of the 32nd Youth Academic Annual Conference of Chinese Association of Automation, Hefei, China, 19–21 May 2017; pp. 1255–1259.

MDPI
St. Alban-Anlage 66
4052 Basel
Switzerland
Tel. +41 61 683 77 34
Fax +41 61 302 89 18
www.mdpi.com

Applied Sciences Editorial Office
E-mail: applsci@mdpi.com
www.mdpi.com/journal/applsci

www.ingramcontent.com/pod-product-compliance
Lightning Source LLC
Chambersburg PA
CBHW061556210926
51850CB00009B/2122
9783036506401